PRAISE FOR
GLOBAL WASTE MANAGEMENT

'The book provides a brilliant description of the mechanisms behind the waste crisis, one of the biggest challenges for creating a more sustainable future. Kamila Pope offers a sharp analysis of how we arrived in this situation and shows how adequate governance could lead us out of it. A paradigm shift in the current dominant economic model and legal order for the use of a more systemic view is wisely proposed. This is an essential read for all working in sustainability issues and environmental policy.'
Dr Michelle Bonatti, Deputy Head, Sustainable Land Use in Developing Countries Department, Leibniz Agricultural Landscape Research Centre (ZALF), Germany; and Lecturer on Environmental Policy, Humboldt University of Berlin

'The issue of global waste management is ripe with challenges of social-ecological justice. This applies to national jurisdictions, but increasingly also to cross-border waste transfers to the Global South. This book casts a sharp eye on the striking absence of systemic legal control instruments. Against this backdrop, Kamila Pope suggests the creation of international regulation oriented towards ecological integrity and dignity of life in an intra- and inter-generational perspective. This book is an alert of a global problem and a handbook for its solution; a must-read for decision makers and scholars alike.'
Christina Voigt, Professor of International Law, University of Oslo

'International trade of useful and valuable products is a well-studied subject. But international trade of substances and objects which are not anymore, or never were, interesting for the owner, is a phenomenon which deserves scientific attention and critical analysis. This is why the reflection on the complex causes and consequences – for people and the environment – of global waste management is so important.

Following exhaustive research, Kamila Pope has gone beyond the mere legal approach in the framework of the *Basel Convention on the Control of Transboundary Movements of Hazardous Wastes and their Disposal*, and has presented a multidisciplinary approach to the global problem of waste management. The visualizing data methods – graphs, charts and diagrams –

included in the book enhance effective communication and contribute to a thorough understanding of its central message: the equity or inequity of international relations among world regions, as a consequence of waste management options.

Reading *Global Waste Management* could help change the mindset of policy makers, decision makers, journalists and public opinion leaders, social and environmental NGOs, companies and all those who are concerned with the effects of current production and consumption patterns and waste management systems between the so-called Global North and Global South regions.

Global Waste Management provides clear insights on the best waste management options necessary to trigger a shift from a linear economy based on a throw-away society, towards a new circular economy paradigm, based on a frugal, responsible and more sustainable society.'

Alexandra Aragão, Professor of Environmental Law, University of Coimbra

'This timely book shows the depth and urgency of the global waste crisis and critiques the liberal, anthropocentric model underpinning current waste management. It makes a well-reasoned case for laws based on eco-centric ethics. Indeed, nothing short of a paradigmatic shift to ecological law and governance is needed to solve the problem of waste and, for that matter, of disintegrating Earth systems (climate, oceans, biodiversity).'

Klaus Bosselmann, Professor, University of Auckland; and Chair, Ecological Law and Governance Association

'This is a book worthy of being read because it brings a fresh perspective to a long-standing problem, the continuously but silently increasing waste crisis. Adopting a socio-ecological perspective and digging deeper to material flow analysis, urban metabolism and environmental justice narratives, the book not only presents the ongoing waste crisis but documents the need for a systemic shift towards a new scientific paradigm that will start from governance. The need for a new global governance pattern that will include not only waste but mainly material flows is one of the key elements of the systemic shift required, and its necessity is one of the most important conclusions of this book. *Global Waste Management* includes valuable suggestions that bridge the gap between the mainstream circular economy discussion and the role of waste management. I strongly recommend this book to any professional, researcher or decision maker who deals with waste management, especially to those who are looking for a fresh perspective and are dealing with global waste flows.'

Antonis Mavropoulos, President, International Solid Waste Association (ISWA)

'This original and far-ranging book shows the reader how waste is both a problem of justice and a legal challenge, especially as it crosses national boundaries from the Global North to the Global South. Pope offers careful insights into the critical case of Brazil while never losing sight of how it fits into larger patterns of the global waste management system. Highly recommended!'

Kathryn Hochstetler, Professor of International Development, London School of Economics and Political Science (LSE)

'*Global Waste Management* is an absolutely timely book on the international waste challenge and governance. I believe that Chinese policy makers, researchers and citizens will find this book thoughtful and helpful for three reasons: First, the book focuses on the hot issue of cross-border waste transfer that China has been fighting against. Second, the book develops a theoretical framework that combines social justice and ecological sustainability and calls for a paradigm shift in this field. Third, the book proposes a new model of global waste governance that will help to tackle the international waste crisis more reasonably and effectively.'

Professor Zhu Dajian, Director, Institute of Sustainable Development and Governance, Tongji University

'Thought-provoking and altruistic, *Global Waste Management* takes a fascinating angle addressing the waste management debacles and waste emergencies that are encountered by many countries. In this highly researched book, Dr Pope provides her insights and models to deal with the waste crisis – something that is highly critical in recent times. She supports the theory that waste is a justice problem, which is spot on and yet often neglected in mainstream schools of thought.

It is well known that there are recurrent transboundary movements of hazardous and other waste, and while the typical discourses usually revolve around technical matters, this book delves deeper to understand the root of the waste problem. As such, it is a quintessential reference for researchers, students, government officials and consultants working in the waste sector around the world. The knowledge gained from this book may provide a base for critical thinking to deal with the waste issues in the readers' respective countries.'

Dr Aretha Aprilia, Environmental Specialist, CDM Smith Europe GmbH, Jakarta

'I first read a work written by Kamila Pope in 2017, when she wrote a book on planned obsolescence. It was highly informative and became required reading for my environmental law students. I approached this new book, on global waste management, with high expectations – especially as waste management is now a thorn in the flesh for national and global policy makers. Kamila explores the problem of global waste and examines the enduring challenge of waste management while our production seems designed to generate continuous waste flows. With in-depth skill and evident knowledge, she provides models that could help policy makers. She also discusses important tools, such as the management of materials flows, which embrace prevention of waste generation in the first place as the most effective way to avoid waste at all. Clearly, Dr Pope is an authority in this field, and in this book has provided valuable conceptual and practical solutions that will remain relevant for a long time. It is a text I highly recommend to students of environmental law, policy, management or governance, as well as to academics and practitioners concerned with waste management across the world.'

Dr Robert Kibugi, Senior Lecturer, School of Law, University of Nairobi

Global Waste Management

Models for tackling the international waste crisis

Kamila Pope

KoganPage

First published in Great Britain and the United States in 2020 by Kogan Page Limited

2nd Floor, 45 Gee Street
London
EC1V 3RS
United Kingdom
www.koganpage.com

122 W 27th St, 10th Floor
New York, NY 10001
USA

4737/23 Ansari Road
Daryaganj
New Delhi 110002
India

Kogan Page books are printed on paper from sustainable forests.

ISBNs

Hardback 978 1 78966 079 1
Paperback 978 1 78966 077 7
eBook 978 1 78966 078 4

British Library Cataloguing-in-Publication Data

A CIP record for this book is available from the British Library.

Library of Congress Cataloging-in-Publication Data

Cataloging-in-Publication Data is available. Library of Congress Control Number: 2020002982

Typeset by Integra Software Services, Pondicherry
Print production managed by Jellyfish
Printed and bound by CPI Group (UK) Ltd, Croydon CR0 4YY

To my mother, Dilcéia, a true source of inspiration. To my husband, love and best friend, Richard. And to our children, Thomas and Luca, light and joy of our days, for whom I hope we can leave a better world.

CONTENTS

Foreword by José Rubens Morato Leite and Melissa Ely Melo xi
Acknowledgements xiv
List of abbreviations and acronyms xvi

Introduction 1

How to read this book: define your own route 7

PART ONE
The socio-ecological problem of waste 11

01 The waste crisis: origins and causes 13
Introduction 13
Waste: a contemporary problem 14
Social metabolism and the transfer of socio-ecological costs 28
Modern Western thinking: the mechanistic paradigm 40
Final considerations 74

02 Waste as a problem of justice 78
Introduction 78
The environmental movement and its schools of thought 81
Some challenges and limitations of the schools of thought 92
The 'what' and the 'who' of justice for socio-ecological issues 106
Final considerations 138

03 Models to overcome the waste crisis 142
Introduction 142
Epistemological and geopolitical paradigm shifts 143
Paradigm shifts in economics 157
Paradigm shifts in law 174
Final considerations 190

PART TWO

A new waste management system for a sustainable and just future 195

04 Waste as a legal problem 197

Introduction 197
Waste: legal concepts and classifications 200
Waste management: legal regimes 227
Final considerations 259

05 Legal regulation of cross-border waste transfer 264

Introduction 264
International regulations: the Basel Convention and the OECD Decision 265
Regional and national regulations: the EU and Brazil 293
Final considerations 321

06 Global governance and the management of waste 325

Introduction 325
Governance of material flows 326
Global waste management: strategies to overcome the global waste crisis 336
Final considerations 367

Conclusion 369

References and further reading 375
Index 386

FOREWORD

by José Rubens Morato Leite and Melissa Ely Melo

It is with great joy and immense satisfaction that we present this work by Kamila Pope. The author began her research for this book perceiving the insufficiency of international legislation currently regulating cross-border waste transfer to promote just and ecologically sustainable global waste management. With the aim of investigating the issue, she developed and used a theoretical framework that combines the notions of social justice and ecological sustainability through (and promoting) the plurality and dialogue between different theories already consolidated. Kamila named this theory 'socio-ecological justice'.

The audacious path taken by the author has as its main goal to propose a model of global waste management, based on the assumptions of socio-ecological justice, and using a critical view of the mechanistic paradigm and the current international legal framework for cross-border waste transfers. Kamila Pope achieves this objective with triumph, firstly presenting a detailed survey of data and statistics, making it possible to understand the causes of the waste crisis. She then indicates the probable biophysical, geopolitical, economic, legal, and epistemological origins of such a crisis, showing that models originating from the mechanistic paradigm have deepened social and geopolitical inequality. As a result, a decrease in the quality of life and disrespect of the biophysical boundaries of the planet could be identified.

Subsequently, she explores the diversity existing within the environmental movement to identify which discourses and strategies show the greatest potential for managing the waste crisis. One highlight of the book is, without a doubt, the analysis of theories of justice carefully selected for the proposition of a theoretical framework of socio-ecological justice that is effective in dealing with socio-ecological issues, including those related to waste. The general assumptions of an *a posteriori* and pragmatic definition of justice are established. In response to the diagnosis of the probable origins of the waste crisis, epistemological, geopolitical, economic, and legal models for overcoming the ecological crises, which encompass the waste crisis, are outlined. That is just the first part of this high-quality book.

The propositional normative analysis of the subject can be found in the second part of the book, based on the theories previously presented and proposed. The book houses a comparative description of waste law of Brazilian, European, and international standards instruments selected by the author. The same legislation is then assessed focusing on cross-border waste transfer to ascertain whether they establish a true global waste management for socio-ecological protection. As a final contribution, strategies for the legal regulation of a system of global waste management are developed from the socio-ecological perspective, that is, a system that complies with the assumptions of socio-ecological justice. The main strategy is the innovative creation of a Governance of Material Flows as a broad and comprehensive notion to contain global waste management.

The research carried out by Kamila leading to this book already holds a quality seal, as a Law Experts Commission evaluated and selected it as the best Brazilian legal thesis of 2019. This prize was awarded by CAPES from the Brazilian Ministry of Education, and Kamila's thesis competed with all PhD theses in law submitted in postgraduate programmes in Brazil. However, this book is essential reading not only for jurists but also for academics and practitioners, from private and public sectors, in areas such as economics, sociology, political science and philosophy, among others.

There are several reasons for claiming the relevance of this book, for making it a must-read for those who are dedicated to solving socio-ecological problems, mainly those related to waste management. Among these reasons are the urgency of the chosen subject (the waste crisis) and the highly original presentation of the facts and the knowledge by the author. In this outstanding book, Kamila Pope has been able to face the immense challenge of transdisciplinary dialogue with unparalleled depth while making the reading both extremely instructive and enjoyable. The charts and diagrams presented throughout the book make it much easier to understand the author's complex diagnoses and propositions.

We also take the opportunity to congratulate Kogan Page for the publication of this brilliant book, whose subject is one of the most relevant in contemporary times. We invite everyone who cares about sustainability and a better future to read it!

Professor Dr José Rubens Morato Leite,
Professor of Environmental Law at the Federal University of Santa Catarina (UFSC),
Coordinator of Environmental Law in the Risk Society Research Group (GPDA)

Professor Dr Melissa Ely Melo,
Professor of Commercial Law at the Federal University of Santa Catarina (UFSC),
Senior Researcher at Environmental Law in the Risk Society Research Group (GPDA)

ACKNOWLEDGEMENTS

Among all the people and institutions that directly and indirectly helped the development of this book, from the research stage to the final production, to which I register all my gratitude, I would like to highlight some of them here. A special thanks to:

My Brazilian University, Federal University of Santa Catarina (UFSC), for the free and quality education, as well as to UNIEDU Program for the fellowship that made it possible to do the research for this book;

My advisor, Professor Doctor José Rubens Morato Leite, for kindly sharing his knowledge, his support, guidance and confidence in my work, and my co-advisor, Professor Doctor Melissa Ely Melo, for her friendship, encouragement and careful work in reviewing and following the development of the contents;

All the professors, professionals and researchers who accompanied me in my journey as a researcher, for the valuable and critical contributions, especially Professor Doctor Maria Alexandra Aragão, whose generosity, care and availability in sharing her ideas and knowledge were invaluable;

My dear friends of the Environmental Law and Political Ecology in the Risk Society Research Group (GPDA), for the rich debates and collective learning;

All the Kogan Page team, particularly Julia Swales and Adam Cox, for their competence in the completion of this book;

My families – Guimarães, Moraes and Popes, for their love and support;

My sisterhood – Renata Guimarães Reynaldo, Michelle Bonatti, Larissa Hery Ito, Patrícia Biava, Patrícia Minck, Luana Debatin, for their sorority;

My dear friend Cilene Silva, for her help in making possible the writing of this book and for devoting a lot of love and care to my most precious people, Thomas and Luca;

My mother, Dilcéia Gonçalves Guimarães, for her example and always being by my side;

My husband, Richard Pope, for working with me as a team, bringing lightness and joy to my days, inspiring me with his strength, resilience, intelligence and humility and, above all, for his unconditional love;

And, finally, to my treasures, Thomas and Luca Pope, for making my journey more meaningful, allowing me to experience the greatest and most intense love of my entire life and making me happy every day.

To all these people and institutions, my enormous gratitude!

LIST OF ABBREVIATIONS AND ACRONYMS

ABNT	Associação Brasileira de Normas e Técnicas (Brazilian Association of Technical Standards)
ABRELPE	Associação Brasileira de Empresas de Limpeza Pública e Resíduos Especiais (Brazilian Association of Public Cleaning and Special Waste Companies)
BAT	Best available technology
CONAMA	Conselho Nacional do Meio Ambiente (Brazilian National Environmental Council)
COP	Conference of the Parties
CSWM	Community Strategy for Waste Management
EC	European Community
EEA	European Economic Area
EEA	European Environment Agency
EFTA	European Free Trade Association
EMF	Ellen MacArthur Foundation
EU	European Union
EWC	European Waste Catalogue
GDP	Gross domestic product
GNP	Gross national product
GWMO	Global waste management outlook
HELP	High level of environmental protection
HS	Harmonized Commodity Description and Coding Systems
IBAMA	Instituto Brasileiro do Meio Ambiente e dos Recursos Naturais Renováveis (Brazilian Institute of the Environment and Renewable Natural Resources)
IBGE	Instituto Brasileiro de Geografia e Estatísticas (Brazilian Institute of Geography and Statistics)
INMETRO	Instituto Nacional de Metrologia, Qualidade e Tecnologia (Brazilian National Institute of Metrology, Quality and Technology)
ISO	International Organization for Standardization
ISWA	International Solid Waste Association

ISWM	Integrated Solid Waste Management
IUCN	International Union for Conservation of Nature
LCA	Life cycle assessment
LDCs	Least developed countries
LULUs	Locally unwanted land uses
MFA	Material flow analysis
NCM	Mercosur Common Nomenclature
NGO	Non-governmental organization
NI	Normative Instruction
NIMBY	Not in my backyard
NPP	Net primary production
NSWP	National Solid Waste Policy
OECD	Organisation for Economic Cooperation and Development
PET	Polyethylene terephthalate
POPs	Persistent organic pollutants
UN	United Nations
UNEP	United Nations Environment Programme
USA	United States of America

Introduction

In 2009, a substantial part of the Brazilian and the international press covered the discovery of at least 89 containers with more than 1.62 tons of hazardous waste (UNEP, 2009). The Brazilian authorities found, among other items, syringes, used disposable nappies, chemical toilets and batteries, all coming from the UK. That was the first time I had heard of cross-border waste transfer.

From that moment, I started to be aware of other cases, such as the illegal transfer of toxic waste carried out by a European multinational company to the Ivory Coast three years earlier. This case of cross-border waste transfer resulted in at least 17 deaths and thousands of cases of poisoning. The health problems affecting the local communities, as well as the irreversible ecological impacts on local ecosystems of that region, were explicit. Other cases of illegal waste transfer will be described more fully in Chapter 1.

To me, it has become clear that the rise in global waste production has generated serious ecological, social and public health problems, challenging the existing methods of waste management hitherto adopted, and representing a real waste crisis.

For various reasons, waste is increasingly being transferred to final destinations other than its place of origin. Such transfers are creating two problems: they are deepening the waste crisis, and generating new social injustices. As we will read in the course of this book, the first issue arises because the disposal of waste in areas remote from its origins makes people and organizations lose any concept of the real impacts and consequences of the high production of waste.

This lack of perception, coupled with the logic of infinite economic growth, reinforces the model of linear production and consumerism, aggravated by the ever-increasing generation of hazardous waste. These mountains of waste are accumulating in the ecosphere in a way never before experienced, and our inability to manage it in an efficient and sustainable way becomes more and more evident.

Just as in the overall ecological crisis, the waste crisis affects different communities and countries in distinct ways. Factors such as economic and political power as well as 'level of development' influence the way the waste problem impacts different populations. In addition to deepening the waste crisis, waste transfer generates social injustices arising from the fact that many of these waste movements are destined for communities, regions or countries with low economic, political and/or informational power. They therefore cause serious social and ecological problems at those receiving sites which carry the negative costs of current development and economic models disproportionately. These countries will be referred to in this book as 'Southern countries' or the global 'South'.

The adoption of the global North–South approach is mainly inspired by the theoretical work of Boaventura de Sousa Santos and, as will be seen in Chapter 1, it is more of a geopolitical approach than a geographical one. While the term 'Northern countries' will be used as a synonym for 'developed countries', the term 'Southern countries' will represent countries classified as 'developing'. The decision to use this approach is justified by the fact that this book was written by a Latin-American researcher, located geopolitically in the global South. Thus, the book is written from the global Southern point of view, to which increasingly large amounts of waste are destined and where important ecological and social impacts have been identified.

Historically, this book has been written at a moment in which international power relations, for various reasons, are still unbalanced. The global South currently occupies a position of vulnerability and submission in relation to the North, disproportionately bearing the negative consequences, whether ecological, social or economic, of current development and economic models.

This is the context from which this book started. However, the models and strategies that will be developed and proposed in the following chapters will take into account the possibility of changes in the current geopolitical global equilibrium (or imbalance). It will thus consider the possibility that countries currently classified as developing could assume roles of domination in the future. It is believed that, with due contextualization and recognition of the place occupied by the author of this book, it would be possible to think of long-term proposals that could be applied against the background of possible global geopolitical changes.

Regarding current regulations on the subject of waste, the Basel Convention is the most comprehensive in the international context. It seeks to control the cross-border movement of hazardous waste. Yet Southern countries, whose national legislations are usually more flexible and/or whose inspection structures are more precarious, continue to receive large quantities of waste that do not fall into the category of 'hazardous', as well as hazardous wastes camouflaged as other categories. In light of these findings, the critical need for investment in studies that help to overcome the current waste crisis is clear. A more detailed investigation of the operation of cross-border waste transfer is much needed.

The decision to focus on this type of operation may give the impression that an 'end-of-pipe' approach will be adopted in this book, ie that it will present solutions solely to define the best destination, in ecological and social terms, for the waste already generated. However, it is important to clarify this will not be the approach adopted. The current waste crisis cannot be overcome only by finding the correct destination for the waste already generated. On the contrary, the prevention of waste generation must always be at the centre of any proposal that dares to present possible ways of overcoming the crisis.

On the other hand, it is also known from the teachings of the law of entropy and of social metabolism that there will always be some waste production, given the biophysical impossibility of complete closure of the production cycle. In this sense, the theoretical developments that will be investigated in this book, as well as the strategies presented in its final chapter, will always have as their top priority the prevention of waste generation and, when this is not possible, the best possible destination for the waste in socio-ecological terms. Thus, this book will adopt an eminently preventative approach to the promotion of socio-ecological justice for present and future generations, of both humans and non-humans.

Unfortunately, there are few academic studies addressing the issue of cross-border waste transfer which link the perspectives of justice and sustainability (to be realized here through the notion of 'socio-ecological justice'). This is even more evident when we discuss waste transfers which are considered legal, such as those of waste classified as non-hazardous for recovery. This particular issue shows how relevant this investigation is, as it raises a number of important questions. Among them, one in particular has generated great concern, and represents the central problem to be answered by this book: does current international legislation to regulate the cross-border transfer of waste promote a global waste management system that is just and ecologically sustainable?

The research conducted for this book started from the initial hypothesis that international standards and regulations have proved insufficient to promote the establishment of a just and ecologically sustainable global waste management system. It was soon realized that for such a hypothesis to be properly tested, then confirmed or rejected, it would be necessary to use a theoretical framework that aligns the notions of social justice and ecological sustainability. From the outset I was faced with the difficulty of locating such a theoretical framework. In fact, I realized that, although some social movements and even certain authors have worked in conjunction with justice and sustainability, they did not present sufficiently advanced developments for the theoretical structuring of this framework.

This was the first great challenge of the book. The strategy used to address the challenge was to search for a way, via discussion of and engagement with the many different and already established theories, to construct the theoretical framework presented here, called 'socio-ecological justice'. As a result, an overall objective was established for the book: to propose a global waste management model based on a theory that unites social and ecological demands. This model will also be constructed from critiques of the mechanistic paradigm and the current international legal framework that regulates cross-border waste transfer.

Within this scope, specific objectives were established. In Part One of the book, the following objectives were pursued:

- collection of data and statistics to understand the origins of the waste crisis;
- investigation of possible biophysical, geopolitical, economic, legal and epistemological causes of the waste crisis;
- study of the diversity existing within the environmental movement to verify which discourse(s) and strategy(ies) show greater potential to deal with the waste crisis;
- analysis of selected theories of justice to propose a theoretical framework of socio-ecological justice that is adequate to deal with the complex socio-ecological issues, including waste problems;
- search for possible epistemological, geopolitical, economic and legal models to overcome the ecological crises, in which the waste crisis is contained.

For the development of the second part of the book, focused on normative analysis and with a propositional bias, the following objectives were pursued:

- a comparative description of selected Brazilian, European and international legal instruments on waste;
- evaluation of these same instruments in terms of cross-border waste transfer to ascertain whether they establish a true global waste management system for socio-ecological protection;
- development of strategies for the legal regulation of global waste management that is more appropriate from the socio-ecological point of view, fulfilling the assumptions of socio-ecological justice.

This book intends to go beyond a mere analysis and understanding of the subject, as it will seek, through theories already developed together with those proposed here, possible legal solutions to rethinking and changing the current practice of cross-border waste transfer. The main focus will be on the protection of ecological integrity and of dignity of life for current and future generations of humans, non-humans, nature and its inter-systemic relationships. The proposals presented in the final chapter will offer guidance on their practical application, which may provide inspiration for the negotiation and drafting of regulations on global waste management.

Waste, as well as being a socio-ecological issue, represents a complex theme, involving multiple levels of reality, governed by different logics. In this sense, it is believed that an approach based merely on individual disciplines – in the case of the author, the law – would be insufficient and too reductionist to deal with all the nuances of the subject. On the other hand, neither a multi-disciplinarity nor an inter-disciplinarity approach would comply with the proposition of this book, since it does not aim at simply analysing the waste problem from different points of view (multi-disciplinarity), or using a method of analysis that has been transferred from another area of knowledge (inter-disciplinarity).

Although it is quite challenging, this book will seek to adopt a transdisciplinary approach, which, according to the Charter of Transdisciplinarity, is multi-referential and multi-dimensional; it contextualizes, at the same time that it does not exclude, the existence of a 'trans-historical' horizon. In addition, rigour in debate, taking into account all the data collected, openness to accept the unknown, the unexpected and the unpredictable, as well as tolerance of opposing ideas and truths, are fundamental characteristics of the transdisciplinary attitude and vision (Morin *et al*, 1999).

Assuming these elements, this book will investigate the waste crisis, especially the issue of cross-border transfers, through the study of, discussion about and engagement with theories from different fields of knowledge (such as law and economics), to identify what connects (ie themes of common interest), crosses (ie ideas applicable to all) and surpasses (ie concepts generated from cross-boundary dialogue) them. In order to contribute to the advancement of the debate on waste, notably regarding the establishment of global waste management systems, strategies will be developed based on the new data and the possibilities that emerge from the different knowledge bases presented in this book – particularly in the final chapter, where the transdisciplinary and transnational bias will become even more evident. It is important to note that, according to Cilliers and Nicolescu (2012), neither complexity nor transdisciplinarity are neutral; they will always have an underlying ethical dimension. It is therefore important to clarify that the transdisciplinary and complex approach adopted in this book will come from a socio-ecological ethic.

Since many of the cross-border waste transfers flow from the global North to the global South, the scope of the methodology, in addition to the study of international legal regulations on the subject, also includes an analysis of the legislation established both at the European level, as many members of the European Union are waste exporters (representing the global North), and also in Brazil, a relevant recipient of waste in the international context (representing the global South). This book will therefore discuss local, regional and international legislation for a better understanding of the need to establish new strategies for global waste management.

In addition, recognizing the existence of different classifications of waste (hazardous and non-hazardous, urban, industrial, hospital, etc) in the collection of data and analysis of selected legislation, it is important to highlight that the proposals to be presented in this book will have a wide applicational scope, leaving space (openness) for the debate on the specifics of different types of waste in individual cases. It is not intended to establish absolute truths in the proposed strategies, but to present them with theoretical rigour, while at the same time remaining open to debates on areas of uncertainty.

The transdisciplinary and transnational elements are of fundamental importance for this book to make sure it deals in sufficient detail with the complexity of the subject. In terms of methodology, the research approach will be based on the principles of complexity, such as transdisciplinarity and transnationality. The procedural method will be the monographic one.

In terms of research techniques, the investigation will initially take the form of a literature review aimed at compiling the theoretical framework for the discussions on justice, environmental and ecological justice, Southern epistemologies, complex thought, ecological economics, productive ecology, circular economy, ecological law and waste law.

The bibliographical research will involve a literature review of national and international books and journals, as well as the gathering of documents, treaties, conventions, directives, regulations, laws, normative instructions and other sources that address the subject of waste at the international as well as the European and Brazilian levels.

In terms of theoretical frameworks, this book will have as its main foundations the theories of justice of Nancy Fraser and Martha Nussbaum, the theories of ecological justice of David Schlosberg and Klaus Bosselmann, as well as the theory of materials flows law of Maria Alexandra Aragão. It is important to emphasize that these will be the guiding threads of this book, but the work of other authors with different areas of knowledge will also be discussed – such as Southern epistemologies of Boaventura de Sousa Santos, complex thought of Edgar Morin, ecological economics, productive ecology of Enrique Leff, circular economy, as well as ecological law of Fritjof Capra, Ugo Mattei and François Ost – to ensure the holistic treatment of the theme.

How to read this book: define your own route

The book will be divided into two parts, each with three chapters. Part One will have a more theoretical bias, while Part Two will take a more legal/practical perspective. All chapters contain discussions and investigations on the subject of the global waste crisis and its management from different perspectives.

Chapter 1 will set the problem in context, dealing with the waste crisis, its possible causes and origins. Firstly, it will seek to provide empirical data on the generation and cross-border transfer of waste, and which is capable of demonstrating the existence of a waste crisis. The chapter will also explore the biophysical causes and epistemological, geopolitical, economic and legal origins of this crisis.

Chapter 1 then presents a more detailed, abstract and theoretical debate on the possible roots of the ecological and waste crises. If this is the first time you have come across these issues, the second part of Chapter 1 might be a more challenging and slow read for you. However, if you are interested in exploring these roots to better inform your thinking on the subject's big picture, you should begin reading from Chapter 1 and take your time to process all the information presented there.

If you feel too pessimistic after having read Chapter 1 and cannot wait to read about some possible solutions, you may choose to move straight to Chapter 3, which presents the main theoretical models developed to overcome the problems outlined in the first chapter. Chapter 2, especially its later sections, brings a more abstract, detailed and conceptual debate that can be more challenging for those not familiar with such arguments. However, it must be said that all theoretical models presented in Chapter 3 are directly related to the critical debate featured in Chapter 2, so if you prefer to follow the reasoning going through the sequence of chapters of this book, you are advised to read Chapter 2 before Chapter 3.

In Chapter 2, we will have the opportunity to understand why waste has become a problem of justice. Attention will be given to the diversity of the environmental movement in some of its streams and schools of thought, in order to highlight its potentialities and limitations to deal with the many ecological crises (or 'polycrisis' as Morin refers to them) in general and with the waste crisis specifically. The choice of 'justice' as the topic of discussion for socio-ecological issues does not mean other topics will be excluded from consideration. Rather, they will be addressed in order to develop a theoretical framework, here called 'socio-ecological justice', that seeks to emphasize the most promising strategies of different environmental streams. Theories of justice selected for their potential application on socio-ecological issues will be evaluated, pursuing the definition of the scope of application of this justice (the 'what' of justice), as well as the subjects of this justice (the 'who' of justice).

Finally, a model of socio-ecological justice will be proposed in Chapter 2 as a theoretical framework for the models of global governance and management of waste developed in the last chapter of the book (Chapter 6). Hence the importance of reading Chapter 2 if you would like to better visualize where this model of socio-ecological justice comes from and how it was theoretically built. However, if you are happy to check how it can be applied to waste governance and management alone, you can move directly to Chapter 6.

Chapter 3 presents some of the main theoretical models developed to overcome the problems raised in the first chapter. Such theories have as a shared goal the aim of overcoming the mechanistic paradigm and, ultimately, the ecological crises, in which the waste crisis is contained.

Part Two of the book has a more digestible content for readers focusing on the legal regulation of waste, either for research or practical use purposes. Therefore, if you are more interested to get an overview of how waste has been regulated with all its complexities, you can start your reading from Chapter 4, which focuses on the reasons why waste has also become a legal problem. The regulation of waste will be studied in Brazilian and European Union legislation, and at the international level, the Basel Convention and the OECD Decision will be analysed. It will investigate, in a comparative way, the legal concepts of waste, the classifications regarding the hazardousness of waste, the concepts, subjects, responsibilities, control instruments and hierarchies of waste management in all these legal instruments.

In Chapter 5 these legal instruments will be analysed in the context of cross-border waste transfer, a key operation for the classification of waste management as global. These instruments will be evaluated to verify whether they adopt an economistic perspective, treating the theme as international waste trade, or adopting a socio-ecological perspective, establishing rules for a global waste management system. If your reason for reading this book is to get a critical assessment of the current international regulations on waste transfer, this is your target chapter, together with Chapter 6, which proposes an idealized model of international regulation that would place ecological sustainability and social justice at the core of its objectives.

All the discussions presented throughout the book will finally come together in Chapter 6. Using the theories proposed in the first part of the book, this final chapter will present possible strategies for overcoming the specific problems pointed out in earlier chapters, for developing regulations that are appropriate to the establishment of a global waste management system as a legal instrument that will help overcome the current waste crisis from a socio-ecological perspective.

The strategies presented in Chapter 6 can be read and understood in isolation. However, regardless of your interests, background or reason for reading this book, they will make more sense if you read and absorb the entire book.

PART ONE

The socio-ecological problem of waste

01

The waste crisis

Origins and causes

If waste is the physical memory of the people, what people of sad memory we will be – vast and long memory, a heavy heritage of a lifestyle of use and abuse of Nature. If, as archaeologists say, waste is a sign of social content, what kind of society would this have been that, instead of producing waste to live, it seems to live to produce waste?

ALEXANDRA ARAGÃO

Introduction

At the end of 2012, the United Nations (UN) declared we are living through a genuine 'global waste crisis'. Not long ago, however, waste was not a matter of concern. The international debate only began to highlight this crisis from the 1990s. Questions arise from this observation from the UN, such as 'Why is humanity believed to be facing a waste crisis?' as well as 'What are the possible origins and causes of this crisis?'

The main objective of this chapter is to collect data, statistics and commentary on waste in recent decades. It is important to stress that the figures presented are only examples. An exhaustive survey of all the data of all types of waste is not the aim of this book, mainly due to time and spatial limitations. As will be seen from the data on waste generation, for example, the main focus will be on municipal solid waste figures, since they can be more easily accessed and verified. The numbers and information presented are nevertheless sufficient to demonstrate the waste crisis, even if they reveal only the tip of the iceberg.

Following the data collection, through the social metabolism approach, the biophysical reasons explaining the waste crisis will be explored. Finally, the dominant epistemological, geopolitical, economic and legal bases will be evaluated, which, at a deeper level, may have generated and deepened the overall ecological crisis, of which the waste crisis is a part.

Waste: a contemporary problem

Waste generation

From the outset of our existence, humans have generated and needed to dispose of waste. However, waste was not a significant source of damage or social and ecological conflicts and claims through much of our history as a species (Aragão, 2006). According to Aragão, in the Palaeolithic period, about 20,000 years ago, human populations lived exclusively from hunting, fishing and collecting plant foods. Known anthropologically as 'hunter–gatherers', the waste they generated was small in quantity and innocuous in quality, since the materials used by these people were essentially natural and rarely processed.

Hunter–gatherers were typically nomadic and used to move away from the waste they generated. This logic only changed about 10,000 years later, with the Neolithic Revolution. From this point, populations started to gradually settle in fixed places, mainly as a result of the development of agriculture and the domestication of animals. In addition to other important consequences, sedentism and a rapid increase in human populations occurred in the Neolithic period, giving rise to the need to create specific sites for waste disposal, away from human households (Aragão, 2006).

In the last phase of the Neolithic period, known as the Metal Age, approximately 6,000 years ago, the discovery of metal led to the development of new instruments. This caused not only the diversification of human activities but also the increase in waste generation and its durability. Fast forward 4,000 years or so and the conditions for the beginning of urbanization arose with the increase of agricultural productivity, the creation of surpluses and the resulting trade and professional specializations. The population density and type of consumption of the cities of the classical world, in the Greek and Roman Empires, around 2,800 years ago, finally turned waste into a subject of social concern (Aragão, 2006).

In the Middle Ages, the emergence of the artisanal economy, the development of trade, and the drastic increase in population made waste disposal a bigger issue. Coupled with the absence of any waste management system and the habit of throwing waste into the streets, it led, 650 years ago, to the spread of the Black Death in Western Europe and the death of approximately half the population within four years. With the Industrial Revolution and the phenomenon of rural exodus, there was an intensification of urbanization and the problems of space scarcity, conflicts, and damage caused by waste. Waste management had become increasingly challenging, and waste law emerged as urban law (Aragão, 2006).

Finally, about 60 years ago, from being an issue resolved by urban law, waste became a problem of environmental law. The different types of waste coming from different sources were now transferred and deposited together in places increasingly distant from their sources, and with no indication of their origins. Consequently, pollution and cumulative damage got worse, and those responsible for this damage and its victims could not be determined. The damage was now being mass produced and increasingly difficult to prove, no longer fitting into the civil legal schemes of establishing causation. Its current characteristics are typical of a new type of society: the Risk Society (Aragão, 2006).

On the Risk Society theory, see the works of Ulrich Beck.

From the second half of the 20th century, the global ecological crisis has been facing a new issue: the waste crisis. The constant and exponential increase of the world's human population is a decisive factor for this problem. However, as will be seen in the course of this chapter, it is not the only one. Humanity is currently faced with its own inability to sustainably manage waste generated in ever-increasing quantities and of increasingly hazardous quality.

In 2012, the world generated more than 1.3 billion tonnes of municipal solid waste, and future forecasts are not encouraging. The World Bank argues that by 2025, more than 2.2 billion tonnes of waste will be generated per year (Hoornweg *et al*, 2012). On the other hand, statistical studies conducted by the European Union show the average amount of municipal solid waste generated per person in European countries in the last decade has been relatively stable: in 2007 this average was 523 kg per capita. By 2017, this number had dropped to 486 kg (Eurostat, 2019a).

> All official numbers on waste generation and management in the EU are available on the Eurostat webpage: ec.europa.eu/eurostat

These figures indicate that the increase in waste generation predicted by the World Bank will occur more intensively in developing countries. By analysing the data for each European country separately, it remains clear the average waste generated by Western Europe is much higher (in many cases double) than in other European countries (Central and Eastern Europe), as well as most developing countries. For instance, while Denmark had a small decrease in waste generation from 790 kg per capita in 2007 to 781 kg in 2017, Poland, which generated 322 kg of waste per capita in 2007, reduced to 315 kg in 2017. Switzerland showed a fall in waste generation from 720 kg per capita in 2007 to 706 kg in 2017. These figures are halved in Latvia, which generated 391 kg per capita in 2007 and 438 kg in 2017 (Eurostat, 2019a).

In Brazil, surveys indicate in the year 2000, around 161,827 tonnes of municipal solid waste were generated each day, about 0.95 kg per capita per day, resulting in the disposition of just over 59 million tonnes of this type of waste in the year. These surveys report that Brazilian cities with up to 200,000 inhabitants collected between 450 and 700 gm of solid waste per capita per day, while in cities with more than 200,000 inhabitants this amount varied between 800 gm and 1.2 kg per capita (IBGE, 2001).

These numbers demonstrate that large urban areas generate a more significant amount of waste when compared to smaller cities or rural areas. Many variables can provide reasonable explanations for this, such as:

- i higher processing of products marketed in these large urban centres (mainly through overpackaging);
- ii higher levels of consumption due to greater purchasing power, among others.

Research conducted by the IBGE and ABRELPE show that, in 2016, the amount of municipal solid waste generated in Brazil increased to more than 78 million tonnes, ie 214,405 tonnes per day; around 1.04 kg per capita per day (ABRELPE, 2017). The country's municipal solid waste generation, therefore, had an increase of more than 19 million tonnes each year from 2000 to 2016. The annual generation per capita increased from 346.75 kg to 379.6 kg, ie each Brazilian generated about 33 kg of waste more than the previous average.

These numbers are a matter of concern and represent only part of the problem since they report only the municipal waste collected. From the extraction of natural resources, through to the production process, transportation, and other links in the chain, a vast amount of waste is generated across diverse sectors (industrial, real estate, transportation among others). Thus, the numbers highlighted can be considered quite literally the 'tip of the iceberg' in terms of waste generation.

In addition to the larger quantities of waste, humanity faces another new reality: the generation of increasingly hazardous waste. As mentioned, the waste generated by the risk society presents new elements that magnify the damage and risks to the environment. Currently, nuclear waste, electronic waste, hospital waste and other types of toxic and infectious waste are generated, such as those from pesticides.

Take electronic waste for example. It is estimated that around 44.7 million metric tonnes were generated worldwide in 2017, with projections for this number to grow to 120 million tonnes annually by 2050 in the worst-case scenario (PACE, 2019). Statistical data raised by the studies on the subject make clear the relationship between high generation of electronic waste and the increased gross domestic product (GDP) of the country. For instance, in 2014, while 1.9 million tonnes of electronic waste were generated across the African continent (1.7 kg per capita), Europe generated 11.6 million tonnes of waste (15.6 kg per capita) (Kumar *et al*, 2017).

The increasingly hazardous waste has been generated and disposed of at a growing rate. Together with other types of waste, it contains high levels of pollutants and toxins, causing soil, water and air contamination, among other types of damage to public health and the environment (Slade, 2006). These numbers are worrying and illustrate why waste is a contemporary problem and has been a recurring topic in international debates and agendas over the last few decades.

Waste transfers

Faced with these exponential rises, both quantitative (larger amounts of waste) and qualitative (greater variety of waste – hazardous and non-hazardous), waste has been transferred to final destinations increasingly distant from its origin, and cross-border waste transfer has become a reality. According to Kellenberg (2015), the global waste trading market has seen dramatic growth between 1992 and 2012, rising from 45.6 million tonnes in 1992 to over 222 million tonnes in 2011; an increase of over 500 per cent in two decades.

In that period, not only did the amount of waste transferred increase but also its destination underwent major changes. During the 1990s, up to 1997, when volumes of waste transferred internationally were still relatively low, much of the waste was destined for developed countries, with developing countries receiving only 18.7 per cent. Since 1998 however this trend has changed, and the amount of waste sent to developing countries has started to increase substantially, growing more than 40 per cent by 2009. Not only has there been an extraordinary increase in the amount of waste transferred to different countries in recent decades, but also a constant expansion of these transfers to developing countries, ie to Southern countries (Kellenberg, 2015).

Perhaps the (now) famous confidential memorandum sent in 1991 by Lawrence Summers (the chief economist of the World Bank at that time), defending cross-border waste transfer (especially to developing countries), may offer some explanation to this trend. He said:

> Just between you and me, shouldn't the World Bank be encouraging MORE migration of the dirty industries to the LDCs [least developed countries]? [...] I think the economic logic behind dumping a load of toxic waste in the lowest wage country is impeccable and we should face up to that. [...] I've always thought that under-populated countries in Africa are vastly UNDER-polluted [...].
>
> SOURCE Summers, 1991

It seems the logic of 'economism' used by Summers prevailed. So much so that many cases of cross-border waste transfer from countries of the North to countries of the South, and the damage they were causing, began to be reported in the late 20th century. In the 1980s, one of the most famous cases was the *Khian Sea* ship incident, which carried more than 14,000 tonnes of toxic ash from incinerators located in the United States. Several countries (Bahamas, Dominican Republic, Honduras, and Panama, among others) rejected such waste, which ended up stranded, sailing for months across the Atlantic Ocean (Greenpeace, 2010).

> The term **economism** (or 'economicist') is used here in a sense of criticizing economic reductionism, that is to say, the reduction of all social facts to economic dimensions, not taking into consideration other important elements, such as environmental protection and social justice.

Finally, in 1988, the ship's crew unloaded 4,000 tonnes of the waste on a Haitian beach, falsely classifying it as fertilizer. Subsequently, *Khian Sea* attempted to unload the rest of the waste in different North African and Asian countries, and the remaining 10,000 tonnes of incineration ash mysteriously disappeared on the route between Singapore and Sri Lanka at the end of 1988. The case of the *Khian Sea* shocked the world and was one of the leading cases that drove the start of negotiations for the creation of the Basel Convention (Greenpeace, 2010).

In addition to promoting the negotiation and approval of such a Convention, the United Nations, through its Human Rights Council, in 1995 designated for the first time a Special Rapporteur to examine the human rights implications of management and disposal of hazardous substances and wastes, including cross-border transfers. Such special rapporteurship still exists and has accompanied, through visits, events and the use of information and public participation instruments, cases of human rights violations involving hazardous waste. It has also been collecting information, advising and pressuring the countries involved in such cases to take the necessary measures.

With the entry into force of the Basel Convention in the early 1990s (which will be further explored in Chapters 4 and 5), cross-border transfer of hazardous waste is now regulated, with the legal definition of illegal traffic of waste. Unfortunately, however, in almost 30 years, only seven cases have been officially confirmed by the Convention Secretariat. It is interesting to note that Brazil was the destination in three of the seven confirmed cases, a fact that demonstrates the importance of the topic for the country, which seems to be a significant part of the international waste trade route.

> For updated information on officially confirmed cases of illegal traffic of waste, consult the Basel Convention webpage:
> www.basel.int/Implementation/LegalMatters/IllegalTraffic/CasesofIllegalTraffic

The first case reported by Brazil was in 2009, involving three loads of waste illegally transferred to the country, captured in the ports of Santos and Rio Grande. Although the exporter described them as 'ethylene polymers and plastic wastes intended for recycling', a total of no less than 89 containers were identified, loaded with approximately 21.6 tonnes of municipal solid waste mixed with plastics and other polymers in the process of decomposition (UNEP, 2009). This waste was exported from the United Kingdom, with transit through Spain, Belgium and Germany, to Brazil, its final destination. As a result of this transfer being caught by the Brazilian authorities, all the companies involved were notified of the immediate return of the waste to the United Kingdom under the Basel Convention terms. They were penalized, receiving fines of US $419,000 under Brazilian legislation (UNEP, 2009).

The second case reported by Brazil was in 2011 as a result of the discovery by the Brazilian authorities of six containers containing approximately 61,000 kg of PET plastic bottles mixed with organic materials in an advanced stage of decomposition. However, in the trade description, the material transferred was classified by its exporter as plastic bottles for recycling. The cargo was exported from Spain directly to Brazil and was identified at the Port of Itajaí. Following the rules of the Basel Convention, the companies involved were notified of the immediate return of the waste to the exporting country, Spain, as well as being fined 2 million Brazilian reais, under Brazilian legislation (UNEP, 2011c).

The third and final case reported by Brazil occurred in 2012 with the discovery, again in the Port of Itajaí, of an export from Canada, with transit through the United States, of two containers with approximately 41 tons of used plastic bags and scraps, all contaminated with decomposed organic matter. The exporter had described these wastes as being used supermarket plastic bags destined for recycling. The Brazilian authority informed the Secretariat of the Convention that they would notify the exporter of the immediate return of waste to Canada and would impose a fine under Brazilian legislation (UNEP, 2012a).

Although only seven cases have so far been confirmed by the Secretariat of the Basel Convention as cases of illegal traffic of waste, many others have been and continue to be reported from different sources since the Convention was adopted. In Brazil, although there is no official compilation and systematization of the data, it is estimated that, by 2014, Brazilian official bodies identified at least 54 incidents of waste traffic, encompassing illegal imports and exports by sea and land (Öberg, 2016).

Of the 54 cases, 17 were of waste imported by sea (shipped to Brazil), 4 exported from Brazil to different countries and 34 imported by land to Brazil. All waste imported by land was identified in the south of Brazil, coming clandestinely from Uruguay, though the origin of the waste was omitted. In turn, 15 of the 17 identified maritime import cases came from developed countries (Europe, USA, and Canada). The two remaining cases of illegal transfer came from Mexico, although there is no information on the complete route of the waste, which means they may have originated in other countries (Öberg, 2016).

Öberg (2016) points out that the first case of illegal import of waste reported by the Brazilian authorities occurred only in 2001, ie almost a decade after the signature and internalization of the Basel Convention by the country. It does not mean there were no cases before that year – on the contrary, many cases were identified by different Brazilian agencies and even by the media. However, the country did not have clear rules and the operational infrastructure to cope with such occurrences.

Besides, even after 2001, underreporting of cases of waste traffic is evident in Brazil. It was only in 2010, with an increase in inspection efforts, that more cases were officially identified. Of all the cases reported by the Brazilian authorities, only 11 were returned to their country of origin, meaning the rest were disposed of in Brazil (Öberg, 2016).

After the first identified cases, the Brazilian authorities concluded that the procedures indicated by the Basel Convention were slow and ineffective. In one case, the company importing the waste was prepared to return the cargo to the exporting country but was prevented from doing so due to the lack of response from the Secretariat of the Basel Convention. As a result, the illegal waste loads were not returned and, after 11 years of waiting at the Port of Santos, were disposed of in Brazil, with the costs borne by the Brazilian government (Öberg, 2016).

The analysis of the cases officially identified by the Brazilian authorities shows the return of waste to the exporting countries became more effective with the decision of the Brazilian government to carry it out without the officialdom of the Basel Convention. By 2014, for example, eight reports of illegal traffic of waste were made to the Secretariat of the Convention, but only three have been officially recognized. It illustrates the problems highlighted by the Brazilian authorities regarding the procedures of the Convention (Öberg, 2016). Other limitations can be identified, such as the impossibility for Brazilian authorities to punish exporters from other countries who transfer illegal wastes to Brazil and the lack of a joint international effort to identify illegal waste flows and to take measures to curb them.

Not all wastes illegally transferred to Brazil are classified as hazardous. By 2014, for example, Brazilian public authorities identified the illegal entry of almost 540 tonnes of hazardous waste in Brazil. However, in the same period, more than 1,745 tonnes of plastic wastes were illegally transferred to the country (Öberg, 2016). Until April 2019 plastic waste was not classified as hazardous by the Basel Convention even with all the negative ecological impacts it has been causing worldwide.

PROBLEMS OF THE PLASTIC ERA

Plastic waste has quickly become a global problem that has generated unprecedented impacts. Thousands of marine animals are dying because of the immense amount of plastic in the oceans. In addition to islands of plastic waste floating for more than dozens of years in all the oceans of the planet, plastics and microplastics have been found in the most unlikely places, such as in polar glaciers, in drinking water consumed by humans and non-humans, as well as in fish and other marine animals consumed by humans. It is estimated that by 2050 there will be more plastic than fish in the oceans of our planet. This plastic crisis has led many international environmental organizations, such as UNEP and IUCN, to launch campaigns to reduce and ban plastic use. Faced with strong evidence of the magnitude of the problem, in 2018 *National Geographic* magazine released a specific issue on the subject.

It was only in May of 2019 that some categories of plastic waste were added to the list of waste requiring special consideration ('other waste' – Annex II) and to the list of hazardous waste (Annex VIII) of the Basel Convention. However, such amendments will only enter into force (become effective) from 1 January 2021. Still, plastic waste that is 'almost' free from contamination and is destined for recycling in an environmentally-sound manner is still listed as non-hazardous. Apart from these amendments, COP14 also created a partnership on plastic waste and decided to update the technical guidelines for the identification and environmentally-sound management of plastic wastes and for their disposal.

For more information on the latest actions taken by the Basel Convention on plastic waste, see: www.basel.int/Implementation/Plasticwastes.

The data recorded by the Secretariat of the Basel Convention does not reflect the reality of cases of waste traffic. Focusing on Brazil alone, the Secretariat of the Basel Convention recognized only 3 of the 54 cases officially identified by the Brazilian authorities. The number of officially reported cases made by Brazil also represents numbers far below reality due to the lack of infrastructure and efficient inspection systems in the country. It is believed the situation is replicated in most countries around the world, particularly Southern countries.

With few exceptions, most of the occurrences of illegal trafficking of waste identified in Brazil have not reached the public's attention, either nationally or internationally. In terms of illegal trafficking of waste, perhaps one of the most internationally famous cases has been the case of the Ivory Coast, Africa. In 2006, the ship called *Probo Koala*, coming from the Netherlands, chartered by Trafigura Beheer BV, unloaded toxic waste at the port of Abidjan in the Ivory Coast. The toxic waste importer, a local company called Tommy, disposed of the waste in 12 different locations across the city, killing at least 17 citizens and directly impacting the health of more than 100,000 people. It is believed that the catastrophic outcomes may have been even worse, yet no health monitoring or epidemiological studies were conducted in the city to assess the medium- and long-term impacts of this crime (Amnesty International; Greenpeace, 2012).

The case has not been confirmed as illegal trafficking of waste by the Secretariat of the Basel Convention. A report resulting from an international investigation published by the non-governmental organizations Amnesty International and Greenpeace disclosed Trafigura had deliberately disposed of toxic waste. The report showed that such a case happened because of flaws in international law that regulate the issue (mainly the Basel Convention) and because of bad decisions made by the countries involved, which have also failed to provide appropriate support to victims after the occurrence of the socio-environmental damages (Amnesty International; Greenpeace, 2012).

Data published by the European Union Environmental Agency indicates that, between 2001 and 2005, Member States of the EU reported the occurrence of illegal waste transfers estimated to be between 6,000 and 47,000 tonnes. Such numbers represent the equivalent to 0.2 per cent of all the waste notified as transferred from the EU. One-third of these transfers were classified as non-hazardous waste, most of which was destined for non-OECD countries for recovery (EEA, 2009). The European Environment Agency (2009) indicates that these figures represent only a small portion of the illegal transfers made from the European Union since many of the reports submitted by the EU members do not contain information on amounts sent to other countries illegally. Furthermore, the Agency found that the number of illegal transfers increased during the period analysed, considering it only a small fraction of the actual number of cases due to underreporting by the Member States.

However, as will be seen in Chapter 5, not all cross-border waste transfers are considered a crime or illegal trafficking. Most of these transfers are currently permitted and regulated by international, regional and national legislation. Although not as evident as the problems generated by illegal waste traffic, cross-border transfers considered to be legal can also be a source of ecological, social and public health damage in importing countries, as well as indicating the absence of sustainable waste management.

Thus, once again, cases of illegal waste traffic are only the 'tip of the iceberg' when it comes to cross-border waste transfer. In terms of figures, legal transfers point to extraordinary and worrying numbers. In Brazil, for instance, no official figures are reporting the amount of waste transferred to the country legally. However, given the quantities announced by the waste recovery industry, it is assumed that these numbers are quite high.

In 2011, for example, it was reported that, since the recyclables collection levels were still deficient in Brazil, the country imported, in 2010 alone, more than 201,000 tonnes of scrap iron, steel and aluminium, as well as polyethylene terephthalate (PET) in flakes and paper chips (Ninni, 2011). That means that, instead of using the waste materials generated in Brazil to supply the waste recovery industry, these materials are imported from other countries, wasting valuable materials and putting pressure on local ecosystems with a large amount of waste disposed of in landfills and dumps.

In the European Union, the number of transfers of hazardous waste increased by 53 per cent between 2001 and 2015, from 3.99 million tonnes in 2001 to 6.1 million tonnes in 2015. The peak of such transfers happened in 2007, with 8.1 million tonnes transferred to other EU states and outside the EU. According to Eurostat (2019b), 23 Member States increased their waste transfers – notably France, Germany, Italy, Sweden, and the United Kingdom. The Low Countries, on the other hand, showed a significant drop in these figures, mainly due to changes in the classification of certain wastes, which are now classified as non-hazardous (Eurostat, 2019b).

The term **Low Countries** generally refers to the territory of the Netherlands, Belgium, and Luxembourg as well as parts of northern France.

In total numbers (hazardous and non-hazardous waste), the amount of cross-border waste transfers from the EU has tripled between 2001 and 2015: from 6.3 million tonnes in 2001 to 19.3 million tonnes in 2015. The number of waste transfers notified to the EU by the Member States has steadily increased since 2005, with only one decrease in 2012 because Finland reclassified its iron oxide flows to China as products and no longer as waste (Eurostat, 2019b).

> For more information on the data about waste transfer from the EU, see the Eurostat webpage: ec.europa.eu/eurostat/statistics-explained/index.php/Waste_shipment_statistics.

Therefore, in addition to the increase of hazardous waste transfers, the number of transfers of wastes classified as non-hazardous (part of the EU Green List – such as paper, plastics and metals) has also grown significantly in recent years, with a sharp rise of transfers from the EU to the Far East, particularly to China. The quantities of non-hazardous waste exported to Asia increased 10 times for paper, 11 times for plastics and 5 times for metals (EEA, 2009).

It is interesting to note, for example, that while the EU exported 17,000 tonnes of waste (hazardous and non-hazardous) for recovery in OECD member countries in 2001, 36,000 tonnes were exported to non-OECD countries. A decade later, in 2011, 50,000 tonnes of all reported wastes were transferred for recovery in OECD member countries, and over 1.2 million tonnes went to non-OECD countries (EC, 2015a).

Regarding hazardous waste, in 2001 the EU exported just over 1,000 tonnes for recovery in OECD countries and almost 12,500 tonnes for non-OECD countries. In 2006, these figures reached their peak, with 6,100 tonnes and 127,000 tonnes, respectively. From 2010, however, the number of transfers of hazardous waste for recovery in non-OECD countries reported by the Member States to the EU has zeroed (EC, 2015a).

From the data highlighted, it is noted that, although still significant, the amount of hazardous waste transferred to outside the EU by the Member States is much smaller than the amount of waste classified as non-hazardous. Moreover, the most substantial transfers out of the EU have been towards non-OECD countries, in other words, to developing (or Southern) countries.

The compilation and systematization of the numbers of cross-border waste transfers (legal and illegal, hazardous and non-hazardous) are well advanced in the EU. The same cannot be said about Brazil and many other developed and developing countries around the world. If figures related to cross-border waste transfers considered to be legal are challenging to access, those relating to illegal waste traffic are even more nebulous.

There is a clear underreporting of cases of illegal traffic of waste since the vast majority of these cases are not identified due, among other reasons, to the lack of preparation and the precarious inspection structures of the recipient countries. Despite the efforts made by different international organizations, because of the nature of these activities, there is also no systematic collection and archiving of the cases of illegal transfers discovered.

Accurate or not, the existing data on cross-border waste transfer translate into very significant numbers, generating many controversies. In addition to more fundamental factors, which will be further analysed later in this chapter, it is believed that increasing rigidity and bureaucratization of environmental standards in Northern countries initially drove such transfers (in many cases, there is excessive or hyper-regulation of the theme, making it even more difficult to comply with the rules). Added to this is the economic and political vulnerability of Southern countries. These two factors have made (and continue to do so) the immediate economic costs for the final waste destination (whether for recovery or disposal) cheaper in developing countries than in developed countries (Anand, 2004).

In this sense, Kaza *et al* (2018), in a paper published by the World Bank, presented the differences between costs for different types of final destination of waste according to the country's GNP (see Table 1.1).

It is evident therefore that waste treatment in developed countries is much costlier than in developing countries. That is due to several factors,

TABLE 1.1 Comparison between costs of different types of final disposal of waste according to the country's GNP

US$/tonne	Low-income countries	Lower-middle-income countries	Upper-middle-income countries	High-income countries
Collection and transfer	20–50	30–75	50–100	90–200
Controlled landfill to sanitary landfill	10–20	15–40	20–65	40–100
Open dumping	2–8	3–10	–	–
Recycling	0–25	5–30	5–50	30–80
Composting	5–30	10–40	20–75	35–90

SOURCE Kaza, Silpa, Lisa Yao, Perinaz Bhada-Tata, and Frank Van Woerden (2018)

such as the strictness of environmental control, the mechanization or not of the waste management processes, and the value of land, among others.

This situation led to the beginning of a cost-reduction strategy for waste management: the creation of a global waste trading market has become a very profitable business. In 2013, it was reported that this market moved around US$443 billion per year, with a steady increase in export volumes. Only the trade in plastic waste, for example, has generated an estimated value of more than US$5 billion a year, with waste flowing mainly from developed countries to developing countries such as China (Moses, 2013).

In addition to promoting economic benefits, international waste trade can also generate ecological benefits. The reintroduction of these materials into the production cycle can lead to a reduced extraction of new natural resources, as well as a reduced wastage of such materials disposed of as waste. It is undeniable, therefore, that international trade of waste presents immediate economic advantages and potential ecological advantages. However, when evaluated in a more systemic and long-term way, such advantages cannot always be identified in ecological, social or even economic terms.

All of the risks and damages to Nature and to public health that can be caused by waste are widely known: contamination of soil, water, and air, the spread of diseases, genetic defects, infant mortality, among others (Cointreau, 2006). Besides increasing the processing of matter and energy with the inclusion of new steps in the waste management process, cross-border waste transfer adds new risks and damages to this management, especially when destined for developing countries.

Cointreau (2006) points out that, in general, while in developed countries the legal regulation and the physical and economic infrastructure of waste management have high standards of environmental quality and public health, developing countries are marked by major socio-ecological problems related to the theme. Much of the waste ends up having an environmentally inadequate final destination in these countries, and most of the individuals working in the sector, formally and informally, do not have the support and infrastructure required to meet minimum environmental and safety standards.

Waste transfer to developing countries (whether legally or illegally, hazardous or non-hazardous) further presses and negatively impacts ecosystems and local communities by creating ecological and public health problems. Moreover, it is true that these costs are not accounted for by the international trade of waste, let alone the costs that will be borne by future generations with a decrease in the environmental quality of these regions.

In this regard, it is worth noting that China has recently banned the import of 24 types of waste to the country, many of them non-hazardous, such as plastics and textile refuse. This new Chinese law, which entered into force in early 2018, will have a massive impact on the international waste market, as it received about 70 per cent of the waste transferred mainly from developed countries. According to the Ministry of Environment of China, this was a necessary measure for the country's environmental protection, since much of the imported waste was being processed in the country in an extremely polluting way (Fontdeglória, 2018).

The figures and surveys highlighted here regarding waste and its transfer reveal how important and problematic this issue is. It is evidence that we are currently experiencing a true waste crisis. In the next section, in which the 'social metabolism' approach will be adopted, we will try to understand the biophysical reasons for the occurrence and risks presented by the waste crisis and by its cross-border transfer, as well as some of its possible socio-ecological consequences.

Social metabolism and the transfer of socio-ecological costs

Social metabolism

The word metabolism comes from the Greek language and means 'change' or 'exchange'. In biology, metabolism is regarded as the set of transformations within living organisms. According to Alves *et al* (2013), the three main objectives of metabolism are:

- **i** the conversion of fuels into energy for the realization of the main cellular processes;
- **ii** the conversion of fuels to the construction of matter;
- **iii** the disposal of waste.

Biological metabolism is mainly composed of two groups of transformations:

- the anabolic, which allows the synthesis (or production) of substances and matter; and
- the catabolic, responsible for the decomposition and degradation transformations.

SOURCE Alves *et al*, 2013

Both groups, however, are interdependent: the anabolic processes use the energy created by the catabolic processes to create matter, and the latter uses the matter produced by the anabolic processes to create energy. When there are more catabolic than anabolic activities, the body loses mass (decreases), and when the opposite occurs, the body gains mass (grows). When both processes are in equilibrium, the organism will be in dynamic equilibrium (Alves *et al*, 2013).

The biological concept of the metabolic process of natural systems and organisms is applied by analogy to the concept of 'social metabolism': to the physical transfers of materials and energy carried out between socio-economic systems and the natural system (ecosystem). The term social metabolism encompasses all the flows of materials and energy necessary for the sustenance of all human economic activities, from nutrition to all the artefacts required by the economy as a whole, such as instruments, equipment, buildings, infrastructure works, among others (Fischer-Kowalski *et al*, 1999).

The social metabolism theory proposes to study the extraction of primary materials, water and air from Nature, as direct or indirect inputs to the socio-economic system. In such a system, natural resources (materials and energy) are transformed into products and finally returned to the ecosystem as waste and emissions. Fischer-Kowalski *et al* (1999) point out that, unlike the terms 'industrial metabolism' or 'urban metabolism', which deal with material and energy flows that pass through an industrial system, 'social metabolism' or 'socio-economic metabolism' is not restricted to industrial societies; it also refers to non-industrial modes of subsistence, being, therefore, a broader concept.

Braungart and McDonough (2009), highlight the existence of two great metabolisms on the planet:

1 The first of these is the biological metabolism, or the biosphere, represented by the cycles of Nature. In practical terms, the biosphere is a closed system, ie nothing can enter or leave the planetary system except heat and occasional meteors. It is composed of two essential elements, mass and energy, which are both valuable and finite.

2 The second metabolism is the socio-economic one, called by Braungart and McDonough 'technical metabolism' or 'technosphere', represented by the cycles of the industry, including the stage of extraction of natural resources.

Fischer-Kowalski *et al* (1999), after an extensive literature review on the subject, concluded that some 'waves' of research on social metabolism have appeared in recent decades. In the late 1960s, the first wave of conceptual power emerged to shape the environmental debate, especially in the United States, where the subject was more advanced in political terms. This first wave established milestones for strategies developed in subsequent research. The second wave started in the early 1990s in Europe, with most of the publications by authors from Scandinavia, Germany and the Netherlands.

The analysis of social metabolism can be done by different means, levels, frames of reference, and flows under consideration. Currently, various instruments are used to measure energy and material flows, such as material flow analysis (MFA), mass balance, life cycle assessment (LCA), network environmental analysis, among others. Industrial ecologists are leaders in urban metabolism studies (Chrysoulakis *et al*, 2013).

Regarding frames of reference, there are two main perspectives of the analysis of anthropogenic material and energy flows known as:

a socio-economic system perspective; and

b environmental (or ecosystem) perspective.

The first focuses on a socio-economic system, treating it as an organism. The second perspective looks at this same system from the larger system in which it operates and is inserted, the ecosystem. It relates the inputs and outputs of the analysed socio-economic system to the stocks and flows from and to the broader system (Fischer-Kowalski *et al,* 1999).

Initially, the first perspective was developed as a theoretical framework of analysis for the social sciences, while the second perspective was developed in the natural sciences. Both of them, however, are concerned with the fate of socio-economic systems and with the sustainability of their metabolism, using different frames of reference. Although from different points of view, the socio-economic system and the ecosystem perspectives can be linked employing mutual feedback loops. Such a link can be achieved by developing a general formal model of interaction between socio-economic and natural systems or by combining analytical and normative modelling techniques often used in projections related to the notion of 'sustainable development' (Fischer-Kowalski *et al*, 1999).

For Chrysoulakis *et al* (2013), the analysis of the metabolism of human socio-economic systems must be done from an ecosystem perspective. That is because current societies are much closer to the notion of an ecosystem (encompassing a complex group of multiple individuals located within the same environment, interacting with each other and with the environment) than to the notion of an organism as a singular individual.

Regarding the flows under consideration, when analysing the metabolism of a socio-economic system or the material and energy flows between that system and its environment, either can be considered:

i the total volume of transfers of matter or energy (or both); or

ii select specific flows of materials or chemical substances.

These same two options apply to the ecosystem approach (or analysis of feedback loops) (Fischer-Kowalski *et al*, 1999).

Typical research topics of the ecosystem perspective include:

i the comparison between the rates of production of specific resources by the reference ecosystem with the rates of consumption of these same resources by a particular social system; and

ii the relationship between the outputs of a socio-economic system (waste and emissions) with the absorptive and stocking capacities of the ecosystem that embraces such system; or

iii simulating the ecosystem reactions to such outputs (such as global warming studies).

SOURCE Fischer-Kowalski *et al*, 1999

In terms of urban metabolism, there are currently two predominant schools. One describes the metabolism in terms of solar energy equivalents (or emergy); another one, broader, expresses the energy and material flows of cities in terms of mass flows. However, there is more research dedicated to energy flows, mainly with the aim of developing technologies and alternative strategies for achieving energy eco-efficiency. Research on material flows, on the other hand, is incipient. Even rarer are the studies focusing on the materials outputs of socio-economic systems, in terms of waste and ecosystemic absorption and stock capacities (Chrysoulakis *et al*, 2013).

The establishment of boundaries that separate social and natural systems (technosphere and biosphere) is still a matter of academic dispute, but a consensus on the boundaries between different socio-economic systems seems to exist. In general, studies involving socio-economic metabolism distinguish 'extractions from the environment' or 'primary inputs' from 'imports' from other socio-economic systems. The same applies to outputs, since 'depositions', 'emissions', 'dissipative uses' and 'losses' are distinguished from outputs in the form of 'exports' to other socio-economic systems (Fischer-Kowalski *et al*, 1999).

From these definitions arise other questions that impact the social metabolism analysis of particular social systems, such as the definition of the amplitude of the material flows induced by an economic activity. Approach options range from a narrow definition of the boundaries of the socio-economic system (only materials with economic value are considered inputs and outputs) to a fairly broad definition, including all materials transformed in some way by an economic activity or process (Fischer-Kowalski *et al*, 1999).

The more encompassing the definition adopted becomes, the more materials with economic value will be diluted in the total of materials considered. On the other hand, if the definition of the amplitude of these material flows is very restricted, a large number of materials tend to be neglected (a parcel known as 'ecological rucksack' or hidden flows), among them: the environmental burden associated with the extraction of primary materials, imports, disposal of waste and exports (Fischer-Kowalski *et al*, 1999).

Alier (2003) defines ecological rucksack as the quantity of raw materials that intervene and must be handled in the life cycle of a product, as well as returned to the environment as waste. Such a concept reflects the hidden flows of matter and energy required to manufacture and to dispose of a product without, however, being part of its direct physical composition (thus, they are not valued by the economic system). The diagram presented in Figure 1.1 seeks to synthesize the notion of socio-economic metabolism, including its hidden flows (ecological rucksack).

FIGURE 1.1 Synthesis of the notion of socio-economic metabolism

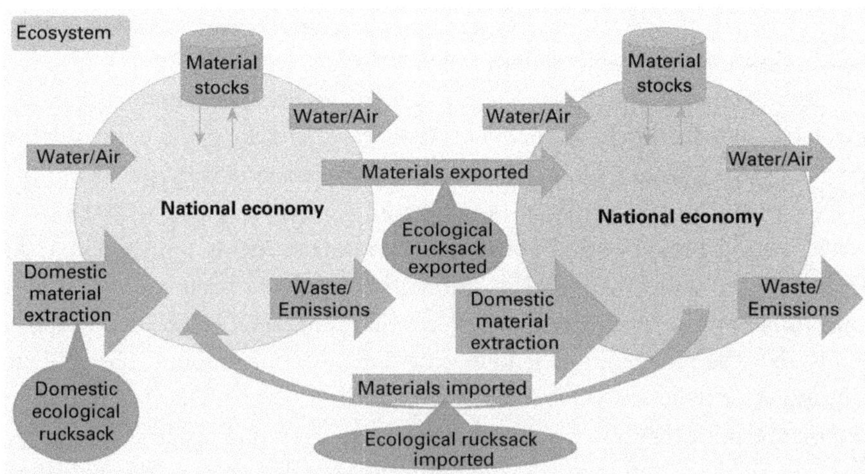

SOURCE Adapted by the author from Fischer-Kowalski *et al*, 1999

For Muradian *et al* (2012), the study of social metabolism is a very appropriate conceptual tool for understanding the biophysical size of the economy. It provides more precise distinctions between stocks and flows as a prerequisite for verifying empirically whether a socio-economic system is still 'growing' (in physical terms), whether it is steady or shrinking. The contributions of economic processes to the consumption and depredation of natural resources can thus be evaluated.

Another significant contribution of this type of study is the demonstration that, no matter how sophisticated the products and services generated in industrial economies, there is a very limited number of inputs material for these production processes. Also, social metabolism studies show that the waste generated by these same production processes (the outputs of the socio-economic systems to the ecosystem) tend to be chemically and physically complex and challenging to systematize quantitatively. That makes studies on waste even more challenging (Fischer-Kowalski *et al*, 1999).

Fischer-Kowalski *et al* (1999) emphasize that social metabolism has proved to be fundamental for studies on the exhaustion of natural resources, pollution, entropy, the inefficiency of economic services, the closure of production cycles and the growth of metabolic transfers. Based on the elements highlighted by these authors, it is possible to affirm that the approach of social metabolism is exceptionally relevant for studies on waste from different perspectives.

> **Entropy** refers to the second law of thermodynamics, which states that in the processing of matter and energy, a part of them will always be irreversibly lost. The lost matter occurs in the form of waste, and energy, in the form of heat and emissions. The subject of entropy will be studied in more detail in Chapter 3.

The study of social metabolism brings essential elements, notably:

a The systemic demonstration of the interrelations between natural resource extraction, production processes, and waste generation.

b A significant change of focus on wastes considered to be hazardous to the ecological effects of large (hazardous and non-hazardous) waste flows, as occurred in studies on pollution (emissions).

c The incorporation of the entropy law in all stages of social metabolism, with the adoption of the notions of 'dissipative losses' and 'dissipative use of materials' in the analysis of issues involving recycling processes, for example.

d Understanding that the satisfaction of a large part of human needs is given through the provision of services and not of energy and matter, which stimulates the dematerialization of these services with the consequent reduction of waste generation.

e The search for the closure of production cycles, preventing new extraction of natural resources and new eliminations of waste in the environment, understanding that this option does not apply to all materials and processes.

f The understanding that, independently of the specific additional effects generated by socio-economic systems in the environment, the large volume of transfers between such systems (high metabolic flows of energy and matter) is an essential factor for the generation of significant ecological damages. Thus, the current dominant model of socio-economic systems (based on the idea of infinite growth) must be questioned under penalty of irreversible overloading of the ecosystem in a short period.

Studies of social metabolism have shown that, while natural ecosystems are, as a rule, self-sufficient in energy or subsidized with sustainable inputs, and also able to conserve mass through recycling processes, urban ecosystems, on the other hand, have a sizeable linear metabolism with high flows of energy and materials. In other words, these studies demonstrate that the metabolic cycles of urban ecosystems are open (trading with the environment) and unsustainable due to the rates of material consumption and the high rates of the waste generation that do not correspond to the rates of assimilation of the environment (Chrysoulakis *et al,* 2013).

On this point, Alier (2003) clarifies that, according to thermodynamics, systems are classified as 'open' when they allow the entry and exit of energy and materials (such as the socio-economic system). There are also 'closed' and 'isolated' systems. Closed systems are only open for energy input and output (such as the Earth), while isolated systems do not allow any input or output, either of energy or materials. As in natural processes of the ecosystem, the availability of energy and materials allows the development, organization, and increase of the complexity of economic processes.

Because socio-economic systems are open systems, there is energy dissipation and waste generation in economic production processes. The study of social metabolism shows that some of the waste can be recycled (hence, preventing new materials flows from and to the environment). However, when this is not possible, the socio-economic system needs the input of new natural resources and sends waste to the environment. When the economic model prioritizes growth (and at an excessive speed), natural cycles fail to reproduce resources or absorb and assimilate waste at the pace required by the economy, often generating irreversible environmental burdens (Alier, 2003).

Haberl *et al* (2011) also point out that simple calculations of social metabolism demonstrate the problems (and the biophysical impossibility) of the globalization of the current Western economic model and its patterns of production and consumption. According to the authors, assuming the world population growth rate will remain the same, it is estimated that about 8.5 billion people will inhabit Earth by 2050.

Supposing the total energy use would be equal for all people of all countries (with the total globalization of current patterns of production and consumption), according to the average value of today's industrial societies, there would be a total energy use three times greater than today's usage worldwide. That means that humankind's energy consumption would then be roughly equal to all terrestrial net primary production (NPP), ie to the total amount of biomass that green plants produce annually on the Earth's surface by photosynthesis (Haberl *et al*, 2011).

Many authors argue that humanity has exceeded the 'carrying capacity' of the planet, which means the maximum size of a species population in a given territory able to live sustainably, ie without depredating its resource base. Not denying this possibility, Alier (2003) emphasizes that, regarding human populations, some additional points must be taken into account for the analysis of the carrying capacity. One is the consumption pattern since the definition of a maximum population in a given territory will depend on its levels of consumption.

The transfer of socio-ecological costs in waste transfers

Another critical point to be considered, in particular for this book, is international trade. For Alier (2003) this type of trade (or cross-border transfer) resembles the notion of 'horizontal transfer' of ecology, and its execution by current human populations causes the intentional regulation and alteration of the carrying capacity of their territory. The carrying capacity of a territory can be increased, for example, when a country imports materials from another country or when it exports waste to another country. The occurrence of ecologically unequal exchanges will inevitably lead to an increase in the carrying capacity of certain territories while decreasing that of others.

> **Horizontal transference**, in biology, means the transmission of infectious agents among contemporaries of a population. It is, therefore, an intra-generational transfer. Vertical transference refers to the transmission of infectious agents between generations (from father/mother to child), representing an inter-generational transference.

Many territories, both urban and rural, have moved (transferred) socio-ecological problems to a larger spatial scale and to a wider timescale. Urban areas are a focus of special attention since large and prosperous urban centres are always based on the intensive use of fossil fuels and the externalization of environmental costs (Alier, 2003).

In addition to the flexibilization of the carrying capacity of different socio-economic systems, international trade has also helped to externalize the socio-ecological costs of these systems. The trend of importing energy and materials and exporting waste and contamination occurs not only at the global level, with the dislocation of environmental pressure from metropolitan countries to the global periphery (North-South direction), but also at regional and national levels. Consequently, large metropolitan regions are improving their environmental quality by transferring their negative externalities to increasingly distant peripheries (Alier, 2003).

The concept of 'externalities', widely disseminated by the English economist Arthur C Pigou in the 1920s, refers to the consequences of the production process that are not accounted for by the socio-economic system and are borne or enjoyed by the society as a whole (Montibeller-Filho, 2008). Thus, externalities can be positive or negative. The exhaustion of natural resources and the environmental impacts caused by production process are examples of negative environmental externalities, whose costs are borne by the society as a whole and not by the economic subjects who obtained the profits of such processes.

The idea of 'environmental externalities' does not consider the negative consequences generated by the production processes as problems to be solved by the economy. The term 'externality' represents the exclusion and non-consideration of Nature (ecosystem) as a fundamental element for the inputs and outputs of the socio-economic system to enable production processes. It generates, according to Leff (2006), processes of ecological destruction and environmental degradation.

What orthodox (neoclassical) economists regard as 'externalities' or even 'market failures' is understood and termed by ecological economists as 'success in transferring costs' (Alier, 2003). The latter statement remains even more evident when seen through the lens of social metabolism. International trade of natural resources and waste, when carried out without accounting for all the real ecological and social costs (when it ignores the hidden flows of energy and materials, the ecological rucksack, of such transactions), ends up facilitating the displacement (transfer) of social and ecological costs to other socio-economic systems and, ultimately, to Nature and future generations.

The study of social metabolism, with its perspective on energy and material flows, is an essential tool for describing, analysing and understanding environmental and sustainability problems at micro and macro levels. As early as 1999, Fischer-Kowalski *et al* affirmed and defended the enormous potential of this approach for the integration of several concerns within a transdisciplinary perspective on ecological issues, whose simplicity ensures the achievement of political relevance for possible socio-ecological changes.

The potential impact of these social metabolism studies will largely depend on the political context in which they are conducted and reported. In this sense, Fischer-Kowalski *et al* (1999) noted that environmental indicators based on the analysis of material flows in developed countries, such as the United States and EU countries, led to the development of policy goals focused on ecological modernization based on, for instance, recovery processes and greater control of hazardous substances. However, the reduction in the global quantity of materials of national economies (degrowth) was not (and still is not) a goal addressed by these environmental programmes.

The ecological modernization approach will be discussed in more detail in Chapter 2.

Moreover, the study of social metabolism has been relatively successful in the operational definition of policy goals, but not as efficient for the development of concrete strategies to achieve these goals (Fischer-Kowalski *et al*, 1999). Perhaps this is due to the fact that creation and concrete intervention of public policies in socio-economic systems are generally seen as functions of political and economic actors (and not of researchers/scientists).

The study of social metabolism is not as popular in social sciences as it is in exact sciences. Cultural anthropology studies are an exception since they traditionally use concepts related to metabolism. Social science analyses, however, are fundamentally important for the definition of a sufficiently complex image of the social systems analysed, as well as to enable concrete changes in intrinsic forms of regulation of socio-economic metabolism (Fischer-Kowalski *et al*, 1999).

Muradian *et al* (2012) argue the power of the study of social metabolism to search for solutions to complex socio-ecological problems is extended when combined with studies of economics and political ecology of resource extraction, its use, generation and final disposal of waste. For the authors, the combination of social metabolism elements with political ecology allows the establishment of connections between materials and energy used by some social actors and the socio-ecological impacts experienced by others, often in geographically distant locations. It also highlights the resistance of the impacted actors, usually in the form of environmental conflicts.

The elements highlighted through the study of social metabolism (such as ecological rucksack, carrying capacity and the transfer of socio-ecological costs) lead to a new perspective in the analysis of international trade. Understanding that socio-economic systems are open and thus exchange energy and material flows with the environment and with other socio-economic systems has made many (ecological) economists oppose 'free trade' due to the occurrence of 'ecologically unequal exchanges' (Alier, 2003).

According to Alier (2003), ecologically unequal exchange means the export of materials from poor countries or regions, disregarding the externalities involved in their production and the depletion of natural resources, unequally exchanged for goods and services of the richest regions. Reflecting on the issue of international waste trade, we can define ecologically unequal exchange as the cross-border transfer of materials in exchange for compensations that disregard socio-ecological externalities (ecological rucksack) of the whole life cycle of such materials.

Cross-border waste transfer has the potential to generate significant ecologically unequal exchanges between the countries involved in such transactions. The notion of 'ecologically unequal exchange' refers once again to the ecological rucksack concept (or hidden flows) and allows us to reflect on the local socio-ecological externalities (in this case, the place of treatment and final disposal of waste) not accounted for, which, for this reason, are not considered in the decision-making processes on waste transfer. This problem is even worse in cases where such transfers occur in the North–South direction. Most of the natural resource exporting countries are economically and politically vulnerable in global terms (global South). These are the countries that have been receiving extensive, hazardous and non-hazardous, waste flows from the Northern countries in recent decades, once again shaping ecologically unequal exchange.

Hornborg (1998) uses the analysis of social metabolism to explain the economic growth and development of the global economic centres (North). He argues that market prices constitute a mechanism by which these centres import exergy (available energy) from the peripheries and export entropy (dissipated energy) to them. The global market allows and provides the transfer of available energy and matter to Northern countries and of their waste and emissions to Southern countries, generating ecologically unequal exchange (Alier, 2003).

For Alier (2003), Southern countries, faced with their economic and political vulnerability, are incapable of curbing such transfers, internalizing externalities in the adjusted prices, applying the precautionary principle or diversifying their trade. In order to maintain current patterns of production and consumption, the dominant Western economic system seeks new territories, and thus modifying spatial relationships. In this globalized market, the relative carrying capacity of developed countries increases, while that of developing countries decreases, a process which also generates a change in temporal relationships, since the new territories incorporated into this global trade are no longer governed by Nature's time of reproduction, regeneration, and absorption.

The new territories incorporated into international trade thus allow the supply of materials and energy and the elimination of waste to occur at an ever-increasing and uninterrupted pace imposed by capital circulation. The irreversible destruction of ecosystems and cultures that valued Nature in different ways evidences the antagonism between this economic time and the geochemical and biological time controlled by the rhythms of Nature. This destruction is deepened whenever the unequal exchange between North and South intensifies, generating the need for more commercial transactions to boost Southern countries' economies (Alier, 2003).

The ecological crisis and the waste crisis therefore find biophysical explanations in the study and understanding of social metabolism. Although such biophysical foundations are discussed and accepted in certain scientific areas, such as physics and ecology, they are still unknown to society as a whole and to other areas of knowledge that insist on ignoring or moving away from their ontological bases.

The next section of this chapter will seek possible answers for the occurrence and perpetuation of such crises. First, a deeper analysis will search for the epistemological and geopolitical foundations of these crises. Then the repercussions of these foundations will be verified in the current economic model, patterns of production and consumption and legal model.

Modern Western thinking: the mechanistic paradigm

Mechanistic epistemology

Throughout most of our history, humankind had (and continues to have) the need to construct frames of reference for organizing life's daily activities, which is an essential cultural ingredient of every society. These frames of reference, known as 'worldviews', are considered successful when they are internalized in individuals, members of society, to the point of becoming unquestionable. A worldview, when fully internalized by society, will cause its members to lose awareness of how much it affects the way they act and how they perceive the reality around them (Rifkin *et al*, 1980).

According to Rifkin *et al* (1980), modern Western hegemonic (dominant) thinking genuinely believes the world is progressing towards a state of higher value as a result of a stable accumulation of human knowledge and techniques. It also considers that (among others):

- **i** individuals exist as autonomous entities;
- **ii** Nature has an order;
- **iii** scientific observation is objective;
- **iv** people have always wanted private property;
- **v** competition has always occurred.

All such beliefs are considered by the Western worldview as part of 'human nature' and therefore to be immutable (Rifkin *et al*, 1980).

However, other present-day societies, and societies from the past, would be unable to comprehend many of these beliefs described as part of 'human nature'. That is precisely the power of a worldview internalized in the collective unconscious: it makes it impossible to imagine another way of seeing the world. Indeed, the modern Western worldview, here named as a mechanistic paradigm, took shape around 500 years ago, and even though it has been refined and modified during this time, its essence remains and strongly influences current Western and Westernized societies (Rifkin *et al*, 1980).

According to Capra and Mattei (2015), it was in the 16th and 17th centuries that the vision of an organic, living and spiritual world, typical of Antiquity and the Middle Ages, changed radically, being replaced by a view of the world as a machine – a dominant metaphor in Modernity to the present day. The emergence of the mechanistic paradigm occurred due to revolutionary changes mainly in physics and astronomy. They became known as the Scientific Revolution, inaugurating Western Modernity.

Although the beginning of the Scientific Revolution is attributed to Nicolaus Copernicus and other scientists, the real paradigm shift in sciences occurred with the astronomical discoveries of Galileo Galilei, who scientifically validated Copernicus's hypothesis on the centrality of the sun (and not of the Earth) in the solar system. This new perspective freed humankind from the natural moorings coming from the idea of having a fixed and unchanging place in the universe. It initiated an untiring process of seeking understanding and domination of the planetary movements (Ost, 1995).

Galileo undertook experiments with bodies in motion, describing them in mathematical terms. He advocated that scientists should restrict themselves to the study of the properties of material bodies that could be measured and quantified (shape, size, and motion, for example). He therefore relegated other properties (such as colour, sound, and smell) to subjective projections, which should be excluded from the science domain. He reduced all quality to quantity (Capra and Mattei, 2015).

For Rifkin *et al* (1980), a range of works by a number of scientists and researchers brought about the Scientific Revolution, inaugurating the mechanistic paradigm. In the middle of the 18th century, all the critical elements for the wide diffusion of this new paradigm had already been integrated into a unified scheme. According to these authors, the mechanistic Western worldview had as its principal founders three men: Francis Bacon, René Descartes and Isaac Newton, great disseminators of the new paradigm, who influenced Western thought and whose ideas still reverberate in the current dominant thinking.

Francis Bacon, opposing the Greek method of contemplation for understanding and harmonious co-existence with Nature, sought a methodology to actively understand, dominate and control Nature. Bacon developed an empirical scientific method, formulating a theory of inductive method. It is an approach that separates the observer from the observed, providing a 'neutral' environment for the development of an 'objective knowledge'. It calls for the conducting of experiments to reach conclusions, which can be tested in other experiments. Bacon profoundly changed the nature and purpose of the scientific quest and believed that objective knowledge would enable people to take charge over natural things (Rifkin *et al*, 1980; Capra and Mattei, 2015; Bacon, 2012).

According to Ost (1995), Bacon's approach was much more than a scientific method: it proved to be a real scientific and political programme for the understanding and domination of the universe. First, Nature should be understood; then imitated and perfected to, lastly, create the artifice, the automaton, the supra-Nature. Coincidence or not, these were precisely the steps taken by humanity over the course of centuries, especially through technoscience.

Continuing the work begun by earlier philosophers, Descartes reaffirmed the idea that the key to understanding the world, the way to decipher its hidden secrets, to control and to mould it to satisfy human goals, would be found in mathematics, the ideal representation of total order. For Descartes everything in the world had its place, and all relations were harmonious, eliminating the Western antiquity view of a chaotic and alive world (Rifkin *et al*, 1980; Descartes, 1998). Descartes argued the antiquity worldview, which understood reality as chaos and decay, was not mathematical and therefore was false. The Christian worldview of the Middle Ages, in turn, did not bring the predictability necessary to be considered scientific since it depended on the commands of a hidden divinity (Rifkin *et al*, 1980).

Descartes developed the analytical (Cartesian) method for pursuing scientific truth. It consists of breaking down thoughts and problems into parts, rearranging them later in their 'logical order'. He founded his view of Nature on a fundamental division (dichotomy) between two separate, independent and fundamentally different realms: that of the mind (*res cogitans*) and that of matter (*res extensa*). Humans would be the only ones containing both elements and Nature would be a machine composed of parts (living or non-living) with no intrinsic relation between them. For Descartes (1998), Nature would work according to mechanical laws and could be explained in terms of the organization and movement of its parts (Capra and Mattei, 2015).

The Cartesian method consists of three main elements:

- **i** intuition, which provides the intellectual vision through clear and distinct representations;
- **ii** deduction, which implies the logical passage from one magnitude to another;
- **iii** division, which presupposes that all things must be divided into equal and comparable parts.

The method is then applied to explain the most complex problems, reducing Nature to a total sum of matter, divisible into specific parts and in constant motion (Ost, 1995).

Descartes proposed three principles on the movement of material bodies, which he called 'Laws of Nature'. The first and second laws state that a still body will remain stationary and that a moving body will move in a straight line and at a constant speed unless something external interferes (Capra and Mattei, 2015; Descartes, 1998).

These laws were later reaffirmed by Newton (2016), who discovered a mathematical method to describe mechanical movement, subjecting Nature to the laws of mathematics (Rifkin *et al*, 1980). Newton represented the culmination of the mechanistic paradigm and the 'Laws of Nature' concept, synthesizing the works of Copernicus, Galileo, Bacon, Descartes, and others. Newtonian physics, the crowning glory of 17th-century science, provided a consistent mathematical theory of the world that provided the solid foundation of scientific thought until the 20th century (Capra and Mattei, 2015).

The conception of the universe as a living whole was, in Western culture, replaced by the vision of a universe made up of parts that could be separated, studied, quantified and rearranged according to human needs. Accuracy, efficiency, and speed are central values of the mechanistic paradigm, which views the world as a great moving machine made up of all sorts of parts that need to be put together for the system to work. Human work is an endless engineering activity, and progress is achieved through machine improvement (Rifkin *et al*, 1980).

As Capra and Mattei (2015) point out, the mechanistic paradigm, while useful in many ways, has caused the fragmentation characteristic of modern Western thinking and academic disciplines. It has thus led to the widespread diffusion of a reductionist attitude in science, resulting from the belief that all aspects of a complex phenomenon can be understood by reducing it to its smallest constituent parts. These beliefs further alienate science from its ontological foundations.

These characteristics of the mechanistic paradigm have led, according to Morin *et al* (1999), to profound blindness about what should be relevant knowledge. Relevance, according to this paradigm, increases with specialization and abstraction. Specialized knowledge, in itself a form of abstraction, extracts the object from its context, rejecting its links and intercommunications with the environment. Then, it inserts this object into compartmentalized disciplines, whose boundaries break the systemicity and multi-dimensionality of phenomena. Such a paradigm leads to mathematical abstraction and separates the research from reality by prioritizing what is calculable. It ignores the context necessary for the understanding of its objects of study.

The Western mechanistic paradigm, applied to the conduct of disciplinary research that isolate problems, has allowed specialists to achieve high performance in their fields of knowledge and to cooperate in non-complex sectors, especially those linked to artificial machines. However, by ignoring the inter-feedbacks and the circular causality of systems, by considering the different ecological and social phenomena according to linear and deterministic causality, this paradigm has been insufficient to perceive and conceive the global and the fundamental, and the complexity of the current problems which are interdependent in time and space (Morin *et al*, 1999).

> The concepts of retroactive and recursive circles will, along with other notions of complex thinking, be explored in more detail in Chapter 3.

The fragmentation, compartmentalization, isolation, and reductionism of the mechanistic paradigm break up the complexity of the world and fractionalize problems, separating what is connected, and uni-dimensionalizing the multi-dimensional; thinking becomes short-sighted, if not blind. This dominant Western thinking ends up destroying the possibility to change and improve current models and take a long-term view. The organization of closed and isolated scientific knowledge into 'disciplines' makes it impossible to think about complexity. The more multi-dimensional the problems, the greater our inability to think of their multi-dimensionality; the greater the crisis, the greater our inability to think about it; the more global the problems become, the less they are considered (Morin *et al*, 1999).

Boaventura de Sousa Santos (2010b) defines the present moment of human history as times of 'strong questions and weak answers', a witness to the crisis of the dominance of this modern Western paradigm, times of paradigmatic transition. Today's strong questions go beyond the discussion on the options for individual and collective life, addressing the epistemological foundations that create such options available by the system. The weak answers, in turn, try to answer without questioning these foundations, creating a virtual reality in the attempt to respond to all questions.

This discrepancy between questions and answers gives rise to a certain confusion, which results from the fact that current complex problems no longer fall within modern science's remit as it has moved away from the complexity of its foundations in the nature of being. For Santos (2010b), the process of Western modernization of knowledge led to two significant contractions. Firstly, Western science became the only type of valid and rigorous knowledge, a form of privileged knowledge by comparison to all others. At the same time, the science of the mechanistic paradigm filters out and seeks to solve only those problems it can answer; it is a science that deals exclusively with problems posed by itself. That is what Santos (2010b) calls 'orthopaedic thinking', a type of thinking constrained and impoverished by the contraction of problems it tries to solve to fit into analytical and conceptual frameworks alien to such problems. The vastness of the existential problems underlying science is then ignored.

Secondly, this process of modern Western thought moving away from the complexity of its ontological foundations resulted in the scientification of the humanities, submitting them to the dominance of science and extending orthopaedic thinking to other areas of knowledge (Santos, 2010b).

Today, therefore, academia is dominated by research into theoretical problems that are increasingly far removed from the very real problems facing humanity, and the resulting answers and responses are increasingly irrelevant to the problems of existence that originated them. This process of knowledge monopolization by modern Western science becomes apparent with the growing discrepancy evident between the strong questions and the weak responses characteristic of the present times (Santos, 2010b).

The fragmentation, isolation, dichotomy and contractions generated by the mechanistic paradigm do not end here. The epistemological monopolization of modern Western science in relation to other areas of knowledge is only the visible separation created by this model. Boaventura Santos (2010a) emphasizes that this paradigm makes invisible other separations established through radical lines that segregate social reality into two distinct universes: the universe 'on this side of the line' (North) and the universe 'on the other side of the line' (South).

In this sense, modern Western dominant thinking is also what Boaventura Santos (2010a) calls an 'abyssal thought', a system of visible and invisible divisions. The 'other side of the line' in this system disappears as a reality, becomes non-existent in the sense of not being considered relevant, comprehensible or commensurable. The fields of knowledge and law represent, for the author, the best manifestations of this epistemology.

In the field of knowledge, the mechanistic paradigm has granted the monopoly of the production of acceptable knowledge to modern science 'on this side of the line' (in the North) over alternative forms of knowledge such as philosophy and theology. It generated the well-known dispute between scientific and non-scientific forms of truth (Santos, 2010a).

These disputes (or distinctions) are the visible part of this system based on the invisibility of forms of knowledge that do not fit into any of these ways of knowing (such as traditional or indigenous knowledge (see Figure 1.2). That is because they do not meet the scientific criteria of truth or the criteria of the forms of knowledge recognized as alternatives. On 'the other side of the line' (South), therefore, there is no real knowledge, but only beliefs, magic, intuitive understandings that, at best, can become the subject of scientific inquiry (Santos, 2010a).

According to Grosfoguel (2010), the dominance of knowledge produced 'on this side of the line' (North) and the invisibility of the knowledge coming from 'the other side of the line' (South) are mainly created by the 'neutrality' and 'objectivity' of the mechanistic paradigm in the production of knowledge. According to Grosfoguel, the Western modern dominant paradigm replaced God, the foundation of European knowledge in the Middle Ages, by (Western) man, the foundation of knowledge in modern Europe. This Western man now has privileged access to the universal laws, to the ability to produce scientific knowledge and theories; he is the owner of the universal truth that is beyond time and space.

FIGURE 1.2 Abyssal lines in knowledge

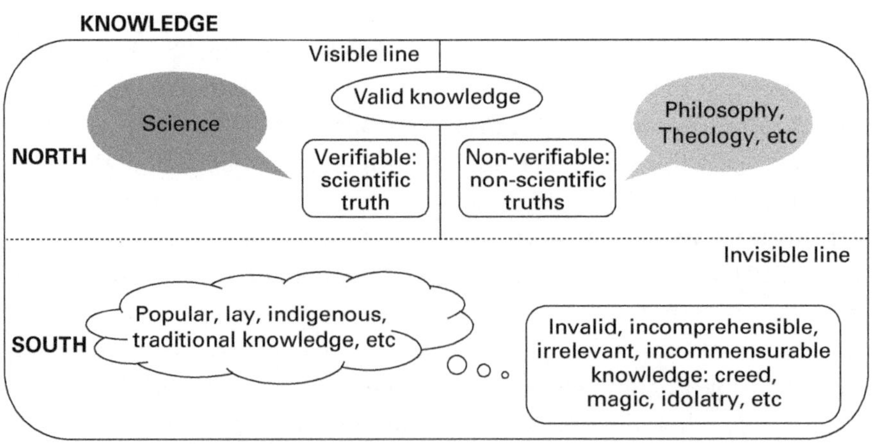

The mechanistic paradigm has broken the link between the scientist or researcher and their ethnic–racial/sexual/gender 'epistemic place' (ie the dominant method, knowledge and reasoning on which they base their research), generating the myth of true universal knowledge. However, universal knowledge hides not only the researcher but also ignores their epistemic place and the geopolitics of the structures of power/knowledge from their own location. This 'neutral' and 'objective' view conceals the local and concrete perspective of the researcher under an abstract universalism, enabling Western man to represent his knowledge as the only one capable of achieving a universal consciousness and dispensing other types of knowledge because they are 'particularistic' (Grosfoguel, 2010).

Mechanistic geopolitics

It is this feature (universalism) of modern Western thought (or what it obscures) that highlights the importance of the notion of spatiality for understanding the extension of the mechanistic tentacles to global geopolitical relations. As Maldonado-Torres (2010) points out, the concept of Modernity (temporal) hides the essentiality of the imperial–colonial relationship (spatial) for the success of the propagation and dominance of the mechanistic paradigm in a global context.

The reduction of the social (and global) hyper-context to a supposed universality of modern Western thought has occurred (and continues to occur) thanks to the epistemological intervention that was (is) only possible with the use of political, economic and military interventions of modern colonialism and capitalism to non-Western peoples and cultures. The resilience of the current dominant models (epistemological, developmental, economic, legal, among others) over the centuries has only been possible due to the repeated use of an instrument that seemed to have been historically neutralized: colonialism (Santos, 2010a).

In fact, the end of political colonialism – defined by Aníbal Quijano as a structure of domination/exploitation where the control of political authority, production resources and labour of a given population dominates another of different identity and whose central headquarters are also located in another territorial jurisdiction (Quijano, 2010) – did not mean the end of the extremely unequal social relations that it generated. On the contrary, colonialism continued (and continues) to propagate itself in the form of coloniality of power and knowledge (Santos, 2010a).

It means that colonized territories may even be defined as post-colonial after their political independence and liberation from the constraints between empire and colonies but not that the coloniality of power and knowledge to which they are subjected have been superseded (Santos, 2010b). Coloniality of power was built within colonialism and is understood as a constitutive element of the world's power pattern, based on the imposition of a racial/ethnic classification (and hierarchy) of the world population as the cornerstone of this power (Quijano, 2010).

According to Maldonado-Torres (2010), many authors consider Modernity to be responsible for the end of colonization and dependence as if Modernity had not been intrinsically associated with the colonial experience. The complex matrix of the coloniality of power has established a new pattern of domination and exploitation with the creation and expansion of new trade routes with the South. Thus, although much of the analysis about Modernity neglects the constitutive role of coloniality in the idea of what is modern; in fact Modernity never existed, nor can it exist without coloniality. It is in this sense that Grosfoguel (2010) affirms that the mechanistic paradigm represents a patriarchal, capitalist, colonial and modern world-system that calls itself universalist, neutral and objective.

The disinserted and non-situated neutrality and objectivity of the geopolitics of knowledge are Western myths since all knowledge is epistemically situated either on the dominant (North) side or on the subservient (South) side of power relations. The success of the mechanistic paradigm is to make subjects socially situated on the oppressed side (their social/global place) think epistemically (from their epistemic place) as those who are in dominant positions (Grosfoguel, 2010).

This epistemic strategy has been crucial for Western epistemological hegemony, since, in concealing the place of the subject of enunciation, the mechanistic paradigm has built a hierarchy of higher and lower knowledge and peoples throughout the globe. From the 'people without writing' of the 16th century it moved to the 'people without history' of the 18th and 19th centuries; to the 'people without development' of the 20th century; and, more recently, to the 'people without democracy' of the 21st century (Grosfoguel, 2010).

The mechanistic paradigm of Western Modernity, as seen, generates a system of visible and invisible separations which, in the field of knowledge, has privileged modern Western science as the only source of truth opposed to other forms of knowledge, visible (such as philosophy) and invisible (such as traditional knowledge). In geopolitical terms, the visible separation (North) of this system is characterized as a paradigm based on the tension and dichotomy between regulation and emancipation through social conflicts.

Underlying this distinction lies the invisible dichotomy of appropriation–violence applicable to the South (Santos, 2010a). While the visible dichotomy of Northern regulation–emancipation does not apply to the 'other side of the line', the invisible dichotomy of appropriation–violence of the South is inconceivable to the North (Santos, 2010a). Figure 1.3 illustrates the distributions of this geopolitics.

Regarding environmental issues, the separation between mind and matter (or between the subject enunciator and observed object) generated a double hierarchical dichotomy in global terms. The dichotomy that has been revealed (made visible) is the one between humans and Nature; the separation and the hierarchy that places human beings in a position of domination over Nature, allowing the excessive appropriation of the latter by humans. There is also a second hierarchical division (dichotomy): between Western (North) and non-Western (South) human beings, according to which the former is placed in a superior position in relation to the latter.

> Many other intersections can be added to these dichotomous relationships, such as those highlighted by feminist movements between men and women, white and black women, Western and non-Western women, and so on. Such intersections, although of great relevance, are not the focus of this book and will therefore not be developed here.

FIGURE 1.3 Abyssal lines in geopolitics

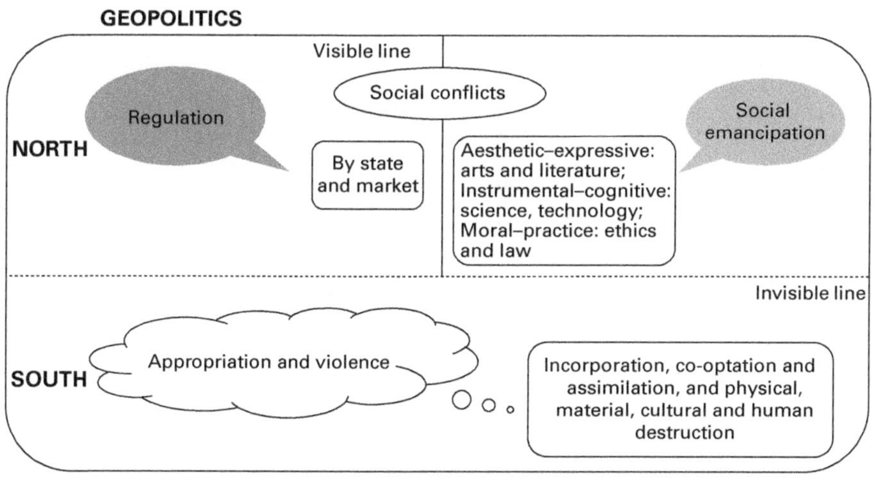

Between these two dichotomies, there is a hierarchy where the first (visible) has been prioritized in relation to the second (invisible) by many environmental studies. It means that, even if the human–nature dichotomy is overcome to end the logic of appropriation–violence imposed on Nature (including it in the regulation–emancipation logic), there will still be the non-Western human being and his Nature in the invisibility of the appropriation–violence logic (Figure 1.4).

That is not just a theoretical inference, but a concrete reality invisible to dominant Western academic, political and economic eyes. Despite being subject to criticism, over the last 40 decades the North has increasingly seen and regulated Nature through socio-environmental conflicts in its territories. As a result, the North has been improving its own environmental and life quality. However, this has not been brought about by removing the abyssal line between humans and Nature but by displacing this line to the South, generating the appropriation–violence of Nature of the South by the North.

The transfer of socio-ecological costs, from the exploitation of natural resources to the sending of waste to the South, is a clear trend and explains the possibility of continuous economic growth in the North associated with the improvement of its environmental quality. Hence the terms 'waste imperialism' and 'toxic colonialism' have emerged, used by social movements in opposition to the transfer of waste from the North to the South (Alier, 2003). The understanding of the existence of these abyssal lines between North and South explains not only the fact that these transfers of environmental costs occur in the North to South direction but also their invisibility.

FIGURE 1.4 Abyssal lines in global socio-ecological problems

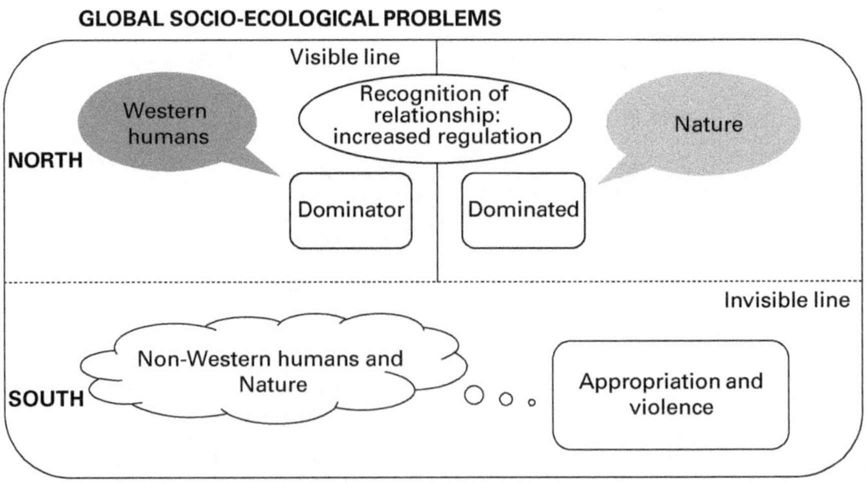

Proof of this is the difficulty in accessing reliable data on cross-border waste transfers (even legal ones) that have been made in increasing quantities in recent decades. Even more difficult is access to data on the many cases of waste traffic to the South and the disastrous socio-ecological consequences generated with more or less intensity and speed by both legal and illegal transfers. It is worth noting that the most famous cases of waste traffic have only become visible to the world due to denunciations made by non-governmental organizations and journalists of the North (such as the *Khian Sea* and Ivory Coast cases). It leads to reflection on how many thousands of other cases remain invisible due to the logic of appropriation and violence applied to the South (like the Brazilian cases described earlier in this chapter).

> The case of the *Khian Sea* only gained prominence internationally thanks to the campaign launched and headed by Greenpeace. The Ivory Coast case became notorious due to an investigative report made and published by the British mainstream media, BBC, followed and endorsed by the British *Guardian* newspaper.

It is important to reiterate that the terms North and South here employed are more political, economic and epistemological than geographic. It can be considered as a geopolitical division between the countries known as 'developed' (most of them located in the global North) and the so-called 'developing' countries (most of them located in the global South). However, the differences are more complex than they may appear. Countries geographically located in the global South can be understood as belonging to the North (like Australia) and vice versa. Also, parts of the South can be found within the North and of the North within the South in terms of internal power relations between different classes, races, and even species.

> In global terms, this book will use the term **North** as synonymous with so-called 'developed' countries and the term **South** as a synonym for 'developing' countries.

Boaventura Santos (2010a) emphasizes that the geopolitical conceptions of North and South overlap the geographical North and South only partially. In geopolitical terms, the South represents all the countries in regions of the world that have been subjected to European colonialism and which, with exceptions (such as Australia and New Zealand), have failed to achieve levels of economic development such as those of the global North. Also, there are relations of domination both within the North and in the South, where there have always been small local elites benefited by the implemented model.

According to Miller (1995), the terms North and South reflect the experiences shared by certain people as a result of historically determined social and economic conditions resulting from the colonial and imperial past, which position a qualitative dividing line between South and North. Anand (2004) stresses that an essential aspect of relations between the countries of the North and the South is that the latter group is economically weaker and more vulnerable to the global economy. Because of this subservient position, the countries of the South are also politically weakened.

Of course, there are many political, cultural, economic and social differences between countries in the North and between countries in the South that cannot be ignored. Nevertheless, what these countries have in common transcends differences and gives them a shared identity and a reason to work together for common goals. The similarities of the powers between the countries of the North and between the countries of the South (and their implications) in the international system are the crucial criteria for global environmental policies and regulations (Anand, 2004). That is the focus adopted here.

According to Anand (2004), although 'the South' is not a homogeneous category, there are many similarities between the countries composing it. Some have many of these shared characteristics, others only a few of them, but generally speaking they are countries typically characterized by (among others):

- low per capita income;
- low levels of education;
- poor health;
- low life expectancy;
- limited infrastructure;
- economic vulnerability, especially in times of global crisis;

- large sectors of agriculture;
- dependence on the export of primary products;
- weak governments;
- lack of governmental responsibility;
- corruption;
- lack of public participation;
- fragility of the rule of law.

It is possible to identify differences in the development and ecological agendas of the North and the South. Regarding development, while the countries of the North focus on the protection of their monetary interests, on the affluence of their lifestyles and the social welfare state, the countries of the South aim, above all, to overcome poverty and to reach a more reasonable quality of life (Anand, 2004).

There is also a discrepancy between their environmental agendas. While the environmental problems of the North are the result of overdevelopment, extravagant consumption and unrestricted demand for ever greater quantities of goods and services, the environmental problems of the South are more closely linked to the poverty issue. Northern environmentalists focus on the prevention of global environmental disasters, while Southern environmentalists seek to protect and sustain their local environment, given their limited financial resources (Anand, 2004). Even more complex, between these two realities are the countries of the emerging economies. Although they are still geopolitically located in the global South, adding to the pending poverty problems, they start to face typical Northern ecological problems.

These incompatibilities between the environmental and development agendas of the countries of the North and South are due to (and are accentuated by) the global economic system in which they are positioned. With the decolonization process of the Southern countries, complex social, economic and political problems arose. It was from this position of vulnerability that these countries were incorporated into the globalized economy. There are many elements that help to explain why the Western hegemonic economic model was adopted by the countries of the South, such as the long-standing connection between these countries and their former colonizers or the perception that the economic strategies used by the North were successful and, therefore, should serve as models for the South (Anand, 2004).

The central fact is that the South's entry into the global economic system was based on unequal economic and political powers. As a result, Southern countries are unable to exert a significant influence on the international economic and environmental systems and its many institutions to, for example, intervene in the operation of legal regimes to monitor trade and the environment. This unequal relationship of power in the global context (or the 'coloniality of power') reflects the epistemological, political and economic hegemony of the North, and grants privileges to the demands and interests of the Northern countries, including the design and operation of international law (Anand, 2004).

The modern Western epistemology has generated blind knowledge, unable to account for the context and the planetary complexity, and which produces unconsciousness and irresponsibility (Morin *et al*, 1999). This mechanistic theory of knowledge has eliminated ecological, cultural, and political contexts, generating implications not only in the production and reproduction of knowledge but also in geopolitical and social relations. Colonialism, in addition to all the dominations by which it is widely known, also operates epistemological domination, maintaining not only political and economic unequal relations but also of knowledge–power (Santos, 2010a).

In order to find strong answers to the strong questions of our times, it is necessary to understand the epistemological and geopolitical bases from which the current complex problems come. According to Morin *et al* (1999), the intellectual solutions used to solve the current planetary problems are themselves the most urgent and most serious problems to be solved. The reductionist, fragmentary, allegedly universal and objective approach of the mechanistic paradigm has proved to be insufficient and incapable of responding to the ecological (and waste) crisis, constituting less a solution, and more the problem itself.

Bringing to light the reductionisms and dichotomies generated by the mechanistic paradigm has been a first step in the search for stronger responses to the waste crisis. From an understanding of the foundations of dominant epistemological and geopolitical paradigms, we can now proceed to higher and more visible levels of study. The next section will thus be devoted to a brief analysis of:

- how these foundations affect the economy, production and consumption patterns and the law; as well as
- how they contribute to the ecological crisis and the waste crisis currently experienced.

Mechanistic economics

GROWTHIST ECONOMY

The mechanistic paradigm is based on the study of matter in motion because it is the only element of the real world that can be measured mathematically. Everything composed of matter can be divided, understood and reordered to overcome natural chaos. The simplification, predictability and functionality offered by this paradigm were also irresistible to the (now known as) social sciences, which sought to understand and reorder the behaviour of people, governments and the economy still in a chaotic state because they were not following the laws of Nature that ruled the entire universe (Rifkin *et al*, 1980).

Searching for ways for the laws of Nature to apply to human beings and social institutions in order to achieve a society in perfect order, the journey from a world of pure matter to a world of pure materialism was rather short. The history of humankind has since come to be seen as a progressive path from a state of disorder and confusion to a well-organized and entirely predictable state represented by the machine world. For Rifkin *et al* (1980), the application of the mechanistic paradigm to the functioning of society was initiated mainly by two authors:

- John Locke, on the functioning of government and society;
- Adam Smith, concerning economics.

Just as the founders of the mechanistic paradigm did with the hard sciences, Locke (2014) excluded God from social issues because it was not a verifiable foundation, and therefore unable to form a basis of public activities. Locke developed the notion of a 'natural' basis of society, observing men and women as mere physical phenomena interacting with other matters in a mechanical universe. He also conceived the idea of a society whose sole purpose would be to protect and allow the enhancement of the properties of its members. Such a society would be formed by individuals who create their own meanings (Rifkin *et al*, 1980; Locke, 2014).

Self-interest (or individualism) would be the solid foundation for the establishment of the state in Locke's (2014) formulation. He argued that society becomes materialistic and individualistic because reason leads humankind to conclude this is the natural order of things. The laws of Nature, then, call upon each individual to act in accordance with their role as a social atom, seeking personal prosperity, which will ultimately lead to social prosperity. The government's aim would be to guarantee the freedom necessary for social individuals to use their power over Nature to produce wealth (Rifkin *et al*, 1980; Locke, 2014).

For Locke (2014), human beings should emancipate themselves from bonds with Nature. Besides, the social role of the state should be to promote Nature's subjugation so social members could appropriate the material prosperity necessary to satisfy their needs without value judgements. In this context, the progressive increase of social wealth would lead to the improvement of social harmony and private property would not only be a right but an obligation for the necessary generation of wealth. With this theory, Locke became the philosopher of unlimited expansion and material abundance, reducing social individuals to the activities of production and consumption, to individual and material interest (Rifkin *et al*, 1980; Locke, 2014).

In economics, Adam Smith (2014) was one of the main writers responsible for the process of scientification of this field in mechanistic terms. The ideal of economic growth has since become a key theme for economics. For Smith, productivity increases originated with the division of labour and with the search for progress and prosperity (Leff, 2006). Smith argued that, if the economy obeyed the laws of Nature unveiled by the mechanistic paradigm, economic growth would occur with the progressive division of labour combined with the use of a maximum proportion of productive labour (Rifkin *et al*, 1980; Tamames, 1983).

Smith's theory (2014) defended the view that government regulations and controls on the economy violated and interfered with these immutable laws of Nature (in this case, represented by the 'invisible hand of the market') by directing economic activities artificially. In this way, he believed in the inefficiency of guiding the 'natural' economic forces, advocating the most efficient method to achieve economic prosperity (in the sense of growth) would be to let the market act without obstacles (Rifkin *et al*, 1980; Smith, 2014).

Following similar steps to those of Locke (2014), Smith (2014) excluded any notion of morality from economics to protect the 'invisible hand of the market' (the natural law that governs the economic system) from any violation. He advocated that individual material interests were the basis of all human activities, and that it would be through the stimulation of these interests that surpluses would overcome shortages, proving beneficial to social well-being (Rifkin *et al*, 1980; Daly, 1993a; Smith, 2014).

The adoption of the mechanistic perspective by economics led to the elimination of the institutional obstacles of the Middle Ages that prevented rapid economic expansion. Mercantilism was, therefore, finally superseded by a new and more flexible international free trade order. According to

Tamames (1983), this new economic perspective inaugurated an era of optimism and confidence in human capacity and infinite growth in a sparsely populated world with large unexplored areas.

According to Rifkin *et al* (1980), the mechanistic paradigm implemented the logic of mathematical order to socio-economic issues, believing in the application of scientific principles for the reorganization of Nature in order to better meet human interests. The operational assumption of this paradigm could be summed up by the logic of accumulation of material well-being by the individuals of a society to achieve a more orderly world. 'Progress' would then represent the accumulation of material abundance, using science and technology as instruments to achieve this goal.

The mechanistic age has thus been characterized by the notion of progress (or of prosperity) reduced to a mere abstraction: a process in which the natural world is subordinated to human beings to create a more orderly material environment and to produce more value from the natural world. Science in this context is the methodology for understanding Nature's actions and then reducing them to consistent rules or principles. Finally, technology is the application of these rules in specific instances to transform natural processes into functional forms of higher value, structure, and order (Rifkin *et al*, 1980).

The mechanistic paradigm brought up profound changes through the revaluation of human rationality, which, freed from the mystical–religious ties of previous periods, produced an anthropocentric and domineering relationship with Nature. It was in this context that new social values were created and strengthened; among them, a characteristically modern faith – faith in progress. The value of 'progress' arose to understand and give ultimate meaning to new notions and institutions, such as science, technology, the nation state, as well as the economic market (Sbert, 2010).

Over time, the term 'progress' has lost prestige mainly as a result of the two world wars and the great economic depression. By the middle of the 20th century, faith in progress had already created and frustrated great social expectations. The term itself had become flagrantly corrupted, being used by the most diverse segments to defend the most varied types of political, social and economic interests. It was then that the word 'development' made a timely entry (Sbert, 2010).

'Development', as a path to be taken, emerged shortly after World War II, with the collapse of European colonial powers, when the United States found the opportunity to give global dimensions to its 'civilizing' mission. It is possible, therefore, to assert the 'Age of Development' began exactly on

20 January 1949, when former US President Harry S Truman, in his inaugural speech, referred for the first time to the South as 'underdeveloped areas' that needed 'help' from the Northern countries to overcome this (new) condition. The term gained wide international acceptance and, as a consequence, provided the cognitive basis for the beginning of a new type of imperialism, both by the uncontrolled interventionism of the North and by Southern self-pity (Sachs, 2010). Sachs says:

> Since then, the relations between North and South have been cast in this mould: 'development' provided the fundamental frame of reference for that mixture of generosity, bribery and oppression which has characterized the policies toward the South.
>
> SOURCE Sachs, 2010: XV

The notion of 'development', posited and diffused from Truman's discourse, far from being just a simple socio-economic enterprise, is a true paradigm that shapes reality according to its precepts; a myth that comforts societies and unleashes passions. At the basis of this notion lies the great mechanistic paradigm of progress (Sachs, 2010). It is necessary to realize and emphasize that this notion of development is homogenizing and partial since it imposes on all the peoples of the planet a single path, with a single direction, to reach a type of 'maturity' of the modern and Western model. From the outset of its use, the ideal of development hides the ultimate goal of Westernization of the world (Sachs, 2010).

The politically correct discourse of 'help', 'aid', 'salvation' of 'underdeveloped' nations was the mask used to conceal the need to create new world markets, to expand the dominant Western economic model and the technology developed by the Northern countries (Ullrich, 2010). To achieve these objectives, this notion of development was constructed from two fallacies: on one hand, the myth that industrialized societies can achieve general well-being, reduce extreme inequalities, and give individuals as much happiness as a society can afford; on the other hand, the reductive conception that economic growth is the necessary and sufficient engine of all social, psychic and moral developments (Morin *et al*, 1999).

> On the subject of 'helping' as a new instrument of power, domination, and maintenance of Northern imperialism, see the work of Marianne Gronemeyer in *The Development Dictionary: a guide to knowledge as power*.

The notion of development came to occupy the centre of a robust semantic network. Undoubtedly, no other concept in modern thought has had its force and influence on human thinking and behaviour (Esteva, 2010). It was precisely because of its strength, linked to blind faith in progress, that blind faith in development allowed to conceal and/or justify great barbarism practised in its name (Morin *et al*, 1999).

This word, in its meaning as understood from Truman, has a particular perception, a point of view: the mechanistic worldview. 'Underdeveloped' is nothing more than a comparative adjective coming from the Western premise of oneness, homogeneity, and linearity of world evolution (Esteva, 2010). There will only be 'underdeveloped' countries if there are 'developed' ones, and these can only be compared if they are considered as hierarchical entities in the same linear process where the arrival point is the (impossible) universal achievement of the dominant modern Western model. In this regard, Morin *et al* argue that:

> The idea of development has always had an economic–technical basis, one that is measurable by growth and income indicators. Implicitly, it supposes that techno-economic development is the engine that naturally brings 'human development' along with it, and that its completed, successful model is that of the supposedly developed countries – in other words, Western countries. This view presupposes that the present state of Western societies is the goal and purpose of human history.
>
> SOURCE Morin *et al*, 1999:76

The result of the propagation of this model was an immense loss of diversities of all kinds. This monoculture, whose main universalizing forces were the market, the state, and science, caused the 'other' to disappear. It corroded many of the viable alternatives to the industrial society, which is geared solely to growth and which dangerously mutilates the human capacity to find answers to the complexity of contemporary problems (Sachs, 2010).

If the concept of development has undergone with Truman the most dramatic and grotesque metamorphosis of its entire history, there is no doubt that it has been further impoverished by its early theoretical advocates, who have reduced it to the notion of 'economic growth'. For such theorists, 'development' would be a simple growth of per capita income in so-called 'underdeveloped' areas. The general and global obsessions with the totalizing process of industrialization and with the absolute goal of GDP growth have obscured any more critical and differentiated view of the notion of development. In the 1950s, optimism prevailed all over the world, and all indices and reports indicating improvements in social programmes considered these results were due to rapid GDP growth (Esteva, 2010).

In the 1990s, the 'economicist' development deepened, reflected in different forms in the global North and South. In the North, development of what was 'poorly developed' or of what was already obsolete was sought. In the South, redevelopment demanded the demolition of what was left of the adjustment process to the modern Western model of the previous years, vacating and opening spaces for (economic, technological and industrial) waste from the North. The obsessive aspiration for overcoming the 'underdevelopment' condition of the Southern countries has destroyed many of the alternatives to this development model (Esteva, 2010).

Even with adjectives such as 'sustainable', the dominant notion of 'development' continues to ignore what is not calculable or measurable, and its only measure lies in economic growth. The mechanistic economic model focuses on economic growth in search of progress and prosperity, which is calculated by the increase of GDP per capita. That is the reason why GDP growth has been the most important political goal around the world for much of the last century, especially in the global South (Jackson, 2011).

LINEAR PRODUCTION AND CONSUMERISM

The growthist model of the economy implanted by the mechanistic paradigm also impacted the patterns of production and consumption. In terms of production patterns, the Industrial Revolution itself was motivated and guided by the logic of material acquisition and accumulation. The secondary sector (manufacturing/industry) thus needed more efficient means of production to manufacture the highest quantity of goods possible for the greatest number of people. That meant the need to migrate from a manual labour system to a mechanized system designed to be practical, cost-effective, efficient and linear (Braungart *et al*, 2009).

In the 19th century, when mass production and consumption practices began, this logic deepened even further and the industrial design model settled on a linear logic, focused on manufacturing products and sending them to consumers in the fastest and cheapest way possible. It is a linear, uni-directional, 'cradle-to-grave' model (also known as the 'take–make–dispose' model), where resources are extracted, shaped into products, sold, used and eventually disposed of in some 'grave', usually landfills or incinerators (Braungart *et al*, 2009).

This production pattern arose when environmental issues were not a source of widespread concern since Nature's regenerative and reproductive capacities were believed to be unlimited, while Nature was seen as a brute force to be controlled and rearranged by humans. Currently, however, it is widely known how Nature, and its processes are fragile, vulnerable, complex and interconnected. Even so, modern industries continue to operate, for the most part, under the logic of the mechanistic paradigm. Many industrialists, designers, and engineers still do not see their work as part of a larger system, beyond the economic system (Braungart *et al*, 2009).

This linear, 'cradle-to-grave' industrial model, derived from the mechanistic paradigm, is based on material extraction, intensive generation and disposal of wastes, and still dominates today's industries. Research shows that in recent decades more than 90 per cent of the materials extracted to produce durable goods in the United States, for example, have become waste almost immediately. In many cases, the said durable product itself has a shorter lifespan than it should or could have. As mentioned earlier in this chapter, what is seen in residential bins is just the tip of the waste iceberg, since the products themselves used and discarded by consumers contain an average of only 5 per cent of the materials involved in the entire production process, transport and commercialization (Braungart *et al*, 2009).

Products manufactured in Europe (from furniture to electronic equipment) have a nine-year average life expectancy. However, ongoing research indicates that the lifetime of products in general, especially consumer electronics, has declined steadily due to consumer preferences and planned obsolescence. This production logic, besides generating overexploitation of natural resources and overproduction of waste, causes the economic value of these products to disappear after their short life due to the losses of matter and energy it generates (EMF, 2015).

About planned obsolescence, see the work *Understanding Planned Obsolescence: Unsustainability from production, consumption to waste generation* (Pope, 2017).

In 2012, for example, only 5 per cent of the value of material and energy used in production processes in Europe were recovered through recycling and energy recovery of waste. The remaining 95 per cent were lost after the first use-cycle of the products. In terms of weight, only 40 per cent of the materials collected by the EU solid waste systems are recycled; 25 per cent are incinerated with energy recovery, and the remainder is disposed of in landfills. Thus, Europe mostly still uses resources only once and then discards them (EMF, 2015).

The industrial infrastructure still prevailing is designed to pursue economic growth without considering other vital concerns, especially regarding human and ecological health and cultural and natural diversities. Except for some positive results, most of the methods and materials used and produced by the contemporary industries generate depletion and contamination of natural resources. If in the pre-industrial period most of the goods produced could be *consumed*, with the industrialization and development of new synthetic materials most of the manufactured products can only be *used*. However, the rapid replacement of products culture still persists (Braungart *et al*, 2009).

> Products that may actually be consumed are the biodegradable ones. In pre-industrial times, most products could readily biodegrade, except metals. These, however, because highly prized, were kept, reused or recycled. With the advancement of industrialization and technological development, most manufactured goods can no longer naturally biodegrade, so they can only be used (rather than consumed). With the decline of material prices and new synthetic materials increasing in the post-war market, it has become cheaper for the industry to make new materials and products and sell them than to create a local reverse logistics infrastructure and reprocessing of such materials. Thus, while in the pre-industrial period and the first decades of industrialization the logic of reuse of consumer goods reigned, today most products are seen as 'disposable', including so-called 'durable' goods (Braungart *et al*, 2009).

Although the Industrial Revolution, driven by the modern Western mechanistic paradigm, has generated many positive social changes, the production pattern it has created (and which still persists) has fundamental flaws that generate crucial omissions and devastating socio-ecological consequences (Braungart *et al*, 2009). To the current production patterns are added, also, unsustainable consumption patterns generated by another revolution, termed by some authors the Consumer Revolution.

The mechanistic economic model based on the need for economic growth led to an acceleration of industrial productivity. However, this model demanded more ferocious consumers, capable of offering a demand compatible with the new supply levels. At this point, profound changes occurred in these societies, resulting in the Consumer Revolution, defined by Bauman (2007) as:

> ... the passage from consumption to 'consumerism', when consumption, as Colin Campbell suggests, became 'especially important if not actually central' to the lives of the majority of people, 'the very purpose of existence'; and when 'our ability to 'want', to 'desire' and 'to long for', and especially our ability to experience such emotions repeatedly, actually underpins the economy' of human togetherness.
>
> SOURCE Bauman, 2007:26

Bustamante (2007) argues that consumerism has its origins in the needs created artificially by the production system, which shapes them as cultural needs through various strategies – such as planned obsolescence, for example. Bustamante defines the phenomenon as typical of the consumer society in places or sectors that do not have economic difficulties of access. It is characterized by the acquisition of unnecessary products and services, by the 'mere act of consuming' constituted into a habit. Consumerism can thus be considered as a significant obstacle to the rational use of limited natural resources.

According to Leonard (2010), Henry Ford played a decisive role in the Consumer Revolution. He developed a strategy to forge a consumer class, who would then work fewer hours, earn higher wages and, consequently, be able to consume more. That was the beginning of the path that would lead to the consumer society, where consumerism, in sharp opposition to previous forms of life, associates the idea of happiness (well-being) with material consumption. However, this happiness would not be realized by the satisfaction of basic necessities, but by the immediate use and by the rapid replacement of goods destined to satisfy an ever-increasing volume and intensity of human desires (Bauman, 2007).

For Baudrillard (1998), the 'abundance' of the consumer society is directly associated with wastage. It means that the representation of abundance in this new type of society ceases to be made by the possession of goods that are useful. It is carried out by the excess and wastage of goods, whether they are useful or not. Thus, the consumption pattern developed from the mechanistic paradigm is further away from the real, from the pure satisfaction of vital and cultural needs. It thrives on the bond of the idea of happiness (well-being, prosperity) to the acquisition of consumer goods (precisely according to Locke and Smith's theories) coupled with the perpetual non-satisfaction of social members (Bauman, 2007). Many strategies have been developed to achieve this formula, such as advertising and planned obsolescence.

The reduction of the notion of development to 'economic prosperity', and of that notion to 'economic growth', is a result of the mechanistic paradigm. The patterns of linear and intensive production along with consumerism have been developed to achieve the growthist goal. Those are all consequences of modern Western logic. However, especially since the end of the 20th century, these models and patterns began to be criticized.

Among such criticism is the one affirming that the economic growth achieved so far has resulted in some benefits, but not in an egalitarian way. According to Jackson (2011), in the first decade of the 21st century, one-fifth of the world's population earned 2 per cent of the global income. The wealthiest 20 per cent, by contrast, earned 74 per cent of that income, proving the vast disparities between rich and poor. Far from raising the living standards of those who need it, the logic of economic growth has left the majority of the world's population out of the prosperity feast for the last 60 years.

This distributive economic injustice is only one of the reasons for questioning the logic of growth. Another reason is based on the fact that economic growth does not help, and to some extent even prevents, human happiness. Almost all indicators trying to measure social 'progress' in the United States (one of the most significant contributors to consumerism) show there has been no change in social well-being, despite continued economic growth in recent decades (Jackson, 2011; Leonard, 2010).

Jackson (2011) and Latouche (2009) suggest the contrary is happening: the progressive decrease in quality of life. That is demonstrated by several factors, such as the increase in obesity, the number of suicides, the use of antidepressants, among others. It seems that materialism promoted by the mechanistic paradigm has been making people unhappy. For Latouche (2009), our societies have become 'toxic-dependent' on growth, where the search for the satisfaction of material desires generates work addiction, maintained by the excessive consumption of drugs (legal and illegal ones).

Finally, perhaps the most obvious of reasons: such economic models and patterns do not take into account the biophysical limits of the planet and, as a consequence, have led to the processing of resources with a higher speed than the population demographic increase (Jackson, 2011). As early as the 1970s, *The Limits to Growth* report warned of this problem by predicting three stages in the process of biophysical limitation for the growthist economy. The first stage was represented by resources scarcity, followed by rising market prices (second stage), and, if material processing were not diminished, the carrying capacity of ecosystems would reach a fatal limit and could prevent the continuation of human life on the planet (third stage) (Meadows *et al*, 1972).

The problem of resource scarcity is therefore only part of the issue of the biophysical limits of the planet. Currently, the debate has been even more intense on the issue of carrying capacity, since it has become increasingly evident how, even before natural resources are exhausted, the planet's carrying capacity will be exhausted (Jackson, 2011).

Many qualifications and refinements have been added to modern economic theories to this day. Still, the economic system continues to be seen as a closed mechanical system in which supply and demand functions continually adjust to give it balance and allow continuous and infinite growth. Likewise, the linear production pattern and the 'consumerism' pattern remain dominant. These models are based on the paradigm of continuous and unlimited material progress, viewing the environment as an inexhaustible source of resources and sinkhole for rejects (Rifkin *et al*, 1980).

However, as will be seen in Chapter 3, the first and second laws of thermodynamics contradict such claims. The first states that all matter and energy are fixed and cannot be created or destroyed, but only transformed. The second states that they can only be transformed in one direction: from available to not available or from usable to not usable. Therefore, whenever matter and energy are extracted from the ecosystems and processed by the socio-economic system, some of them will be dissipated or wasted at every stage, until, in the end, everything, including what is transformed into a product, will end up as waste and as dissipated energy (Rifkin *et al*, 1980).

In biophysical terms, there is no possibility of material progress towards the accumulation of a permanent stock of usable goods. Thus, the increase in industrial productivity means an increase in material and energy flows between the ecosystem and the socio-economic system, increasing the disorder of the ecosystem that will ultimately be endured and borne by society. Furthermore, the complexity and increase in the number of stages of the chain of production processes, characteristic of the current globalized economy, mean that more energy and matter will be dissipated and wasted, and less matter and energy will be available to other living beings and future generations (Rifkin *et al*, 1980).

Currently, it is known that around 60 per cent of the world's ecosystem services have been degraded due to their overuse since the mid-20th century. During that same period, the global economy 'grew' more than five times.

This extraordinary increase in economic activities has no historical precedents and is entirely contrary to current scientific knowledge about the finite limits of natural resources and the fragility of the ecosystems in which humans (and other living things) participate and on which they depend for survival (Jackson, 2011).

The social metabolism logic, associated with the laws of thermodynamics, shows that the continuous and growing processing of the socio-economic system leads to the dissipation and unavailability of matter and energy to other living beings and future generations. Waste produced by this system cannot be 'thrown away' since it will accumulate in ecosystems as degraded matter. If the socio-economic system continues to function with the current patterns, the biosphere will reach the limits of its carrying capacity, and the Earth will become a large 'grave' (Braungart *et al*, 2009).

If the idea of a non-growing economy is seen as an abomination by traditional economists, the idea of a continuously growing economy is an abomination to ecologists. No subsystem contained in a finite system can grow indefinitely in biophysical terms. Thus, the economic model and the production and consumption patterns of the mechanistic paradigm have failed in every way. Besides, the globalization of this model has generated wealth but also increased areas of poverty, especially in the South, provoking a series of crises.

After years seeking this type of development, North–South imbalances and countries' internal inequalities have not only remained but also deepened. While some countries marched towards disaster, others who were able to emerge from economic 'underdevelopment' came to grips with the civilization problems typical of the 'developed' world (Morin *et al*, 1999).

The mechanistic economy presents itself as a mathematically advanced social science, linked to science and technology, supported by the ideals of materialism and progress. Still, it has also proved to be the most backward science in social and human terms since it abstracts from social, historical, political, psychological and ecological contexts and conditions that are inseparable from economic activities (Morin *et al*, 1999).

Even so, the search for economic growth as a path to social progress (welfare) remains the prevailing discourse for public policy around the globe. Such policies are increasingly divorced from reality and less capable of explaining and giving strong answers to the current problems faced by societies (Rifkin *et al*, 1980). The mechanistic economic model, whose patterns of production and consumption caused the current socio-ecological problems, cannot also be the solution to these problems (Santos, 2010b).

The resilience of the mechanistic economy in global terms, as well as the coloniality of Western power and knowledge, has found support in another field of knowledge: law. The mechanistic paradigm has also generated a significant and transformative impact in law, whose effects persist until the current day. Law, as will be seen in the next section, was deeply impacted by this paradigm and has contributed significantly to the perpetuation of the mechanistic view.

Mechanistic law

Following the line of the sciences and economics, modern Western jurists fragmented the medieval legal order – a holistic system which had adapted customary religious norms and Roman law to the practical requirements of human relations. Thus, in the transition to Modernity, Western legal thinkers began to conceive law as an aggregate of distinct parts strictly governed by the natural laws of individual human reason. The old holistic view of the world as a cosmos, of Earth as a generous gift from God to humankind, and as an abundant community that could be accessed by all, was replaced by a humanistic emphasis on individual and human reason. This process resulted in a mechanistic legal vision (Capra and Mattei, 2015).

Modern Western jurists prepared the foundations for a dramatic transformation of legal institutions. They adapted those institutions based on the notion of commons, firstly to form concentrated, legally formalized individual private ownership, and then the capital. Individual private property became the most important legal concept of the dominant legal system. In turn, mechanistic jurists found that the ideal mechanism to govern this new system is what is now known as state sovereignty (Capra and Mattei, 2015).

For centuries, modern Western legal jurisprudence has worked to untangle the strings of solidarity that had bound human beings together up until the Middle Ages. The conquest of individual private property was presented as a claim to freedom. It moved away from the complex network of rights and duties determined by land use to maintain the ecological harmony and interpersonal solidarity existing in previous paradigms. Currently, such rights and duties have been partially reinstated through laws and regulations seen as external constraints (Ost, 1995).

Just as in the sciences, law had its own 'Copernican Revolution', which placed human beings at the centre of the universe in the form of an increasingly vigorous subjectivation. Law began to be created based on the subjective right of the sovereign individual and no longer on common use. Society, in its turn, gained an atomized and dynamic order in whose centre lay the right of individual private property. Moreover, just as Cartesian Nature is wholly fulfilled and can be divided into as many parts as necessary, Mechanistic Law imposes the complete occupation of natural spaces by individual properties, which, besides integrality, also has the characteristic of unlimited divisibility (Ost, 1995).

It is interesting to note that, just as for the sciences, Francis Bacon was also one of the pioneers of Mechanistic Law. He was one of the first jurists to severely criticize the tradition of customary law by drawing an explicit analogy between the laws of Nature and human laws. Hugo Grotius also exerted significant influence on this new law paradigm, founding a legal school based on the mechanistic vision of rational natural laws. The controversial relationship between individual private property and state sovereignty found answers with the works of Thomas Hobbes and Jean Bodin, justifying their co-existence (Capra and Mattei, 2015).

According to Ost (1995), Hobbes's *Leviathan* was for political philosophy what Descartes' *Discourse on Method* was for the theory of knowledge. Both have replaced the view of a solidarity-based, traditional and immutable world by a new and different one. Leviathan was a mortal god, an artificial man who symbolized the state and whose sovereignty served his own production and defence. Like other mechanistic authors, Hobbes liberated society from the divine presence, allowing the positivist law to emerge beside the positivist science.

The mechanistic paradigm has also been extended to international law, simplifying it to a system composed of nation states, completely independent of each other. The two organizational principles of modern Western Law, *individual private property* and *state sovereignty*, were driven by the need to justify the process of colonization of the New World. The concept of legal domination of 'unoccupied' lands inhabited by 'savages', who did not have the Christian God, rationality, and any notion of property, prevailed (Capra and Mattei, 2015).

In colonized territories, the modern Western legal system, freed from holistic concerns of justice and feudal traditions, developed its full exploratory potential. Mechanistic Law defended the fundamental right of human beings and science to dominate both Nature and other human beings in a 'natural state'. It gave a rational justification for the exploitation of new territories and their native communities. Thus, in modern Western law, the visible abyssal line marks the dichotomy between what is regarded as legal or as illegal under national or international law (Figure 1.5). This line stands on the invisible abyssal line that separates the dichotomy of the domain of Law from the domain of Non-Law; the latter applies to lawless or outlaw territories (Santos, 2010a).

Originally, these territories coincided with the colonial zones. In them, everything that could not be thought of in terms of true or false (science), or legal or illegal (law) occurred. Historically, therefore, the invisible abyssal line was the global line that separated the Old World from the New World, which made possible the emergence of Mechanistic Law in the North, and especially in the international arena. Colonial territories, then, did not represent the legal or illegal, but the lawless (Santos, 2010a).

The same logic of abyssal lines can be applied to the legal notion of property. In Mechanistic Law, the property will be divided between public and private. There is also the invisible line separating this dichotomy from 'ownerless properties', *res nullius* (things not yet appropriated) and *res communis* (things that belong to no one and are used commonly by all) to which the logic of appropriation and violence applies. Thus, modern Western property law has the vocation to cover all available space; on the one hand, by the regulation–emancipation of individual private property and state sovereignty in the North and, on the other, by the appropriation–violence of the colonial territories seen as *res nullius*.

The mechanistic paradigm reached a dominant position in the law in France with the triumph of the French Revolution. Numerous mechanistic notions can be found in the Napoleonic Code. Article 5 of this Code, for

FIGURE 1.5 Abyssal lines in law

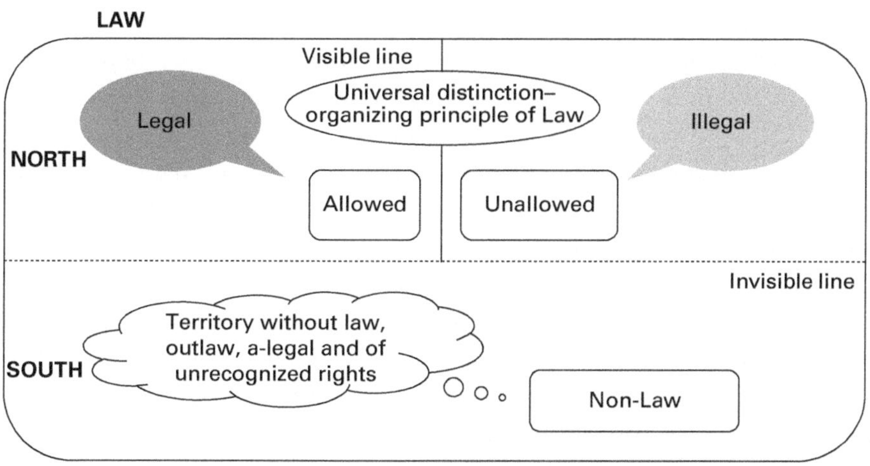

example, limits judges to the mechanical application of the rules to the facts to be judged by them. In other words, the Napoleonic Code, designed as written reason, was developed to be applied by the state (judiciary) with little, if any, interpretation, through pure mechanistic logic. This approach, whether through military imposition or perpetuated intellectual reputation, has become the backbone of professionalized law around the world (Capra and Mattei, 2015).

In the 20th century, the mechanistic paradigm of law became a consensus among jurists around the world. Such jurists agreed that, in advanced societies, law is (or must be) a pure system, ontologically separate and different from politics, morality, justice, religion or cultural norms. Such a positivist and formalist approach to law, which treats it as a purely objective structure and as a system of mastery over facts rather than values, has thus become the only canon of Western law (Capra and Mattei, 2015).

Although he was not the creator of Positivist Law, Hans Kelsen (2009) was undoubtedly one of its great propagators. His Pure Theory of Law is seen as the most influential theory produced during the 20th century in this field, and one of the highest points of modern Western legal thinking. Kelsen devoted himself to the development of the foundations of law as a social (and not natural) science, removing from his analysis all the references that he considered as foreign to the field, especially sociological and axiological references. The notion of justice, for instance, was excluded from the legal field because the imprecision and fluidity of its meaning did not fit into a law that was supposed to be universally valid. Law, therefore, was only a set of hierarchical rules detached from values.

This mechanistic legal paradigm of law remained intact, even with all the legal movements that have emerged since Modernity. That is because all these movements have remained within the limits established by the opposition between private law (market) and public law (state) both in the capitalist and in the communist systems. However, the responses suggested by such movements failed to meet the post-war challenges, such as the problems arising from massive urbanization and industrial production. Instead, they helped alienate Law further from communities, its ontological base, leading to a greater professionalization of the political process of the establishment of law (Capra and Mattei, 2015).

The mechanistic paradigm in law treats reality as a set of distinct and definable individuals, private owners whose individual rights are protected by the state. From this perspective, Law is viewed as an object separate from the subject applying it and to whom it is applied. It bequeaths the imposing top-down view to the construction of the codified Law. By breaking with the medieval pluralistic view, Mechanistic Law sees human laws as a mechanical chain of order transmission from the top to the bottom, to which obedience is due in respect of legality (Capra and Mattei, 2015).

For Capra and Mattei, with the adoption of the mechanistic paradigm, Law became subservient to Economics, the queen of social sciences. It began to be seen as an aggregate of positive scientific concepts from which rules could be deduced as a means to achieving the specific purpose of ensuring market efficiency. This dominant legal view brings the myth of scientific predictability of Law to the efficient organization of the market. It does not open space for the community to give its own meaning to Law, nor for significant local variations, transforming Law from a (local) culture into a (global) technology.

Although widely criticized, the dominant conception of Law is still that of an objective, state-governed system that has the power to decide conflicts between individuals. Law and economics continue to support a short-term view, observing reality in a mechanistic way. Both fields place at the centre of their perspectives the atomized and abstract individual owner. Private owners can exercise their rights to land use, extracting the maximum value from the commons at the expense of others (contemporaries and future generations). Modern Western law embodies the typical mechanistic separation with an objectified legal order. Such a legal system is analogous to Descartes' *res extensa* (the object of thought), which is substantially separated from the domain of human action, the *res cogitans* (the thinking subject) (Capra and Mattei, 2015).

Since Modernity, no truly community-based understanding of law has re-emerged. The mechanistic view has been transformed into an ideological component of the dominant extractive economic model. As a result, legal positivism still governs as the dominant conception of law, treating it as immutable technology that can be accessed only through jurists and legal professionals (Capra and Mattei, 2015).

Although the mechanistic approach has also yielded positive results, its limitations have become increasingly apparent, particularly in relation to Law. According to Capra and Mattei (2015), based on the second law of thermodynamics, an irresistible evolution towards disorder and destruction seems inevitable. Such an image of the world as a machine that deteriorates due to mechanistic and immutable human laws produces disempowerment and distress.

The mechanistic perspective takes law away from communities, expropriating them of their most fundamental common good: control over their own legal order. The top-down elaboration and application of Law and the distancing from its ecological foundations have generated disempowerment of the communities. As a result, its members are discouraged from connecting with the legal order aiming at the transformation (for better) of their own reality. While such expropriation may occasionally produce desirable and positive results, it has often been lethal to the environment and community identities (Capra and Mattei, 2015).

Legal decisions on land use, for example, follow the logic of choice between individual ownership and government interests, which, although apparently conflicting, follow the same logic of extractive values. Thus, issues that have a substantial direct impact on the communities' well-being (such as access to justice) are decided by professionals at levels quite remote from those affected by such decisions: the communities themselves (Capra and Mattei, 2015).

With centralized power in the state and the hierarchy of law courts, the only legitimate choices in Western societies today are between more or less government and more or less property. Any other view of law that was different from this mechanistic, professional and hierarchical legal order was relegated to illegality. For social members, to 'sue' through the filter of legal professionals responsible for the law machinery has become the only 'right of resistance' in the dominant legal system. Law has progressively become a technology to support the constriction of political action (Capra and Mattei, 2015).

These are the visible limitations caused by the dichotomy of the North between the legal and the illegal, where tension between regulation and emancipation reigns. It happens that the mechanistic paradigm, as seen, beyond this visible abyssal line, has established an invisible abyssal line that separates Law from Non-Law, where the logic of appropriation and violence prevails. While in the North, law is unthinkable without the distinction between the rights of persons and the right of things, the logic of appropriation and violence recognizes only the right of things, whether human or not (Santos, 2010a).

Although the political colonialism of Western Modernity has been extinguished, the coloniality of power persists and makes possible the co-existence between Law and Non-Law, between the regulation–emancipation and appropriation–violence dichotomy. The abyssal lines continue to structure knowledge and law and are constitutive of the political and cultural relations and interactions that the West leads. These juridical and epistemological cartographies produce a radical denial and a radical absence: the denial and absence of humanity. The denial of part of humanity is a condition for the other part to affirm itself as universal (Santos, 2010a).

An excellent example of this facet of Mechanistic Law is, especially at the international level, the emergence of Soft Law. Such Law, of voluntary compliance, presents itself as a more benevolent manifestation of the logic of regulation–emancipation but also brings the logic of appropriation–violence whenever very unequal relations of power are involved. According to Boaventura dos Santos (2010a), the plasticity of Soft Law presents intriguing similarities with Colonial Law, whose application depended more on the will of the colonizer than on anything else.

In addition to the denial of part of humanity, Mechanistic Law has also denied Nature. This modern Western legal system, dominated by Mechanistic Economics while at the same time instrumentalizing it, is rooted in the dichotomy between individual property and state sovereignty and fuels the current ecologically destructive practices. Much of the political debate is firmly anchored in economics, which, by calling itself an exact science, determines the design of public policies and legislation (Capra and Mattei, 2015). The economic growthist model and unsustainable production and consumption patterns are then promoted by law, inducing individual extractive behaviours.

Although the social and ecological effects of these dominant models of Economics and Law are quite clear, this understanding has not satisfactorily promoted the development of public policies that escape the mechanistic logic. Fragmentation and linear thinking, typical of the mechanistic paradigm, dominate current political and economic debates under blind faith in technological progress and the infinite growth of the economic system. The prevalent idea of development is fundamentally reductionist, based on the notion of 'progress' of previous centuries and based on the concept of GDP as an index for social welfare measure.

Unlimited extraction and exploitation of natural and human resources underpinned by mechanistic models are at odds with fundamental principles of ecology. The violation of these principles has consequences as serious and lethal as ignoring the law of gravity when climbing a mountain. However, the fact that the consequences of these models are scattered over time and space makes them more difficult to be seen, understood, rejected and overcome (Capra and Mattei, 2015).

Once again, society has been giving weak answers to strong questions. As a source of balance and sustainability to the socio-environmental problems of our times, Economics and Law have given their answer as more science, technology and markets (now environmental markets). The impact of the discrepancy between the strong questions and the weak responses are quite different in the global North and South. Weak responses enable and ensure the continuation of the coloniality of power, so the weak answers in the South translate into impositions and violence of all kinds. What unites, on the other hand, North and South, is the intractable lack of credible and prudent (development, economic, legal) models, capable of guaranteeing everyone, including Nature, a decent existence (Santos, 2010b).

Final considerations

This chapter sought to analyse the possible origins and causes of what the United Nations regards as a 'global waste crisis'. In the first section, data, statistics and literature commentaries showed a quantitative increase in the generation of waste, as well as in the types of waste with increasingly hazardous characteristics in the last decades. In addition to and as a result of such increases, there has also been a substantial rise in the number of cross-border waste transfers, notably in the global North–South direction. This trend can be seen across a number of factors, such as the hyper-regulation of waste in the global North and the ease and cost-effectiveness of waste treatment and disposal in the global South.

The underreporting of waste traffic cases, as well as an absence of registration and systematization of cross-border waste transfers, both illegal and legal, is clear. Still, it is possible to conclude that waste has become a contemporary global problem, the effects of which have disproportionately hit the global South. Cases of waste traffic discovered, whether by public authorities or non-governmental organizations, have generally demonstrated the ineffectiveness of the international law currently regulating the subject, in particular the Basel Convention.

The social metabolism approach confirmed that social and economic systems cannot be treated as closed systems since they exchange matter and energy (flows) with the ecosystem for the sustenance of human activities. It was observed that, based on an environmental approach and with a broad focus on material flows, it is possible to understand:

- the biophysical impossibility of maintaining the model of infinite economic growth and the complete globalization of its production and consumption patterns;
- the relationship between resource extraction and waste generation;
- the need to pay greater attention to large (and not only hazardous) waste flows; and
- the inevitable loss of matter in the form of waste in all socio-economic processes according to the entropy principle.

It was also noted how the social metabolism approach brings to light concepts such as carrying capacity, transfer of socio-ecological costs and ecologically unequal exchange as essential elements of the debate on international trade. That is because international trade has allowed the flexibilization of the carrying capacity of different territories around the globe, as well as the externalization of environmental costs and the realization of unequal exchanges between countries (mainly between North and South) through ecological rucksack (or hidden flows).

These characteristics of current international trade have made possible the continuation of the economic growth of the Northern countries together with the improvement of their environmental quality. The importance of using this approach by social and human sciences (such as law) was stressed. That would allow the development and implementation of concrete measures for a fairer and more sustainable regulation of the social metabolism of the social and economic systems to, ultimately, overcome the waste crisis.

Beyond the biophysical reasons for this crisis, it is necessary to unveil the reasons situated in deeper levels of analysis to bring about paradigm shifts. It was seen how modern Western thought inaugurated an epistemological paradigm, here called mechanistic, marked by reductionism, isolation, dichotomization, determinism and universality, among others. These characteristics meant that knowledge came to be produced increasingly distant from its ontological bases and the complexity of real life, creating general blindness.

This mechanistic paradigm has generated abyssal lines denying part of humanity and also Nature by applying the logic of regulation–emancipation to the North and of violence–appropriation to the South. This paradigm has extended its tentacles to global geopolitical relations, generating domination (first as 'colonial' and then as 'coloniality') of the North over the South, reflected in various forms, including in the development and functioning of international law. Moreover, in addition to the exact sciences, the new

Western paradigm quickly conquered other areas of knowledge, notably Economics and Law.

In economics, the idea of material accumulation by individuals for their well-being in a more orderly world has become its backbone. Since the adoption of this paradigm, theoretical economic formulations have been developed in the search for progress, prosperity, and, finally, development – all notions reduced to the (unrealistic) idea of infinite economic growth. The mechanistic economic logic generated reflections in the patterns of production and consumption of Modernity. The production system inaugurated with this paradigm was designed to function in a linear, uni-directional and intensive way. On the other hand, consumerism became the rule, with the values of speed, excess and wastage guiding the current dominant consumption pattern.

This economic model, however, has not fulfilled its promises of social welfare. On the contrary, in addition to deepening social and geopolitical inequality, the decrease in quality of life and of the biophysical limits of the planet have been witnessed. These impacts are not taken into account by the mechanistic paradigm, which sees the economic system as closed and autonomous, and therefore capable of growing indefinitely.

Law, in the mechanistic paradigm, came to be treated as an aggregate of distinct parts governed by human reason. Private individual ownership has become the most important legal concept in the new legal model, governed and regulated by state sovereignty. This model revolves around the dichotomy of public–private law, so all legal proposals built within this logic are only variations between, on the one hand, more or less government and, on the other, more or less market.

Like in economics, Mechanistic Law has distanced itself from its ontological foundations in the search for a pure and objective structure of the domain of facts, not of values. It generated the separation between the object of study and the subject, and created, as well as in epistemological and geopolitical terms, abyssal lines dividing legal and illegal (in the North) and Law and Non-Law (North versus South). However, the mechanistic view of the world has been and is still so internalized in each of us, that it is hard to question our assumptions and understand criticisms of it, and rendering it successful in its continued domination. For this reason, it prevents the identification and understanding of the dimension and complexity of current social and ecological problems.

The mechanistic paradigm has therefore been blamed for many of today's socio-ecological problems, and even for the current ecological crises (including the waste crisis). As Boaventura dos Santos (2010a) points out, global socio-ecological injustice is closely linked to global cognitive injustice. Thus, in order for the current unsustainable (development, economic, legal) models to be overcome, a new type of thought, of epistemology, is needed. It must be, in Santos' words, a post-abyssal thinking.

02

Waste as a problem of justice

Introduction

The mechanistic paradigm generated (and continues to create) ecological crises whose impact affects North and South differently. The undeniable occurrence of environmental catastrophes around the world since the second half of the twentieth century, the social transfer of socio-ecological costs to the most vulnerable communities and countries, and the inability of current dominant models to provide adequate responses to these social and ecological problems, all indicate this paradigm is collapsing. The visibility of these problems is generating socio-environmental conflicts and the theoretical and political interest to solve such negative consequences of the current development model.

More recently, within these ecological crises arose the waste crisis. According to Aragão (2006), such turmoil is the most evident symptom of unsustainability in the use of natural resources. In reverse chronological order, the enormous generation of waste of increasingly hazardous qualities has caused:

- **a** undesirable land uses (since wastes are solid, voluminous and polluting objects);
- **b** pollution (in all management stages, especially in the elimination); and
- **c** the exhaustion of natural resources with the wastage of matter and energy, affecting all collectively, but especially the most vulnerable communities and countries.

According to Aragão (2006), waste is one of the most complex and relevant issues of today's society, and it implicates:

- **i** the proliferation of legislation, adoption of standards, celebration of contracts and practice of administrative acts all related to waste management at a legal level (for which reason waste has become a legal issue); and ii the multiplication of socio-environmental conflicts related to (poor) waste management at a social level (which shows that waste has also became a problem of justice).

As seen in the previous chapter, the mechanistic legal model, while aspiring for purity, objectivity, and positivism, removed Law from its ontological bases, isolating it from other domains, such as politics, morality, and justice. Thus, concerns about justice have come to be considered as 'pre-legal' issues.

This book intends to analyse how the waste problem has been debated in both fields: of justice and legal. The analysis of the legal regulation of waste will be presented in Part Two. Chapter 4 will give a more panoramic view of the legal regulation of waste within the methodological scope of this book. Chapter 5, in its turn, will emphasize the regulation of cross-border waste transfer, an operation that qualifies waste management as global.

The first objective of this chapter will be the study of the plurality existing within the environmental movement as a form of socio-environmental conflict, with the aim of verifying the potential of the movement's various discourses and strategies to deal with the complexity of the waste crisis. The names given to the streams of the environmental movement highlighted will be:

- cult of wilderness;
- gospel of eco-efficiency; and
- environmental justice.

Their respective schools of thought will be named:

- deep ecology;
- shallow ecology; and
- ecology of the poor.

These terminologies are inspired by the work of Joan Martínez Alier (2003).

It is important to emphasize that any classification has the potential to generate a reductionist approach to the elements analysed. The risk of classifying these streams of the environmental movement is therefore recognized. It is also acknowledged that the theoretical developments made by different schools of thought often do not capture all the complexity and diversity of the environmental movement.

For these reasons, this book intends to point out that all streams are like channels of the same river, whether of the environmental movement or the schools of thought. Theoretical Manichaeism will be avoided, emphasizing that all streams and schools have positive elements running alongside questionable ones. So, a pluralistic and critical approach emerges as a reasonable way to explore and expand theoretical and pragmatic tools that show potential for overcoming ecological crises (including the waste crisis).

After examining some of the proposals of different streams of the environmental movement and their schools of thought, this chapter will then focus on a discussion of 'justice', on the understanding that waste is not only an ecological problem but also an issue of social justice. The second central objective of this chapter therefore will be to carry out an investigation of the selected theories of justice for their potential for proposing a broader and more inclusive theoretical framework of justice.

This framework, here called 'socio-ecological justice', will be developed with the aim of creating a concept of justice that could be more appropriate for dealing with the complex socio-ecological issues of the present time. To achieve this goal, the migration from the mechanistic perspective of justice, represented here by the theory of John Rawls (marked by a liberal–distributive, anthropocentric and individualistic bias), to a socio-ecological perspective will be crucial.

Structures and developments of different theories will be used, which, when placed in dialogue, will allow social and ecological demands to be combined into a single theoretical framework, furthering the construction of assumptions behind 'socio-ecological justice'. In this regard, the theories of Nancy Fraser and Martha Nussbaum will be presented as criticisms of the liberal–distributive bias of mechanistic justice, while the theories of Klaus Bosselmann and David Schlosberg will represent the criticism of the anthropocentric and individualist bias of mechanistic justice.

The task posed is quite challenging, and it should be highlighted that the socio-ecological justice model to be proposed in this chapter does not aim to be finished, unique and immutable. It intends only to offer some possible avenues to debate and to surpass the mechanistic logic in the field of justice. As already mentioned, the investigations into waste as a problem of justice and of Law will be presented in different chapters. The final chapter will then link both analyses with the reintroduction of the assumptions of justice (in particular the ecological) in the international regulation of waste proposed. This approach will be taken in an attempt to reconnect Law with its ontological bases, thus seeking to retrieve its complexity and context.

The environmental movement and its schools of thought

The environmental movement as a socio-environmental conflict

The environmental movement was born as a form of organized representation of socio-environmental 'conflicts of resistance' to the ecological crisis. Muradian *et al* (2012) define such conflicts as social mobilizations carried out by actors (a single social group or a coalition of different groups) who share a negative perception of the environmental effects of economic activities and manifest this perception through visible confrontation. Socio-environmental conflicts do not necessarily represent direct physical confrontations and can occur via legal actions, campaigns and other demonstrations of disapproval of specific activities or impacts generated by these activities (Muradian *et al*, 2012).

The primary objective of the environmental movement is to include environmental issues in the debates and agendas of the most diverse segments (Montibeller-Filho, 2008). This objective stems from the negative perception the movement has of the ecological impacts generated by socio-economic processes. For this reason, in the concept adopted here the environmental movement represents socio-environmental conflicts (described above). The origin of this movement dates back to the mid-1960s when it began as what is known as the North American environmental revolution. In the 1970s, the environmentalist trend expanded to other countries in the North, culminating in the 1980s with its expansion around the world, when it reached Latin America, Eastern Europe, the Soviet Union, and Asia.

Environmentalism has since become a global and multi-sectoral movement, represented by different streams and actors. Alier (2003) synthesizes and classifies in the movement into three main streams, which he names 'cult of wilderness', 'gospel of eco-efficiency' and 'environmental justice'. This classification mirrors various points of divergence between the different types of environmentalism, which does not mean they do not have points of convergence or common elements. Alier describes the streams as 'channels of a single river'.

THE CULT OF WILDERNESS AND THE DEEP ECOLOGY

Chronologically and in terms of self-awareness and organization, the cult of wilderness was the first stream of the environmental movement. It emerged as a reaction to the negative environmental consequences of the dominant development model. This stream defends an untouched Nature, but it does not challenge the ideal of economic growth, focusing its actions on the conservation of remaining spaces of 'original' Nature that are 'outside' the influence of the market (Alier, 2003).

This view of the cult of wilderness is criticized by Emilio Moran (2006) who argues that it is necessary to move away from the myth of an untouched – original – Nature. According to Moran, since our emergence as a species, human beings act, interfere and transform the environment. In this sense, Moran warns: if the goal is also to change the way people relate to Nature to deal with our current environmental crisis, the assumption of a natural world without human beings deviates us from the reality that virtually every ecosystem on the planet has been shaped by human action, from millions of years ago. Our ancestors may not have provoked the conditions we see at the dawn of the 21st century, but they have continuously impacted ecosystems all over the world through hunting, harvesting, using fire, changing the balance between predator and prey and even pressuring herbivorous plant species.

The cult of wilderness originated in the North, in a reality of economic prosperity and global political domination. Precisely for these reasons, some authors interpret this deepening of the appreciation of Nature in terms of a post-materialism culture. From this perspective, this stream indicates a cultural change to adopt new social values, implying a greater appreciation for Nature as soon as the urgent material needs are met (Alier, 2003).

Alier (2003), however, considers the term 'post-materialism' in this case to be extremely misleading. Developed countries only possess and maintain their economic prosperity using an enormous amount of energy and matter from Nature, as well as occupying large areas for disposal of their waste and carbon dioxide. They are thus extremely materialistic. He argues that this environmentalism only grew due to material concerns about increasing chemical contamination and the risks and uncertainties arising from the use of nuclear energy, not because its proponents had already reached a 'post-materialist' stage.

Initially, the scientific basis of the cult of wilderness movement emerged mainly from the biology of conservation. As for its political actions, Alier (2003) states that the central proposals of this stream consist in the maintenance of natural reserves, usually designated national or natural parks. There are degrees of authorization for a human presence in these protected areas, but for the cult of wilderness movement such areas should ideally be free from human interference.

According to Alier (2003), the elements of this stream have been studied and developed in the last 40 years by theorists of the 'deep ecology' school, which proposes an ecocentric vision in opposition to the 'shallow' anthropocentric position. Deep ecology generally advocates the transfer of focus from human beings to the ecosystem, deconstructing the existing hierarchy between the elements of this whole (Ost, 1995). The book *The Shallow and the Deep Ecology Movement: A summary* released in 1972 by the Norwegian philosopher Arne Naess is seen as a milestone in the development of this school of thought.

Naess created the term 'deep ecology', expressing the distinction between the 'shallow ecology' of anthropocentric thinking and the ecological consciousness of the non-separation between human beings and the world: the deep ecology. Shallow ecology sees humans as the source of all values and assigns an instrumental value to Nature. By contrast, deep ecology considers the world as a network of fundamentally interconnected and interdependent phenomena, recognizing the intrinsic value of all living beings without any distinction (Capra, 1996). The school of deep ecology was formed by a number of authors who come from very diverse areas. Ost (1995) argues that, initially, three strands led this vision, namely:

i the passage from objectification to the subjectivation of Nature;

ii the substitution of anthropocentrism for ecocentrism;

iii the migration from individualism to holism.

THE GOSPEL OF ECO-EFFICIENCY AND THE SHALLOW ECOLOGY

A second stream of the environmental movement has developed which is concerned with the effects of the development model not only in 'untouched Nature' areas but also in industrial, agricultural and urban areas. Alier (2003) calls this stream the 'gospel of eco-efficiency', which directs attention to environmental impacts and/or risks to human health from industrial activities, urbanization, and modern agriculture. This stream dominates the current environmental political debates in the United States and Europe, with an anthropocentric bias.

The gospel of eco-efficiency seeks solutions for economic and ecological gains together (a win–win approach), and therefore, it defends economic growth (although not at any cost), sustainable development, ecological modernization and proper use of resources. Specifically, on ecological modernization, Alier (2003) points out that it has two legs:

1. an economic leg, with environmental taxes and an emissions trading market;
2. a technological leg, aimed at decoupling and dematerialization, with the development of measures for reduced consumption of energy and materials.

This stream is represented mainly by engineers of industrial ecology and environmental economists. The primary objective of environmental economists is the incorporation of the environment into the market. To achieve this, they adopt and adapt two theories: the Coase theorem, and the Pigovian tax.

> The formulations of environmental economics will be addressed in detail in Chapter 3. However, its two principal foundations can be summarized as follows:
>
> - The **Coase theorem** applied to environmental issues proposes to transform all commons into private properties, attributing prices to natural resources (well-defined property rights).
> - The **Pigovian tax** focuses on the extension of the economic policy of welfare to the environment, with state intervention for the correction of negative externalities caused by the economic model.
>
> SOURCE Derani, 1997

Environmental economics pursues the internalization of environmental externalities, aiming at the rational use of natural resources. Montibeller-Filho (2008) summarizes its essential content as composed of the following strategies:

- monetary valuation of environmental goods and services;
- internalization of externalities;
- the 'polluter pays' proposition;
- property rights;
- the contingent valuation method;
- the (environmental) cost–benefit analysis.

> The **contingent valuation method** (CVM) seeks to monetarily measure the impact of a qualitative or quantitative variation of environmental goods and ecosystem services to the well-being of individuals. This method uses two value indicators: the willingness to pay (WTP), ie how much these individuals would be willing to pay to improve their well-being; and the willingness to accept (WTA), ie how much they would be willing to accept as compensation for a loss of well-being. Thus, CVM aims to quantify the change in the level of well-being felt by the subjects as a result of the change in the provision of some

environmental good or ecosystem service. This estimation of values by WTP and WTA is based on the creation of hypothetical markets in field surveys, through interviews. With the average WTP or WTA values multiplied by the total number of the population, the total economic value of the good or service will be obtained or, more precisely, the value of the modification of its availability will be defined (Melo, 2016).

For more information about each of the strategies above, see the work of Gilberto Montibeller-Filho.

This reformulation of economics for the consideration of environmental aspects can be easily identified in the sustainability model of the Brundtland Report. The concept of sustainability has been propagated by the official UNEP reports, and is based on three pillars: social, economic and environmental. The model first emerged in the Brundtland Report (*Our Common Future*) and is used by almost all official documents addressing environmental issues. The term 'sustainable' here refers to the co-existence of these three pillars as equivalent entities that, when conflicting, must be balanced (Winter, 2008).

Proof of the propagation of the 'three pillars' concept of sustainability by the UNEP and other international organizations can be seen in the formulation of a '**green economy**', presented in the official speeches of Rio+20, and which, in fact, is environmental economics in a new guise. Green economy shares the same fundamental tenets as environmental economics (maintenance of growth, technological optimism, monetization of natural goods, for example), but also presents some new proposals (decoupling, green energy matrix, among others). On this subject, see the works of José Eli da Veiga and Ricardo Abramovay.

Environmental economics, especially with its new version known as 'green economy', has been gaining ground across many diverse sectors. This is mainly because its proposition of 'decoupling' is based on the reconfiguration of the production process in such a way that supply of goods and services would gain in eco-efficiency, using less natural resources (dematerialization) and less energy, without the economy having to stop growing (Veiga, 2010).

Decoupling proposes a separation between the concepts of growth in the production of goods and services and growth in the use of its material base (natural resources) and energy. Through technological innovations, decoupling is aimed at increasing dematerialization and eco-efficiency in production processes to obtain a reduction in energy and natural resources exploitation. On this subject, see the works of Ricardo Abramovay.

In economic terms, the strategy of **dematerialization** seeks to reduce the use of materials in economic activities. For instance, it promotes the manufacturing of products, keeping their functionality while reducing the quantity of materials used.

Improvements in eco-efficiency at the enterprise level are assessed by product and process life cycle analysis and by environmental auditing. Such tools have therefore been identified as links between industries and sustainable development. However, as Alier (2003) points out, in addition to its use for a 'greening' of industries, eco-efficiency is capable of leading to an extremely valuable research programme on the consumption of materials and energy by the economy and to disconnect the economic growth from its material base (decoupling process).

Although there are other streams of the environmental movement, according to Alier (2003), the two main streams co-existing today in the global context, sometimes intersecting with one another, are the cult of wilderness and the gospel of eco-efficiency. At the theoretical level, while the first stream has been developed by deep ecology philosophers, the second one has been framed mainly by environmental economists, aligned with the perspective of the shallow ecology.

ENVIRONMENTAL JUSTICE AND THE ECOLOGY OF THE POOR

There is a third stream of the environmental movement, which has also been gaining strength in recent decades: the 'environmental justice' stream. As highlighted by Muradian *et al* (2012), most reported socio-environmental conflicts refer to natural resource extraction processes. However, these conflicts occur at different points of the production chain – from extraction, transportation, infrastructure, processing, to disposal of materials in the form of waste.

Aragão (2009) reveals that many waste management facilities, explicitly created in response to the waste crisis, ended up generating several socio-environmental conflicts of great proportions. Such conflicts were called NIMBY (not in my backyard), referring to facilities making LULUs (locally unwanted land uses), of which waste management facilities are the best example.

Such conflicts, initially localized, have been globalized and transferred to other territories through the process of dominant economic globalization. For Alier (2003), as the economy expands globally, more waste is generated, more natural systems are compromised, the rights of future generations deteriorate further, and more genetic resources are lost. Indeed, some groups of the current generation are impacted by all these problems more sharply than others.

Muradian *et al* (2012) reveal the current trend in some countries (especially industrialized ones) to replace the extraction of materials from their territories by imports of materials from other regions of the world. By doing so, these countries transfer the environmental burdens and socio-environmental conflicts associated with them to other areas. For Muradian *et al* (2012) the growing demand for new natural resources, the decline in the quality of natural reserves and the increasing competition for possible land uses, all in developed countries, are some of the causes of such transferences. The same trend can be seen concerning waste. Cross-border waste transfer has led to the expansion of the frontiers of waste disposal (relativization of carrying capacity) to ecologically and socially vulnerable territories, establishing the conditions for the emergence of new socio-environmental conflicts in these territories and at a global level.

From this economic and geopolitical context emerges the environmentalist stream of environmental justice. The source of this stream is identified by Alier (2003) as India in the 1970s as a consequence of the Chipko movement, which sought to defend the forests of Kumaon and Garwhal in the Himalayas. At the same time, in Brazil, the 'rubber tappers' movement headed by the union leader Chico Mendes in the 1980s established the beginnings of the same stream there. Both movements of these two countries were involved in conflicts surrounding the extraction of natural resources.

It is important to emphasize that, currently, at the international level, there are two distinct meanings of the expression **environmental justice**. The first refers to the specialization of justice with the creation of specialized courts and sections to deal with environmental issues.

A second meaning, related to the social movement itself, is linked to social demands in cases of 'environmental racism' or 'environmental discrimination' in a broader sense. In the Brazilian context, for instance, there is important judicial jurisprudence that explicitly addresses the theme of environmental justice with its second meaning. The case of Special Appeal n. 1.310.471-SP (2011/0293295-2), in which Justice Antonio Herman Benjamin was the rapporteur, is a good example. The case dealt with a notorious case of contamination by lead of a vulnerable community in the city of Bauru, São Paulo.

It is to this second meaning that this book refers when using the expression 'environmental justice'.

The environmental justice stream argues that economic growth implies great impacts on the environment, highlighting the geographical displacement of resources (largely from the South to the North) and waste disposal areas (mainly from the North to the South) to meet the increasing demands of industrialized countries. According to Alier (2003), this process of geographic displacement generates impacts not solved by current economic policies or technological innovations, disproportionately affecting some individuals and communities that often protest and resist.

Although they often do not recognize themselves as part of the environmental movement, such social movements undoubtedly act in support of environmental, cultural and social protection of their communities. Some of these groups, especially those located in the global South, call for indigenous land rights and the sacred value of Nature, noting that many indigenous groups and peasants have co-evolved sustainably with Nature, ensuring biodiversity conservation (Alier, 2003).

The central axis of environmental justice was not, originally, the recognition of the intrinsic value of natural elements, but the material interest in the environment as a source of subsistence. Therefore, this stream of the environmental movement does not share the same ethical (or aesthetic) foundations of the cult of wilderness, although they can sometimes use common arguments. The ethics of the environmental justice stream arises from the demand for contemporary social justice among humans, which, for Alier (2003), can be both a positive factor and a weakness.

Parallel to the environmental justice stream which emerged in the global South (usually called 'environmentalism of the poor'), there was a social movement organized in the United States against local cases of 'environmental racism', many of them linked to the waste crisis. This is the North American origin of the demand for environmental justice. It arose in less favoured communities within this developed country or, in the logic of Boaventura de Sousa Santos, of the South located within the North. According to Rammê (2012), the historical roots of this expression are linked to North American social movements in defence of the rights of racially discriminated populations and communities exposed to risks of toxic contamination, who inhabit regions close to huge waste dumps of toxic waste or to large industries emitting chemical effluents.

For Aragão (2006), conflicts generated by waste issues have a possible explanation in the fact that waste is an undeniable source of injustices and, consequently, of conflicts. Such injustices may be expressed or hidden, public or private, inter-subjective or collective, current or potential, and their causes may be related to the damages that waste may cause, the loss of quality of life or the risks it inevitably entails.

Historically, conflicts related to waste were (and still are) mainly headed by the environmental justice movement. The list of cases of such conflicts is extensive, being caused by a variety of reasons: from the generation of waste, its transfer, construction of open dumps, landfills, and incinerators, to their final disposal and its impacts. The social transfer of socio-ecological costs, from which the cross-border waste transfer is a typical example, has undoubtedly led to a spatial expansion of socio-ecological conflicts represented by the environmental justice movement – and providing evidence of the coloniality of power of the North over the South.

The term 'environmental justice' is predominantly seen as a stream of the environmental movement born in the USA. In turn, the 'environmentalism of the poor' originally emerged as a term applied to rural movements of the global South. For Alier (2003), however, both terms are variations of the same stream, here unified as environmental justice. This link between the rural–Southern notion of the environmentalism of the poor with the urban–Northern notion of environmental justice was first suggested by the authors Guha and Alier and further developed by the latter in his work *Environmentalism of the Poor*.

Alier (2003) defends the environmental justice movement as potentially important as long as it is willing to speak on behalf not only of minorities located in the United States but also of majorities existing outside that country (which are not always racially defined). It addresses issues such as biopiracy, biosafety and climate change, in addition to local problems of contamination. What the environmental justice movement inherits from the United States civil rights movement also holds true on a worldwide scale because of its contribution to Gandhian forms of non-violent conflicts.

That said, it is important to note that, although all these movements can be considered part of the environmental justice stream, they are not identical.

Each of them has a different social, political and economic context. For Anand (2004), the environmental justice movement can be described in general terms as a coalition of citizens and groups advocating a broad-based democracy, focusing on larger goals of social justice. In this way, this stream brings elements of an environmental movement alongside other socio-political spheres.

In theoretical terms, the environmental movement has been studied and developed mainly by sociologists and theorists of political ecology, here called the authors of the 'ecology of the poor' school. In this sense, Anand (2004) argues that the theoretical construction of the concept of environmental justice has been carried out simultaneously from two approaches: the bottom-up, through the analysis of grassroots movements (local grassroots groups); and the top-down, through the investigation of political organizations that reflect the terms of such groups. At the same time as acting as a movement, environmental justice is a complex concept concerning the inseparable connections between social, political, economic and environmental issues.

From the 1980s, grassroots activists and academics diagnosed problems and established principles and strategies to tackle socio-environmental issues recognized and identified across the globe. Based on these developments, Acselrad *et al* (2009) state that, in general, the environmental justice movement is based on four principles:

i toxic pollution for no one;

ii creation of a different development model;

iii a just transition to this new model; and

iv democratically instituted environmental policies – politicization versus market belief.

The first of these principles, toxic pollution for no one, shows a clear opposition to environmental conflicts guided by the mechanistic logic of individualism, reinforced by the right to 'individual property' defended and guaranteed by the state, which entails a spatial displacement of pollution. This mechanistic logic generates the transfer of environmental injustice to communities and countries not socially and/or politically well organized, where legislation is less restrictive and/or enforcement is more precarious. The mechanistic individualism ends up reducing environmental conflicts to inter-local disputes caused by the generalized refusal to be close to the sources of environmental damage, the previously mentioned NIMBY movements (Acselrad *et al*, 2009).

Denying the hypothesis of the individualistic stance that seeks to push environmental problems into the backyard of others, the environmental justice movement therefore criticizes the criterion 'always in the backyard of the poor'. It politicizes the debate about coping with pollution and proposes toxic pollution for no one. Such a movement has a solidarity character, bringing to the public debate the socio-political logic that promotes environmental inequalities (Acselrad *et al*, 2009).

The second principle, creation of a different development model, is a logical consequence of the first one. In defending toxic pollution for no one, the environmental justice movement is led to contest the dominant development model. This model, considered to be socio-ecologically perverse, is only maintained because it has set up a geopolitical pattern that puts a strain on the health and well-being of populations deprived of financial and/or political resources in a negative way (Acselrad *et al*, 2009).

The environmental justice movement believes that by protecting the most vulnerable populations from social and environmental risks, it will automatically be fighting against environmental degradation in general. It assumes that by impeding the transfer of negative impacts generated by the development to the least favoured, the main generators of these socio-environmental impacts and risks will be obliged to deal directly with the problem, seeking its elimination and solution. In other words, the propensity to identify and eliminate sources of environmental damage and, consequently, to find new development models, will tend to intensify (Acselrad *et al*, 2009).

Without denying the importance of the debates on the economy of environmental resources, the environmental justice movement argues that to this must be added the question of which purposes these resources are being used for. That is, in addition to the concern about how to produce goods, we must question (and modify) how much is produced and what is produced. This is the heart of the discussion about the need for a new model of production and consumption (Acselrad *et al*, 2009).

The third principle, a just transition, stems from the previous principle. All changes that are necessary to overcome the current development model must always be careful not to generate new environmental injustices. This transition, to be just, must take gradual steps, so, for example, combating unequal pollution will not create massive unemployment of the workers of polluting industries. Acselrad *et al* (2009) argue that the implementation of this principle depends on the creation of alliances with other social movements and syndicalists to obtain more substantive changes.

Finally, the environmental justice movement proposes the principle of democratically instituted environmental policies – politicization versus market belief. Such a principle points to the absence of effective regulation of the great economic agents that generate ecological risks, advocating that this allows them to freely choose vulnerable communities as the preferred victims of their harmful activities. Therefore, the movement believes that environmental injustice will only cease when there is a restraint (by the exercise of politics, within the framework of permanent democratization) of the free will of economic agents with greater power to cause environmental impacts (Acselrad *et al*, 2009).

The environmental movement, seen as an organized representation of the socio-environmental conflicts, is, in fact, a complex and plural movement. The three streams highlighted here in this topic have positive elements. Yet the theoretical developments arising from these streams – here called 'deep ecology' (relative to the 'cult of wilderness'), 'shallow ecology' (or the 'gospel of eco-efficiency') and 'ecology of the poor' (related to 'environmental justice') – while addressing such positive elements, are subject to criticisms that demonstrate theoretical limitations with practical consequences. It is precisely this point that the next section will address.

Some challenges and limitations of the schools of thought

Deep ecology, shallow ecology or ecology of the poor?

The streams cult of wilderness and gospel of eco-efficiency have emerged in the Western context of the North. They are prevalent in the global environmental debate, establishing the theoretical opposition between deep ecology and shallow ecology. Without denying the importance of these two schools of thought, they do not represent a direct or blunt criticism of the current dominant development model based on the idea of infinite economic growth. The understanding that the ecological crisis affects communities and countries in political and economic inequal positions differently is not the object of their attention either.

The most obvious consequence of these omissions is that in some situations, public policies and economic actions proposed on the basis of deep ecology and shallow ecology have resulted in socio-ecological damage to less favoured individuals and/or communities, or in transferring the problem of unsustainability and the inadequacy of current models (economic, development, and legal, for instance) for future generations.

Alier (2003) recalls how many of the national parks created in the United States inspired by the deep ecology thinking could only be built after the displacement or elimination of the native peoples living in those territories. The relationship between society and Nature in the United States, for example, has been seen not as a dialectical socio-ecological history and subject to changes, but as a history of deep and permanent reverence for the wild, so the appreciation for Nature has expanded proportionally to the destruction of landscapes caused by economic growth.

Regarding deep ecology, Ost (1995) argues that the strict monism originally adopted by this school of thought can explain its insufficiency (or its limitations) in the fight against the socio-ecological crisis, given the negative consequences of the public policies inspired by it. For Ost, this would be the most fragile and critical point of this school. Deep ecology arose from the necessary refutation of the mechanistic dualism and led to total assimilation between 'subject' and 'object'. The mechanistic perspective promotes duality without recognition of relationships, with hierarchy and unbridled exploration. Deep ecology, in turn, creates unity without identifying differences, bringing confusion.

According to Ost (1995), the mechanistic dualism and the monism of deep ecology both make the same mistake: one radically opposing subject and object and the other identifying them completely. The author considers both reductionist doctrines that fail to think in complex terms. The only way to do justice to human beings and Nature would be, according to Ost (1995), simultaneously affirming the similarities that bind them together and the differences that separate them.

With regard to shallow ecology, its championing of, for example, the need to close the production cycle, drives the recycling industry, undoubtedly presenting great potential for environmental protection. It faces the linear logic of production patterns and can mitigate the extraction of new material resources and the disposal of more waste in the ecosystem together with the reduction of the use and dissipation of energy in production processes.

This school also seeks solutions to overcome the ecological and waste crisis. The strategies inspired by shallow ecology focus on the inclusion of Nature in the market and on increasing efficiency of production processes. Such strategies are important and essential for the changes we need. However, they are insufficient to deal with all the complex problems generated by the current development and economic models.

As highlighted by Alier (2003), many of the environmental conflicts lie outside the sphere of the market (whether real or fictitious markets) and the arguments debated in such conflicts are therefore unrelated to the dominant economic discourse. The discourses adopted in many of those conflicts encompass arguments such as the intrinsic and ecological value of ecosystems, respect for the sacred, dignity of life, human rights, and environmental justice. For this reason, they are often misunderstood by the state, which operates on market principles.

If shallow ecology does not include criticism of current development, economic, consumer, and legal models, or of the social and geopolitical demands on its investigations, the strategies inspired by this school of thought can generate (and indeed have generated) serious socio-ecological damage, particularly affecting vulnerable communities and countries. Without questioning the mechanistic paradigm and its models, shallow ecology will only spatially and temporally transfer the negative consequences of the ecological and waste crisis to other territories and for future generations.

An example of this is the social and ecological damage generated by waste transfers destined for recycling to regions and countries without support and/or minimum conditions for carrying out such processes. When the transfers are of totally inappropriate or unsuitable waste for recovery, this damage is even worse. The result is severe ecological and public health harm to communities and countries receiving such waste.

Besides, shallow ecology reinforces the unrestricted faith in technological development (ecological modernization) to solve socio-ecological problems, ignoring that the socio-economic system has never been and will never be a closed system. It will always make exchanges of matter and energy with the ecosystem. Even with more efficient production processes, if the economic growth model is not challenged, the dissipation of matter and energy will continue to occur unsustainably and, at best, the socio-ecological collapse to which humanity is headed will only be delayed for future generations.

For Acselrad *et al* (2009), the notion of ecological modernization supposes that environmental degradation reaches everyone equally. It believes that the decoupling strategy (economy of matter and energy with the maintenance of economic growth) is enough to promote socio-environmental well-being. According to Acselrad *et al* (2009), this view focuses on technological solutions and does not take into account the fact that technology cannot solve all problems, thus limiting the space for political debate on the social issues requiring solutions, while focusing on the development of new technologies to solve environmental problems as a great opportunity for economic growth.

Alier (2003) argues that the gospel of eco-efficiency founded on the school of shallow ecology uses the 'ostrich principle', instead of the precautionary principle. It means that this stream of the environmental movement, by closing its eyes (or putting its head in the sand) to other consequences and implications of the development model, produces only temporary relief with its strategies. But the conflict between economy and Nature cannot be permanently silenced by false hopes built on the ideal of a total eco-efficiency. What has been happening is that this conflict has been transferred to more vulnerable communities and countries and to future generations, who are (or will be) facing a disproportionate socio-ecological burden (social transfer of socio-ecological costs).

Deep ecology theorists also criticize shallow ecology for not breaking with the anthropocentric perspective and for its reformist character and economistic focus. For them, shallow ecology only recognizes an instrumental or utilitarian value to Nature, the fundamental flaw of this school of thought and of the public policies and actions it inspires. That is because shallow ecology seeks economic solutions only to the environmental problems that negatively affect human life and health (Ost, 1995).

Theorists of the ecology of the poor school consider that the ecological modernization of shallow ecology can reduce the intensity of energy and matter use. However, until this goal is reached, it causes significant destruction and creates new unknown risks. From the perspective of the ecology of the poor, new technologies not only do not represent a definitive solution to the conflict between Economy and Nature but also generate new environmental conflicts. As typical examples of this assertion, Alier (2003) cites the location of waste incinerators and areas for the storage of radioactive waste, which led to the emergence of environmental justice movements.

These are real issues that escape the reductionism and isolation of the sciences and promote concrete demands. It is in this sense that the ecology of the poor school has been gaining prominence. In addressing the movement of environmental justice, the success of this school lies precisely in its ability to raise fundamental social issues overshadowed in other schools. It also stands out as it seeks geopolitical, spatial and temporal contextualization of environmental problems, demanding other development, economic, political and legal models.

The existence of such positive qualities in this school does not mean that the theoretical developments of the ecology of the poor are not amenable to criticism. Initially, the very definition of environmental justice is criticized since there is no theoretical consensus about its meaning and scope. As a practical consequence, it has been challenging the development of public policies capable of dealing with environmental injustices. A clarification of what exactly environmental justice is and its scope of application is thus fundamental not only from the theoretical–academic perspective but, going beyond, for practical–political concerns (Anand, 2004).

The term 'environmental justice' originated in local social movements against negative environmental consequences disproportionately borne by vulnerable communities. This stream of the environmental movement has however succeeded in connecting different local demands, spatially widening its focus of attention to similar demands at the global level, and becoming a transnational environmental movement. The migration from the original term 'environmental racism' to the now more popular 'environmental justice' demonstrates an expansion of the scope of action of this stream, with the inclusion and connection of different demands, all encompassed by the umbrella notion of 'justice'.

The movement for environmental justice gave visibility to social issues intrinsically connected with environmental issues. Although this movement fights against social problems that also impact the ecosystem and future generations, these two elements do not appear in the original focus of its theoretical development. The school of the ecology of the poor primarily addresses intra-generational justice issues, often not giving due attention to inter-generational and inter-species justice. This characteristic raises criticism concerning the anthropocentric approach of the school. Much of the literature of the ecology of the poor pays no attention to environmental issues that do not directly impact human beings (Schlosberg, 2009; Rammê, 2012). By doing so, this school of thought loses the opportunity to instigate debates as well as to inspire national and international policies that are more revolutionary and less reformist.

Schlosberg (2009), after a detailed analysis of the literature of the movement for environmental justice, emphasizes that, even though this stream uses multiple discursive resources, most of its theoretical developments focus only on the distributive aspect of justice. At most, this literature adopts a slightly broader perspective to discuss issues of participatory equity. A large portion of the ecology of the poor theorists thus takes a liberal view of justice, focusing on the equity element (either of distribution or participation), often neglecting other important dimensions addressed by critical authors of the liberal–theoretical model of justice.

Following this reasoning, Peña (2005) argues that most of the ecology of the poor school do not examine the fundamental ethics of the justice arguments adopted by the environmental justice movement. This can be seen as a weakness of such theoretical developments, leading them to reproduce a notion of justice focused on equity, derived from the liberal paradigm of justice. For Peña, taking a critical stance towards the political values and ideological presuppositions underlying this liberal notion of justice has the potential to clarify how such structures inform and limit the theoretical thinking of environmental justice discourses.

Pellow and Brulle edited the book *Power, Justice and the Environment: Towards critical environmental justice studies,* a critical analysis of the theory of environmental justice to identify its weaknesses and limitations, as well as strategies to strengthen this stream. In the conclusion to the book, the editors summarize the analyses contained in it. They argued that, despite the radical orientation of the environmental justice movement, the focus that its theoretical developments (here called 'ecologism of the poor') have given to achieving justice through the existing (mechanistic) legal system simply reinforces the institutional relationships that created and maintained the environmental injustices the movement is trying to tackle. Such dynamics, according to Pellow and Brulle, demonstrates the need for a theoretical reformulation of the environmental justice movement, in order to align its strategies of action with its radical analysis of the causes of environmental injustice (Pellow and Brulle, 2005).

They go on to explain that while the ecology of the poor questions essential elements of the causes of environmental injustices they seek to tackle (such as the globalized dominant development model), it does not question the epistemological foundations that have created and continue to maintain such a system. In proposing strategies and solutions within the liberal–distributive model of justice, the ecology of the poor school fails to question the mechanistic paradigm underlying the socio-ecological crisis. Therefore, it reinforces the structures of power (social and geopolitical) and 'under-explores' its critical and transformative potential (Pellow and Brulle, 2005).

Directly linked to the focus on the anthropocentric, intra-generational and equity aspect is the fact that most of the theoretical developments of the ecology of the poor work with environmental injustices on which there is some consensus (scientific or not) about its hazardousness, even if all the ecological and public health consequences they may generate are unknown. The ecological burdens over which there is no perceived hazard or immediate threat to humans, usually produced diffusely and cumulatively, are left out of the demands for environmental justice.

Rob Nixon (2011) qualifies such environmental burdens as a type of violence, which he calls 'slow violence'. By slow violence Nixon means that violence which occurs gradually and imperceptibly, causing postponed destruction, dispersed in time and space. Generally, such environmental burdens are not seen as a form of violence, which is usually conceived as an immediate event or action in time, instantly visible spatially – such as the dumping of a large amount of toxic waste in open areas located in vulnerable communities, for example (as in the case of Ivory Coast).

> It should be noted that the examples of slow violence presented by Rob Nixon in his work (such as climate change, deforestation, acidification of the oceans, among others) differ from those highlighted here (mainly the transfer of non-hazardous waste).

These environmental burdens, on the contrary, constitute a different type of violence: gradual, cumulative and exponential, and their repercussions unfold over different timescales. These are atypical environmental catastrophes, which go beyond clear limits in time and space and are marked, above all, by (temporal, geographic, rhetorical and technological) transfers that simplify and underestimate the real human and ecological costs, both in advance and retrospectively (Nixon, 2011).

Nixon's slow violence theory closely resembles Ulrich Beck's 'risk society' theory (1992), according to which the propagation of risks (concrete or abstract) is only possible through institutional structuring (what Beck calls 'organized irresponsibility'). Such structures impede the perception of these risks and dangers using different strategies, among them the diffusion of uncertainties through the issuance of contradictory opinions by specialists who have the information and knowledge power. Thus, organized irresponsibility seems to be one of the possible responses to the absence of social resistance against the practice of slow violence.

> Although they have several points of contact, Beck's theory, unlike Nixon's, does not acknowledge that the environmental damages and burdens of the current ecological crisis impact (politically and economically) vulnerable communities and countries differently.

The greenhouse gas accumulation and the accelerated biodiversity loss are now seen as catastrophic events. Still, the political responses to them have continuously been postponed, often for generations. The scientific and political complexity involving these events could be listed as important reasons for such delayed public reactions. Although all humanity is affected in some way by slow violence, the most direct and immediate victims (human and ecological) are rendered the most invisible. Generally, these victims are the most ecologically, socially and geopolitically vulnerable (Nixon, 2011).

The notion of slow violence poses a series of challenges: scientific, legal, political and representational. In the extended spectrum between the emergence of slow violence and its delayed effects, between the place of its origin and the place of its consequences, its causes and sources disperse, disappear, and are forgotten. An obvious result of this type of violence is the immobilization or social inaction it causes. The temporal and spatial invisibility of these environmental burdens creates barriers of perception and representation of the communities and countries subject to slow violence. As a result, resistance efforts (socio-environmental conflicts) as well as political–legal responses either rarely happen or face severe delays. Meanwhile, the ecological conditions of life support become increasingly degraded (Nixon, 2011).

Muradian *et al* (2012) emphasize that a diverse range of variables is required for socio-environmental conflicts to arise. Such resistance movements are usually determined by (among others):

- the form of land use and appropriation;
- population density;
- the characteristics of the materials flows transferred, and the technologies used in those processes;
- historical traditions;
- formal and informal institutions in action;
- notions of identity and justice;
- local perceptions of socio-ecological risks.

There is no doubt that the global expansion of social metabolism leads to increased socio-environmental conflicts in the world. However, such conflicts have occurred mainly in regions with ecologically fragile ecosystems where there is intensive human occupation of the territory and high levels of social organization (Muradian *et al*, 2012). Although there are some territories with high human population density and regions with ecologically fragile ecosystems in Southern countries, these are not peculiar characteristics of the global South. Besides, even though they are most affected by transfers of environmental burdens, Southern countries normally have low levels of socio-environmental organization when compared to the global North.

Regarding the waste issue, all the known socio-environmental conflicts arising from cross-border transfers were related to hazardous waste, where the environmental and public health impacts were visible and direct. Therefore, the absence of social dispute (socio-environmental conflicts) about cross-border transfers of waste considered as non-hazardous may be related to the lack of perception of the ecological burdens that may arise from these flows.

The fact that most of the transfers of non-hazardous waste are not challenged and do not generate direct socio-environmental conflicts is also reflected in the absence of studies on the subject in general, particularly from the justice perspective. This is an example of slow violence. Most research and reports on waste transfer from the perspective of environmental justice focus on the transfer of hazardous waste, notably toxic waste. Moreover, as will be seen in Chapter 5, this theoretical gap also has legal repercussions: the focus is on the regulation of cross-border transfer of waste considered hazardous, with a lack of regulation on cross-border transfer of waste regarded as non-hazardous.

Ecological pluralism: dialogue and engagement

As a reflection of the plurality of the environmental movement, the ecological schools of thought resulting from this movement are also multi-faceted. They all offer valid proposals as well as having limitations that can be criticized. By proposing an ecocentric vision, adopting teachings of ecology and highlighting the interrelationships among all ecosystemic elements, including humans, the deep ecology school plays a fundamental role in the epistemological questioning of the dichotomized, fragmented and reductionist thinking of the mechanistic paradigm, as well as of the objectification and subjugation of Nature.

On the other hand, there is no doubt that shallow ecology helps decisively in the dissemination of the understanding of the differences between the ecosphere and the technosphere cycles. It brings crucial economic proposals seeking to break with the linear logic of the production systems, aiming to reduce (even relatively) the extraction of new natural resources and waste generation. Though not free of criticism, its focus on practical and economic responses has been successful in terms of political engagement for the development of a more eco-efficient economy.

In turn, ecology of the poor has succeeded in bringing to light the social problems caused by environmental issues, which if not omitted, have been obscured by other ecological schools. The demands for environmental justice addressed by this school have shown that the environmental burdens and the ecological crisis resulting from the current dominant development model impact communities and countries with different political and economic powers in distinct ways, affecting the most vulnerable more deeply.

Regarding the ecologically unequal exchange generated by social transfers of socio-ecological costs, different arguments have been developed to overcome it. Theories of environmental economics (a foundation of shallow ecology) adopt the notions of 'externalities', 'market flaws' or of 'incomplete markets' seeking the inclusion of Nature in the economic system. To do so, they establish new property rights and negotiations in real and fictitious markets (such as carbon markets). Ecological economists, on the other hand, emphasize the powerlessness of those who suffer from those negative environmental externalities, developing concepts such as 'ecological debt' and 'environmental liabilities' (Alier, 2003).

The concepts of ecological debt and environmental liability both come from the idea of environmental injustice. **Ecological debt** represents the ecological debt accumulated by the countries of the North in relation to the countries of the South. Such debt is mainly due to Northern countries' imports of natural resources and exports of waste with an unequal ecological exchange, as well as to the free or cheap occupation of the environmental space of Southern countries. **Environmental liability** represents any ecological impact, caused by public or private entities, that has not been repaired.

Without denying the importance and impact of other perspectives, the justice point of view is preferred here, since it has the potential to escape from the chrematistic (monetary) language conferred by economists by considering other elements that may help in overcoming the liberal mechanistic paradigm (which will be studied in detail in the next section of this chapter). In addition, the adoption of this perspective allows the establishment of criteria to assist both: the negotiation process and drafting of a global waste management regulation; as well as the decision-making process of waste transfer cases (which will be developed in Chapter 6 of this book).

The concept of **economy** can be used either in the sense of 'oikonomia' or of 'chrematistic'. The distinction was made by Aristotle, who stated that the former deals with the art of supplying the goods necessary for life, while the latter refers to the administration of assets for the home or the city. In this sense, in the present-day economic model, chrematistic language refers to monetary terms: a measure of wealth in terms of money (Melo, 2016).

The waste crisis is an issue addressed by shallow ecology, which is concerned with environmental impacts and the wastage of matter and energy resulting from waste. However, the environmental justice movement studied by the ecology of the poor has, since its inception, led the demands of the most vulnerable communities and countries that bear a disproportionate environmental burden generated by waste in contrast to those in a socially and geopolitically privileged position. The adoption of justice as the framework for discussion is therefore quite appropriate for the debate on cross-border waste transfer (global waste management), since waste, besides an ecological problem, is a problem of justice.

Moreover, according to Bosselmann (2016), a discussion based on justice has the potential to lead to better decision-making in ecological matters, since, in terms of principles, it is a discourse resulting from a common social agreement facilitated by certain institutions, such as law and governance. The ethical discourse, on the other hand, reflects rather broad ideals that cannot be directly communicated through these institutions. Hence, ethics, while providing a basis for the concepts of justice, cannot, according to Bosselmann, guide decision-making processes in the same way as justice institutions.

The preference for the discursive resource of 'justice' presented by the ecology of the poor school does not mean the complete exclusion of the other schools of thought highlighted in this chapter. In fact, what will be proposed in the next section is the adoption of an 'ecological pluralism', the synthesis of a kind of dialogue between these 'ecologies', as well as between different perspectives of 'justice'. The result expected from this dialogue is a multi-dimensional and complex approach that can be achieved in a theoretical reformulation of justice for socio-ecological issues, here called 'socio-ecological justice'.

As mentioned before, one of the criticisms levelled at the ecology of the poor is precisely the current difficulty in outlining the meaning of environmental justice and its scope. Bearing in mind that the notion arises from real social movements, it is necessary to avoid universal concepts typical of the mechanistic epistemology. Adopting a universal concept may lead to the weakening or misappropriation of the notion of environmental justice for the maintenance of existing relations of domination. Nor is complete relativism a good path to take since this would result in a rigid, restrictive, particular and unchanging categorization of justice for socio-ecological issues, preventing their appropriation by different movements and situations.

To implement justice in legal and environmental terms without running the risk of inflexibility and the reductionism of a ready and closed concept (general or particular) made *a priori*, the proposal of 'universalism of arrival or confluence' of Joaquin Herrera Flores (2009) seems to be appropriate. Flores' proposal accepts as necessary the prioritization of the development of theoretical assumptions that will enable concepts of justice to be made *a posteriori*.

> Inspired by the work of Joaquin Herrera Flores, the Latin expressions ***a priori*** and ***a posteriori*** are used here in the sense popularized by the philosopher Immanuel Kant. While the former (*a priori* – from earlier) translates a conclusion, concept or knowledge reached before observation by the senses or concrete experience, the latter (*a posteriori* – from later) means the conclusion, concept or knowledge was or will be defined/reached based on observation or on concrete experience.
>
> As it will be developed in this section, although some assumptions can be used as a starting point to guide the development of the concept of socio-ecological justice, in an *a posteriori* approach (pragmatic) such a concept will only be defined after observations and actual experience.

In addressing human rights, Herrera Flores (2009) observes that modern Western (mechanistic) epistemology universalizes a particularism (its notion of what should be human rights, its economic models and social relations) and therefore qualifies it as a 'universalism of departure or *a priori*'. According to Flores, the theories that arose criticizing this mechanistic universalism, seeking a return to the 'local', drowned in the plurality of interpretations they created. Hence, in their way, they also constructed a universalism, called by Flores a 'universalism of parallel lines', in which such lines can only meet (or connect) in the infinity of the field of cultural differences.

'Localism', by resisting universalism *a priori*, ended up despising cultural 'distinctions' aiming to impose a single worldview. In doing so, it fell into the same mistake of creating a dichotomy between 'us and them', and despising 'the other'. Flores proposes a complex view of human rights, accepting the possibility of a universal synthesis of the different options of these rights, but rejecting the universal as a starting point. This is the 'universalism of arrival', which can be achieved after (and not before) dialogue or confrontation, and with the intersection of proposals and not with their overlaps (Flores, 2009).

Similarly, and considering that the environmental justice demands investigated by the ecology of the poor also come from grassroots movements, instead of proposing a ready and closed concept of justice to apply to actual socio-ecological situations, the proposal of a definition established *a posteriori* enables the opposite path to be taken. Thus, based on previously established assumptions, concepts of justice can be developed from a pragmatic perspective for real-life cases. Following this logic, everything that is, in the real situation, outside of what is considered just, will be viewed as an injustice in socio-ecological terms.

With a similar proposal, Schlosberg (2009) states that one of the most evident changes in contemporary social and environmental movements is the existence of different discourses within the same movement. Such differences cannot solely exist side by side in a static state (universalism of parallel lines). There is a need for a dialogue and a real engagement between such discourses so they can be revealed through debate and discussion, creating general assumptions that can be applied to meet the needs of different demands in a coherent and unified, though not uniform way.

David Schlosberg (2009) explains that the concept of unity he adopts is based on the idea of **Unity in diversity**, where differences are recognized rather than ignored. In this type of unity, differences are integrated, not annihilated or absorbed. **Uniformity**, on the other hand, absorbs differences. Schlosberg uses the metaphor invented by Mary Parker Follett, where the unit is represented by a bowl of salad, whereas uniformity would be a pot where all the ingredients are mixed and reduced to a single element.

For Schlosberg (2009), the insistence on singularity and universalism (typical of the mechanistic paradigm) of concepts of 'justice' and 'environment' produces exclusion and the disappearance of other multiple articulations, demands, and movements. Pluralism, on the other hand, is empirically real; it exists in movements for environmental justice, and is pragmatically necessary to avoid the error of exclusion. In considering that these demands and concepts are made in different ways, places, times and contexts, and articulated with the use of different languages, Schlosberg proposes a pluralistic theory of environmental and ecological justice, seeking to incorporate such demands and divergent notions in a concept of justice for socio-ecological issues that is even broader, inclusive and pragmatic (Schlosberg, 2009).

Nancy Fraser (2008) characterizes current times as 'abnormal justice' times, where demands for justice made by social movements generate debates and disagreements about different aspects of justice: on its scope, on who can be the subject of justice, on the necessary procedures for the definition and implementation of justice, among others. All these questionings and the divergence of views challenge the 'normal' (hegemonic) discourse of justice, revealing its structural deficiency to deal with these new demands. Returning to Santos' words (2010b), here, again, there are strong questions for which classical theories of justice have offered weak answers.

Fraser (2008) emphasizes that this is not an unprecedented situation. At other historical moments, the absence of a precise and secure hegemony to deal with new demands has caused conflict between rival paradigms during attempts to normalize new visions of justice. According to Fraser, such cases are no exception: in truth, normal justice is historically abnormal, while abnormal justice is the historical norm. In other words, Fraser argues the notion of justice is continually changing due to the different views that arise from historical specificities never before experienced. Such visions challenge and destabilize established paradigms, and new concepts of justice, separate from the pre-existing perspective considered normal, emerge.

In moments of 'abnormality' (or crisis) the established theories of justice offer little direction and prove to be insufficient. They stem from a shared grammar, not offering strong responses to conflicts that escape the scope of their epistemological foundations. The abnormal demands for justice raise unintelligible questions for theories of normal justice. In this sense, Fraser (2008) proposes an approach of (in)justice in abnormal times, identifying three nuclei of abnormality in disputes about justice (referring to the 'what', 'who' and 'how' of justice) and offering three conceptual strategies for elucidating such abnormalities.

Based on these theoretical developments, to understand more clearly what are (or can be) the assumptions of a 'justice' for socio-ecological issues, the next section seeks to work with two of the three nuclei of abnormality found in disputes for justice, as pointed out by Fraser. Initially, the first nucleus, relating to the scope of action (the objects of justice disputes, the 'what') will be developed. The second nucleus concerns the subjects of justice (the 'who'). The last nucleus of abnormality (the 'how') will be addressed later and developed in Chapter 6.

The following section will be devoted to understanding the extent and the impact that the concept of 'justice' for socio-ecological issues can have, following the critical questioning of its ethical and epistemological bases and its subsequent theoretical reformulation. It is believed that this theoretical development will enable a framework of assumptions of justice to be built, which will serve as guides during the process of designing a concept of justice that is appropriate for the object of investigation of this book: cross-border waste transfer and, ultimately, the promotion of a global waste management system that is just and sustainable. However, it is important to highlight that the model of justice proposed here can also be applied to other socio-ecological issues.

The 'what' and the 'who' of justice for socio-ecological issues

The 'what' of justice: a multi-dimensional concept

Justice is a rather complex and difficult to define notion. The myriad theories with different approaches that exist on the subject are proof of this. These different approaches have challenged the classical (distributive) notion of justice, and one of the nuclei of divergences (abnormalities) refers to the scope of justice, or the 'what' embraced by it. According to Fraser (2008), the dominant notion of the distributive 'what' of justice has been challenged by at least two other views: the politics of recognition and the politics of representation.

There are currently at least three divergent concepts of the 'what' of justice: (re)distribution, recognition and representation. In order to deal with the theme of 'justice' in socio-ecological conflicts, it is essential to begin by making clear which concept of justice will be adopted, its extent and possible dimensions. Following the analysis made by David Schlosberg (2009), among the several theories, of different streams, on the subject, this section will develop the general theoretical design of the distributive–mechanistic approach (or hegemonic–normal, in Fraser's words) of justice, led by John Rawls. In contrast, the critical perspectives of this approach will also be analysed, highlighting the theories of Nancy Fraser and Martha C Nussbaum.

MECHANISTIC PERSPECTIVE OF JUSTICE

Since the 1970s the literature of political theory has defined justice almost exclusively as a matter of equity in the distribution of social goods. This group of theorists was pioneered by John Rawls, with his classic work entitled *A Theory of Justice* (Schlosberg, 2009). In his work, Rawls (2005) argues that the basic structure of society is the primary object of justice, and he develops a theory of justice as equity. His conception of justice generalizes and leads to the maximum degree of abstraction the idea of 'social contract', adopting a contractualist approach.

Rawls (2005) advocates that justice is the fundamental element in achieving a 'well-ordered society' in which all members accept the general principles of justice, which are also met by basic social institutions. For Rawls, these general principles attribute the rights and duties of society and define the appropriate distribution of the benefits and burdens of social cooperation. The notion of justice in this approach provides a pattern by which distributive aspects of the basic structure of society must be assessed, defining the appropriate division of social advantages and costs.

According to Rawls (2005), the definition of the concept of justice is, in fact, the object of the original consensus between free and rational individuals, members of society. Within this reasoning, an appropriate concept of justice can only be developed from the projection of an initial position of equality, hypothetical and ahistorical. This 'original position' proposed by Rawls corresponds to the state of nature of the social contract theory. According to Rawls, the choice that 'rational men' would make in this original position determines the concept of justice in a position of equality, hence the option of presenting justice as equity.

> The hypothetical and ahistorical position proposed by Rawls means that it results from the exercise of rationality not having occurred historically, nor will it occur. Such rational exercise is the basis for an *a priori* concept of justice.

In the original position, no one knows: their place in society, their class or social status; how many benefits would be distributed to them; their abilities, intelligence, strength or other personal and social advantages and weaknesses. The principles of justice are thus agreed in this original position under a 'veil of ignorance'. Rawls (2005) argues that the original position is the ideal *status quo* for an equitable consensus reached by these individuals. It provides the 'impartial' perspective necessary for the agreement of a just concept of justice.

Rawls (2005) presents two basic principles of justice on which he believes there would be a consensus in the original position, and to which a priority order of application applies: ie the first principle must be fulfilled before trying to achieve the second. The first principle of Rawls' theory refers to equality and freedom between individuals. It states that '[...] each person is to have an equal right to the most extensive scheme of equal basic liberties compatible with a similar scheme of liberties for others' (Rawls, 2005: 53). Such a principle requires equality in the attribution of rights (notably freedoms, such as political and private individual ownership) and fundamental duties.

The second principle deduced by Rawls concerns social and economic inequalities. The author considers such inequalities inevitable, due, for example, to social starting positions, natural advantages and historical contingencies. This principle asserts that '[...] social and economic inequalities are to be arranged so that they are both (a) reasonably expected to be to everyone's advantage, and (b) attached to positions and offices open to all' (Rawls, 2005: 53).

According to Rawls (2005), a priority order must also be applied within the second principle. The social and economic inequalities will only be fair if they:

a result in benefits for all (principle of just equality of opportunity); and

b result in benefits in particular for the less favoured social members (the difference principle).

Such principles reflect a general conception of justice in the sense that all social values should be distributed equally unless an unequal distribution brings advantages to all. Therefore, for Rawls, injustice will only happen from inequalities that do not benefit everyone (Rawls, 2005).

Rawls then emphasizes five conditions crucial for the conception of what is right (or fair): generality, universality, publicity, ordering (between conflicting claims) and conclusion (as a definitive, concluding conception of justice deduced through rational consensus). That is, the conception of justice, for Rawls, is a 'set of principles, **general** in form and **universal** in application, that is to be **publicly** recognized as a **final court** of appeal for ordering the conflicting claims of moral persons' (Rawls, 2005: 117, emphasis added).

It should also be noted that Rawls, at the beginning of his work, makes explicit that he intends to formulate a reasonable conception of justice for the basic structure of society conceived as a closed and isolated system of other societies. He develops his concept of justice from the ideal of a well-ordered society without external interactions and interferences. Once formulated, such a 'pure' concept of justice could, with due modifications, be applied to other issues, considered external to the scope of justice as equity (Rawls, 2005).

The concept of justice that Rawls seeks to define is therefore delimited by the attribution of rights and duties and by the proper division of social advantages, providing a pattern for distributive aspects of the basic structure of society. All elements other than distributive and individuals are left out because they do not represent issues of justice, although they are also part of the social ideal. Rawls points out, however, that the principles of justice as equity are the essential part of the social ideal (Rawls, 2005).

Rawls' principles of justice must be accessible to all social members and should guide political institutions seeking the just distribution of primary goods to ensure as much freedom as possible to social members. For these reasons, Rawls' theory is seen by some authors as a validation of the liberal thought. Schlosberg (2009) argues that Rawls represents the source of the liberal justice theory: fair distribution regardless of any substantive agreement on what each individual believes to be 'good' or what constitutes a 'good life'. The notion of justice as equity is then understood as the fair distribution of social, political, and economic benefits and harms.

Schlosberg (2009) stresses that Rawls' and his followers' contractualist approach is only one of the approaches within the spectrum of distributive justice theories. There are other approaches, some more substantive (which start from a specific idea of how a good society should be) or consequentialists (such as utilitarianism, which focuses more on the specific outcomes of the distributive process). What is important to emphasize is that in all these approaches the central conceptual framework of justice is always focused on how and what is distributed in the construction of a just society; and this has been the hegemonic (normal) approach to justice over the last few decades (Schlosberg, 2009).

The word **normal** is used here in the sense given by Fraser, as the hegemonic or dominant definition of justice used in times of normality. In other words, when the concept of justice is not questioned by other demands that do not fit into the dominant discourse.

The fact that the distributive approach to justice is dominant does not mean it has not been questioned by other views, divergent from the normal hegemonic discourse. Since the publication of the work *Justice and Politics of Difference* by Iris Young (and afterwards, of the works developed by Nancy Fraser, Axel Honneth, Charles Taylor, Amartya Sen and Martha C Nussbaum, among others), the materialistic emphasis given by the Rawlsian school to the distributive element, without an examination of the underlying causes of maldistribution, has been challenged by critical theories that also work with post-materialist elements (Schlosberg, 2009).

Peña (2005) criticizes the Rawlsian paradigm, noting that its logic is of a rationalist system and of a 'neutral' and 'objective' positive Law that privileges individual utilitarian rights as a sole precondition for the attainment of justice. For Peña, Rawls' concept of justice privileges a specific class with ownership interests, imposing a restrictive definition of justice as a state enjoyed exclusively by abstract individuals. As a consequence, Rawls reduces his concept of justice to power over space, under the individual private property right without interference from the state and other individuals. It serves thus as an ideological rationality for the establishment of power for the exploitation of the natural and social conditions that support life.

The point raised by Peña is, essentially, a critique of the use of the mechanistic paradigm as the epistemological basis of Rawls' theory of justice, which proposes 'impartiality', 'universality', 'social ordering' and 'objectivity' through a reductionist and exclusionary approach of social complexity. Rawls compartmentalizes the different social values, isolating its object of study (distributive justice) in a system distanced from its context, imagining a basic social structure without external interferences and in an original, hypothetical and ahistorical state.

Peña (2005) argues that a theory of justice for socio-ecological issues that starts from an exclusively distributive concept of justice is not consistent with what the environmental justice movement itself criticizes. Ultimately, this concept seeks the uncertain sharing (fair distribution) of equal parts of a 'rotten and carcinogenic pie'. In other words, by focusing exclusively on the equal distribution of socio-ecological benefits and harms, most of the theorists of the 'ecology of the poor' present solutions that, in practice, reinforce the development model responsible for the socio-ecological injustices they seek to overcome.

One problem underlying the theories of justice for socio-ecological issues exclusively based on the distributional aspect is the unquestioned commitment they assume with the tools of the model they criticize. Theories of justice adopting only the distributive bias are uncritically involved with this model. They limit their strategies of action to a legal and political–administrative universe that privileges interests derived from norms of an individualistic and rationalist project, the mechanistic. This incommensurability between liberal and socio-ecological values requires a reflection on the limits and contradictions of the theories of justice that privilege distributive bias and the exploration of other possible dimensions of justice (Peña, 2005).

Following this line, Schlosberg (2009) argues that one of the main inadequacies of liberal theories of justice is their unique focus on developing and discussing ideal and fair processes for the distribution of goods and benefits. Unlike liberal theories which are developed from the Rawlsian theoretical abstraction of the 'original position', the critiques of the distributive theory are entirely influenced by political injustices of the real world. Focusing on examining real injustices, these critiques argue that there is much more than a maldistribution in the notion of injustice, and this becomes clear especially when who is left out of the real distributive processes is assessed (Schlosberg, 2009).

In general, critical approaches to the distributive perspective of justice argue that, instead of focusing on ideal justice schemes and processes in liberal societies for the improvement of material distribution, any theory of justice that is thorough and pragmatic should explore what are the impediments and social context to explain why such schemes do not materialize in the real world. That is, what are the social, cultural, symbolic and institutional conditions underlying maldistribution? In this way, critical approaches to distributive justice argue that distribution issues are essential for justice, but they are insufficient. It is a mistake to reduce the notion of social justice to the idea of (re)distribution (Schlosberg, 2009).

CRITICAL PERSPECTIVES OF JUSTICE: NANCY FRASER

From the critique on the reductionist aspect of the mechanist perspective of justice, Nancy Fraser (1996) developed her two-dimensional theory of justice, encompassing the elements of 'distribution' and 'recognition'. More recently, her contribution has been updated to a three-dimensional theory of justice, adding a third element, 'representation' (Fraser, 2005). For Fraser, the notion of justice must be seen by a multiplicity of dimensions (as opposed to ontological monism), each associated with an analytically different gender of injustice and revealed in distinct forms of social strife. The author chooses three autonomous and interconnected dimensions as the 'what' of justice, which cannot be reduced to one another, nor legitimately excluded from the contemporary theoretical formulations of justice (Fraser, 2008).

Initially, in the development of her theory, Fraser (1996) criticizes the ontological monism not only of distributive justice theories but also of theories of justice that focus only on the recognition aspect. While the former adopts a reductionist, economistic and legalistic view of social status, assuming that a fair distribution of resources and rights is sufficient to end the lack of recognition, the latter argue that recognition will eliminate all problems of maldistribution. According to Fraser, not every lack of recognition derives from maldistribution and/or from legal discrimination, just as not all maldistribution is a by-product of the lack of recognition.

Many injustices embrace issues of distribution, recognition, and representation simultaneously. Thus, the opposition of such dimensions is, in Fraser's opinion, a false antithesis. She argues for the need to adopt a multi-dimensional concept of justice, capable of providing the necessary levels of social and theoretical complexity, as well as the moral-philosophical discernment. Such a concept of justice must at least encompass distribution, recognition, and representation, without reducing one dimension to the other. It should treat them as distinct dimensions of justice while unifying them in the same broader and embracing theoretical framework (Fraser, 1996, 2005, 2008).

Participatory parity is the normative core of Fraser's thesis. All the dimensions advocated by her (distribution, recognition, and representation) are indicative of injustices when they prevent the full participation of the subjects of justice in public decisions. This unifying norm allows the demands for justice to be normatively binding for all who agree to respect the fair terms of interaction under conditions of pluralism of values (Fraser, 1996).

This normative core of Fraser's concept of justice allows the creation of a broad, comprehensive and unified theoretical framework. The establishment of legal (hypothetical–formal) equity is insufficient for such parity to become a reality, making necessary the satisfaction of other preconditions. A first precondition, called by the author 'objective', tries to ensure that the material distribution of resources is sufficient for the subjects of justice to have independence and voice. It would prevent forms and levels of material inequality and economic dependence that obstruct parity in participation (Fraser, 1996). That is, then, the economic dimension of (re)distribution of Fraser's theory of justice.

Fraser also identifies a precondition she qualifies as 'inter-subjective'. It requires that institutionalized cultural patterns (for example, by law) of interpretation and evaluation express equal respect for all subjects of justice and guarantee equal opportunity for the achievement of social esteem. Such a precondition seeks to prevent cultural patterns from systematically depreciating some categories of subjects of justice and the qualities associated with them (Fraser, 1996). That is the cultural dimension of recognition of Fraser's theory of justice.

On the subject of recognition, unlike Axel Honneth, for example, who understands this as a matter of self-realization, Fraser (1996) considers it unfair to deny some individuals and groups the status of full members in social interactions simply as a consequence of institutionalized patterns of interpretation and evaluation in which they have not participated in an egalitarian way and which depreciate their distinctive characteristics or the distinctive characteristics attributed to them.

With this definition of injustices related to recognition, Fraser made it clear that her view does not focus on the individual psychological aspect of the recognition (as done by Honneth). She criticizes Honneth's approach and turns her attention to the social status of individuals and communities on the need for recognition. Fraser defends the need for a structural understanding of non-recognition more as an institutional practice than an individual experience (Schlosberg, 2009).

For Fraser, the focus on social relations, rather than on the psychological aspect of individuals, assists in understanding non-recognition as a kind of injury to the social status of an individual or community (Schlosberg, 2009). In her view, the focus is on social relations and not on individual psychology. Non-recognition is thus an external manifestation of publicly verifiable impediments to some subjects of justice to participate with parity in public decisions (Fraser, 1996).

To overcome non-recognition, according to this approach, changes in institutional and social practices must take place; more specifically, changes in norms and institutionalized interpretations that create devalued classes of subjects, preventing them from their full participatory parity. Instead of linking to individual and psychological aspects, the social status approach has the potential to demonstrate that certain social acts will be morally indefensible when their institutionalized norms prevent parity of participation, regardless of whether it undermines the subjectivity of the oppressed (Fraser, 1996).

More recently, Fraser (2005; 2008) has identified a third precondition for participatory parity to become a reality. It seeks to ensure fair criteria for belonging to the justice community. That is, this new precondition creates the possibilities to determine who counts as a subject of justice, as well as ensuring fair procedures for filing and resolving disputes of justice. It intends to prevent certain subjects, legitimated to make claims of justice, being unjustly excluded from the community of justice. Such a precondition also seeks to prevent public processes of contestation (ie public debates in the decision-making process) from not allowing the full expression of claims and participation in decisions on disputes by all subjects of justice. That is the political dimension of representation of Fraser's theory of justice. All three preconditions (or dimensions) proposed by Fraser are necessary for participatory parity, and none is sufficient by itself.

Although distribution and recognition are also political factors by nature, Fraser (2005) explains that, in the debate about justice, she considers the political dimension in a more specific and constitutive sense: it concerns the nature of the state's jurisdiction and the decision-making rules by which the state structures social disputes. Thus, the political dimension of representation provides the stage on which the economic struggles for distribution and cultural struggles for recognition are conducted. This dimension focuses on two aspects:

i *belonging* in the establishment of the political boundaries (the 'who' of justice); and

ii *procedures* that structure public decision-making processes (the 'how' of justice).

In both aspects, the question that arises is whether the relations of representation (ie the relationship between representatives and the represented) are just or unjust. The political dimension of representation raises questions about the boundaries of the political community, which can exclude certain subjects even though they hold the right to be represented. The decision-making rules of the justice community should attribute equal capacity for expression and representation in public decision-making processes to all members of that community, but this has not entirely happened in practice. This dimension is irreducible to the other dimensions, meaning that there may be obviously political obstacles to full participatory parity arising from the political constitution of society. These impediments generate conceptually distinct injustices relative to distribution (class structure) and recognition (status order) (Fraser, 2005).

The injustices of maldistribution and non-recognition are then coupled with the injustices of false representation. The latter occurs when political boundaries and/or decision-making rules erroneously deny certain subjects the possibility of equal participation (in public decisions, but not only in political arenas). Although representation is a distinct dimension of justice, it is fully interwoven with the others, guiding not only the decisions on who can make economic and cultural claims for redistribution and recognition but also on how these demands must be debated and decided. Injustices based on a political way of social order can only be adequately understood through a theory encompassing the three fundamental dimensions of justice: distribution, recognition, and representation (Fraser, 2005).

With the addition of this third dimension to her notion of justice, Fraser (2005) reveals that at least two levels of misrepresentation can be distinguished: one political; and one meta-political.

The first one occurs in what Fraser calls the ordinary–political sphere and deals with misrepresentation within the territorial space of nation states. Such territorial space has been the dominant and uncontested grammar of the vast majority of theories of justice, which limit the communities ('who') and the procedures ('how') of justice to within the state borders. However, the claims against injustices arising from the globalization process have challenged the very establishment of the political boundaries (Fraser, 2005; 2008). With this process, a second level of misrepresentation emerges (the meta-political one), called by Fraser 'misframing', where injustices arise when the boundaries of the justice community are set up mistakenly denying certain subjects the chance to participate in the debates and decisions on justice (Fraser, 2005; 2008). Discussion of this topic will be resumed in the next section, where the element 'who' of justice will be addressed.

Fraser (1996) also adopts a pragmatic approach of justice, arguing that the remedy for an injustice should always be defined according to its category. For example, if injustice results from the economic structure of society and economically defined class differentials, the solution will be the material (re)distribution. If injustice is due to the status of social order and culturally defined status hierarchies, the solution will be found in recognition policies. In cases where more than one category of injustice occurs (when they are inter-penetrated), the different dimensions of justice must be used to allow the complete participatory parity of the subjects of justice.

According to Fraser, this pragmatic approach avoids the universalisms of both: by distributional justice theorists, who consider that any lack of recognition can be remedied by the material (re)distribution of goods and rights, and by recognition theorists, for whom the recognition of differences is a universal human need in modern societies. The normative core of Fraser's theory (participatory parity) encompasses all possible dimensions of justice and positions itself as the general pattern to guarantee such demands are met (Fraser, 1996).

Given the general outlines of Fraser's three-dimensional theory of justice, one may question the existence of other dimensions of justice which are not encompassed by the dimensions of distribution, recognition, and representation. On this point, Fraser clarifies that the dimensions of justice are historically revealed through conflicts waged by social movements when they succeed in making plausible claims that challenge the established 'what' of 'normal justice' and render it incapable (or insufficient) to respond to such transgressive demands (Fraser, 2008).

There will therefore always be the possibility that new dimensions of justice emerge in the course of history. Every theory of justice will always be provisional, and it must be open to the inclusion of new dimensions in its theoretical foundations. In exercising this flexible hermeneutics and respect for external visions to the 'what' of the 'normal justice', Fraser (2008) defends the need for validation of arguments that help to ensure the new claims bring forms of injustice that do not fit into any of the three dimensions already existent in her theory.

Multi-dimensional social ontology alone is insufficient to process disputes involving multiple dimensions of justice, requiring a normative principle of unification (not uniformity) to formalize such dimensions within a single measure. Such a principle will enable the evaluation of claims with multi-dimensional implications. It will make it possible to deal with the lack of comparison between different discourses for the idea of justice in real-life cases to materialize (Fraser, 2008). The unifying normative principle proposed by Fraser is 'participatory parity'.

For Fraser, overcoming an injustice of any dimension or multiple dimensions means dismantling the institutionalized obstacles that prevent some subjects from participating in public decisions in conditions of parity with others as integral partners of social interaction. The three forms of injustice defined by her violate this single principle, combining the multi-dimensionality of social ontology with normative monism (Fraser, 2008) (see Figure 2.1).

CRITICAL PERSPECTIVES OF JUSTICE: MARTHA C NUSSBAUM

There is another school of thought of justice theories that deserves attention in this book. It has also sought to expand the concept of justice beyond the single focus on distribution. This school proposes a 'capabilities' approach and gained prominence through the works of Amartya Sen and Martha C Nussbaum. According to Schlosberg (2009), the central argument of this

FIGURE 2.1 Fraser's three-dimensional model of justice

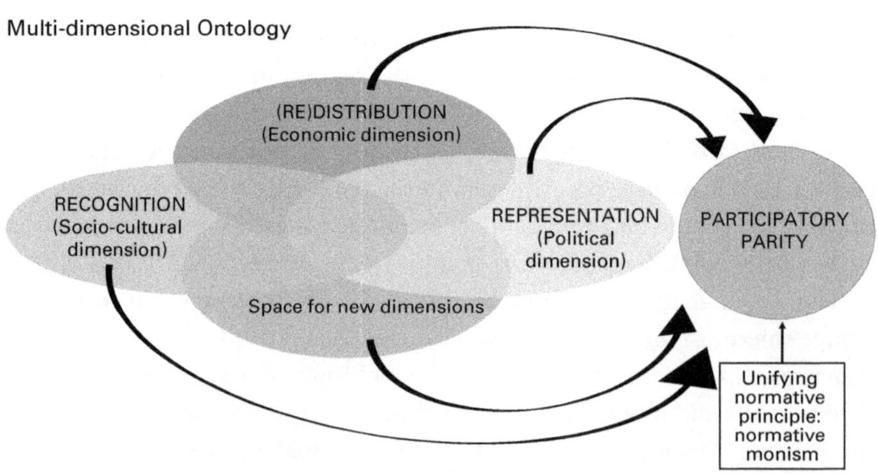

approach is that social justice agreements should be judged not in purely distributive terms, but mainly in observing how such distributions affect the quality and 'functioning' of life of the subjects of justice.

The focus of the 'justice of capabilities' lies in the individual agent (in its functioning and well-being) rather than in traditional indicators of distribution. Its purpose is to examine the specific capabilities of individuals – whatever allows or disrupts a living system's ability from turning basic (primary) goods into the functioning of life. In this approach, the notion of justice moves away from an exclusive concern with the number of goods that each individual receives, to examine what these goods and their absence provide or cause to each individual (Schlosberg, 2009).

Nussbaum (2011) synthesizes the capability approach as a proposal developed both to comparatively analyse the quality of life of different societies and to theorize social justice. Unlike utilitarian perspectives, the justice of capabilities views each person as an end and turns its attention to the opportunities that are made available by the state to each individual. This approach focuses on the element 'choice' or, in other words, on the 'substantive freedom' of each individual. It defends the proposition that the state should offer all subjects of justice a set of opportunities, which may or may not be exercised according to the individual choice.

Nussbaum highlights the importance of using the plural 'capabilities' to emphasize that the most important elements to the quality of life of the subjects of justice are plural and qualitatively distinct. The plurality and the non-reducibility are thus key elements of this approach. The term 'capabilities' refers to the 'substantial freedoms' (opportunities) that each individual must have in order to be able to choose and achieve alternative life-functioning combinations or, in other words, to be and to do what they choose in the context of a given society (Nussbaum, 2011).

Just as important as the term 'capabilities' for this approach is the term 'functioning', which means the actual realization of one or more capabilities. 'Functioning' therefore means states and actions (beings and doings) that represent the concrete result, the realization of capabilities. Although it is understood that functioning gives meaning to capabilities, Nussbaum (2011) emphasizes that capabilities have their own value as freedom of choice. Even if the capabilities are not concretely realized by a choice of the subject of justice, their development is still essential for that individual to be able to choose.

Therefore, it is on the capabilities and not on the functioning itself that the political goals of a state should focus (Nussbaum, 2011). For example, nutrition is a capability, while eating is functioning. An individual should have the opportunity to nourish him- or herself, while being able to choose not to do so (opting for diet or fasting, for example). Literacy could be classified as a capability, while reading would be functioning. The state should thus offer literacy to all subjects of justice, although such subjects will have the freedom not to read if they choose.

This approach focuses on the possibility of different combinations of life functioning and the freedom of each individual to make use of these possibilities or not. The justice of capabilities is grounded in the desire to see everything flourishing to become what it can be. The central measure of justice here is not only how much individuals possess, but if they have the necessary capability to enjoy the life chosen by them, fully functioning (Schlosberg, 2009).

The term 'capabilities' used in this perspective of justice is not restricted to the personal abilities of each individual, but also encompasses the opportunities (freedoms) arising from the combination of these abilities with the proper political, social and economic conditions. To clarify the complexity of the term 'capabilities', Nussbaum (2011) uses different terms for its distinct classes, structuring them within the same theoretical framework. In this theory, 'internal capabilities' are those relating to the personal characteristics of individuals, their personal abilities. These internal capabilities interact with the social, political and economic conditions of functioning, resulting in what Nussbaum calls 'combined capabilities'. Combined capabilities are therefore all 'substantial freedoms', the total opportunities for a person to have the freedom to choose and act within their specific political, social, and economic conditions (Nussbaum, 2011).

The development of both personal qualities and political, social and economic conditions are necessary to enable each individual to choose and put into practice the kind of life they want to live. The creation of contexts of choice (conditions) by the state in different areas (political, social and economic) will be insufficient if it does not educate its citizens, if it does not nourish their powers of mind (personal qualities – internal capabilities). Although the distinction between internal capabilities and combined capabilities is not very evident in practical terms, Nussbaum (2011) argues that it is heuristically useful in diagnosing the achievements and shortcomings of a society.

To these two terms, Nussbaum (2004, 2006; 2011) further adds the 'basic capabilities' as part of the 'internal capabilities', as the internal capabilities that are innate to the subjects of justice: the innate faculties that make possible their development and flourishing. Unlike Sen, Nussbaum addresses this point in more detail, presenting a list of 10 specific basic capabilities that establish the minimum conditions for each individual to flourish to be what they wish to be. This list involves the basic capabilities relating to:

- life;
- physical health;
- physical integrity;
- sense of imagination and thought;
- emotions;
- practical reason;
- institutional affiliation;
- relationship with other species;
- entertainment; and
- control over the environment.

SOURCE Nussbaum, 2004; 2006; 2011

It is important to note that each capability listed by Nussbaum is considered as an independent component. Even if an individual has one of the capabilities well developed, it does not mean that they do not need to develop others. In addition, Nussbaum explains that her list is valid for a certain period and can be modified when facing new conditions. It would always happen in open dialogue with social movements, which can be influenced and influence such presuppositions of justice (Schlosberg, 2009).

Although she considers such capabilities as innate, Nussbaum (2011) emphasizes that her theory is not based on the idea of 'human nature'. Instead, she proposes it as an evaluative theory with preliminary ethical grounds. It is a theory that seeks to understand and analyse, among all the capabilities subjects of justice can develop, which ones are really valuable and which ones a minimally just society must seek to nourish and endure. Nussbaum's theory, in short, seeks to reveal which capabilities provide minimum standards of existential dignity.

On participatory parity, Schlosberg (2009) points out that, under the capabilities approach, public participation is an integral and fundamental part of an understanding of justice. Participation is seen as both freedom (a capability) and a function in itself. For example, the tenth basic capability listed by Nussbaum (control over the environment) has two aspects: one political and one material. The political aspect for humans is defined in terms of active citizenship and political participation rights. For the non-human world, participation must be part of a political concept that will ensure that non-humans are fairly respected. For the latter, the political representation must defend those rights (as with children's rights) (Nussbaum, 2004).

The capabilities approach, adopting a more holistic concept of justice based on the importance of developing basic capabilities for the full functioning of individuals, believes that such development will only take place through the distribution of goods, social and political recognition and other conditions (Schlosberg, 2009). This approach thus establishes a fundamental connection between the economic (distributive), cultural (recognition) and political (representation) components.

The basic capabilities necessary for the full functioning of the subjects of justice are the normative core of Nussbaum's theory. Justice, for her, requires respect for human and non-human dignity, with the guarantee that the means necessary for the complete development of at least their basic capabilities will be provided to the subjects of justice. Accordingly, Nussbaum's list of 10 basic capabilities establishes an existential minimum to guarantee dignity to humans and non-humans. Other capabilities beyond this proposed minimum may be guaranteed by different states, which will be established and specified in their various constitutions or fundamental legal principles (Nussbaum, 2004, 2011).

From everything that has been examined so far, it is clear the 'what' of justice is a complex, plural, heterogeneous and multi-faceted element. It cannot be reduced to a single dimension without suffering the penalty of insufficiency and inadequacy. Justice is a matter of distribution, but also of recognition, representation and other possible dimensions necessary to enable the full functioning of individuals and communities. Although maldistribution is a key element in the configuration of social injustice, it is not the only one. This understanding is only evident with the necessary analysis of the context and the actual causes of such injustice. It represents a critical view of the traditional (liberal and mechanistic) theory of justice, and, within this view, there are different perspectives, which reiterates the impossibility of constructing a theory based on a monological ontology of justice.

From the critical theories highlighted here, it is possible to identify a strong tendency in the literature to move towards a more holistic, complex and less economistic concept of justice. Fraser and Nussbaum's theories, although different in their approaches, both adopt multi-dimensional perspectives of justice, with points of contact on the economic, cultural, and political conditions necessary for the materialization of justice in the contemporary world. The distributive, recognition and representation dimensions are present in both theories, which also highlight the importance of public participation for the concept of justice they advocate.

The study of contemporary critical theories of justice is vital for a better understanding and application of this notion to socio-ecological issues. As argued in the first section of this chapter, the environmental justice movement is critical of both perspectives of traditional environmentalism: the cult of wilderness (represented by the deep ecology school) and the gospel of eco-efficiency (represented by the shallow ecology school). As noted by Schlosberg (2009) and Peña (2005), although the environmental justice movement brings with its empirical experiences a critical view on what Acselrad *et al* (2009) call 'environmental common sense', its theoretical developments (made by the ecology of the poor school) rarely explain which concept of justice they adopt. When this explanation is given, it is clear that most of them are restricted to distributive or, at best, participatory justice.

This is the great paradox – and limitations – of the vast majority of the literature of the ecology of the poor. While there is a heterogeneity of views within the current theories of justice, the ecology of the poor school limits discussions of environmental justice to assumptions and strategies of liberal justice (justice as equity). It leaves out a whole range of argumentative possibilities for expansion, diversification, and strengthening of the notion of justice for socio-ecological issues as a multi-dimensional concept capable of being applied to real-life situations.

Social movements, on the other hand, use a wide variety of concepts of justice, and commonly use arguments for distribution, recognition, representation, and development of capabilities. Notably, the environmental justice movement provides evidence that all such concepts of justice are used in practice, so the theoretical developments that are made about environmental justice must also include these different discourses (Schlosberg, 2009).

The adoption of a multi-dimensional notion of justice that contemplates, at least, the dimensions of distribution, recognition, and representation, seems to be a viable and coherent path along which to search for answers to the complexity of current socio-ecological problems; to search for strong answers to the strong questions of contemporaneity. If the impossibility of constructing a single concept of justice is evidenced by a brief analysis of the multiplicity of theoretical views on justice (not to speak of the infinite empirical views of social movements around the world), it seems the idea of conceptualizing justice *a posteriori*, as well as the use of a pluralistic and pragmatic approach, is appropriate.

But this is just the 'what' of justice. According to Fraser (2008), unless there is an adequate way to understand the 'who' of justice, any method concerning 'what' will be useless. This will be the problem to which the next two sections of this chapter will be devoted.

The 'who' of justice: environmental or ecological justice?

THREE-DIMENSIONAL AND PRAGMATIC DEFINITION OF THE 'WHO': SPACE, TIME AND SUBJECT ASPECTS

Within the boundary of the 'what' of justice, the addition of the dimension of representation suggested by Fraser (2005, 2008) revealed the existence of two levels of justice: a political and a meta-political. The first, called by Fraser the ordinary–political sphere, represents the first-order injustices, referring, at least, to maldistribution, non-recognition and misrepresentation within state territorial spaces. The nation state is, according to Fraser, the referential spatial framework of the dominant grammar of normal justice, having a political community with supposedly stable contours and members.

Fraser calls this frame of reference 'Westphalian', referring to the Peace of Westphalia – the set of treaties that inaugurated the modern international system with the agreed establishment of common notions and principles, such as state sovereignty and the nation-state. In addition to the scope of the notion of justice, the traditional spatial framework of justice has also been challenged with the increasingly evident occurrence of injustices generated by the globalization process. At this point, the meta-political level of justice arises, in which, in addition to the first-order issues of justice, the spatial framework itself is questioned (Fraser, 2005; 2008).

First-order issues of justice are only part of the reality: in addition to ordinary–political injustices (misrepresentation) that manifest themselves within the framework of a delimited political entity (the nation-state), it is possible to see another level of injustice, the meta-political, resulting from the division of political spaces into defined entities. These meta-political injustices, called by Fraser 'injustices of misframing', occur when the boundaries of a political entity are set in such a way as to deny some subjects the opportunity to participate in debates and decisions concerning justice (Fraser, 2008).

Injustices of misframing occur with the mistaken exclusion of certain subjects (which should be recognized) of the political entity that discusses issues of distribution, recognition and first-order representation. According to Fraser (2008), the use of the Westphalian framework as the only possible framework is an example of such injustices, since it divides the political space in such a way as to prevent many individuals, especially the most vulnerable, from facing the forces that keep them oppressed.

This model channels the demands for justice into domestic political spaces of relatively powerless or even failed states, protecting the predatory and foreign state powers, the transnational private powers and the governance structures of the global economy. The Westphalian framework, while permitting a regime of exploitation ruled by these exploratory global state and economic forces, hinders the emergence of a transnational system of democratic decision-making and control over justice issues by relegating this role to nation-states (Fraser, 2008).

In bringing the issue of spatial framing as a matter of justice, Fraser's theory, reflectively, and beyond the 'what' of justice, brings out disputes about the 'who' of justice in spatial terms. She emphasizes that in considering the possibility of other frameworks for an analysis of the first-order injustices, it is necessary to establish a unifying normative principle for the clarification of disputes involving antagonistic concepts of 'who', in the same way the principle of participatory parity is used for disputes between the different dimensions of the 'what' of justice (Fraser, 2008).

Assessing the insufficiency or inadequacy of other theoretical developments for the construction of a concept of transnational justice, Fraser (2008) proposes the so-called 'all-subjected' principle to be the unifying principle, the analysis measure for allegations of misframing and the normative instrument for resolving disputes around the 'who' of justice. This principle provides that all those who are subjected to a governance structure will be subject to its justice.

From this perspective, it is not only the citizenship (the perspective of nationalist affiliation), the possession of abstract human qualities (humanist perspective) or casual interdependence (all-affected perspective) that place a group of subjects in the same level in relation to justice. Fraser's 'all-subjected' perspective defines joint subordination (to any governance structure that dictates the rules on a particular aspect of the socio-economic system) as a critical element for the analysis of who should be included as a subject of justice in claims related to the existing rules, levelling the scope of moral concerns of this subjection (Fraser, 2008).

Fraser goes on to clarify that she understands the expression 'subjected to a governance structure' in a broad way. It encompasses relations between powers of different types, including not only states but also non-state agencies that generate executable rules and define essential areas of interaction such as the World Trade Organization, the International Monetary Fund, the Organisation for Economic Co-operation and Development, and the United Nations (Fraser, 2008).

Fraser (2008) argues these governance structures fulfil functions of regulation of the interaction of large transnational populations subjected to them. However, they do not have any responsibilities concerning those whom they govern (which is a nation-state's responsibility). As a result of this broad view of 'governance structures', the term 'subjection' must also be understood broadly. It is not restricted to subjection to state jurisdiction, including the subjection to the coercive power of non-state and transnational forms of government.

The principle of 'all-subjected' thus determines that an issue will be framed fairly if all those subjected to the regulatory governance structure(s) of one or more relevant segment(s) of social interaction on this issue receive the same consideration (Fraser, 2008). To be entitled to claim for justice, it is enough that this individual or community is subjected to the governance structure that regulates the object of justice (an analysis that applies to the case of waste). For example, all communities and countries involved in cross-border waste transfer operations, whether they are signatories to the international instruments regulating the theme or not, are subjected to such waste governance structures. They are therefore the subject of justice concerning global waste management, even if they are not recognized or duly represented.

Fraser (2008) highlights that everyone is subject to a plurality of governance structures at different levels: local, national, regional and global. Thus, it is vital that contemporary theories of justice open possibilities to different frameworks for distinct justice issues. Accepting the plurality of frameworks to define the 'who' of justice, the principle of 'all-subjected' indicates when and where to apply which framework, defining who has the right to the parity of participation in certain circumstances.

Again, Fraser's approach is reflexive and pragmatic. It considers multiple frameworks of justice with an evaluation principle that helps in determining the appropriate framework of 'who' should be subject to justice in each particular case. Her theory allows, in addition to the first-order injustices, the meta-political injustices of misframing to be debated and overcome by the pragmatic (re)definition of the subjects of justice. The evaluation principle of 'all-subjected' is, therefore, the key tool for such pragmatic redefinition, contesting the Westphalian framework as the only valid one.

Fraser's approach offers viable ways to elucidate disputes between different 'who's in times that she calls abnormal. Her theory provides essential elements for expanding the community of justice beyond state borders in today's globalized world, presenting critical reflexion for the construction of a counter-hegemonic movement of transnational resistance forces. Hence, Fraser's notion of justice is quite adequate for cross-border issues, as are many of the socio-ecological problems (including those relating to waste), and consequently is adequate for the demands of the environmental movement in its different streams.

In addition to being transboundary, socio-ecological problems are also transtemporal and affect not only human beings but also the non-human natural world. Therefore, there are three dimensions of justice issues related to socio-ecological protection to be considered. Two of them are identified by Bosselmann (2016):

i the justice of the environment among people;

ii the justice of the relationship between humans and the rest of the natural world.

To these two dimensions, it is also possible to add a third one:

iii the justice of the environment between generations.

All three dimensions represent relational aspects referring, respectively, to:

i space (the justice of the environment among people);

ii the subjects (the justice of the relationship between humans and the rest of the natural world);

iii the time of socio-ecological problems (the justice of the environment between generations of humans and non-humans).

The first aspect is addressed by the notion of 'environmental justice', which adopts an anthropocentric and intra-generational perspective in most of its theoretical developments (ecology of the poor school). For the spatial issues of socio-ecological problems, Fraser's proposal is an adequate theoretical framework, since it expands the territorial space of traditional theories of justice. It makes possible the inclusion of subjects located outside the state territory in the community of justice when they are subjected to governance structures that have generated the injustices they seek to resist.

The other two aspects, related to time and subjects of socio-ecological problems, are generally not addressed by the 'ecology of the poor' school. They are also largely excluded from the theories of justice in general, whether liberal or critical, since they all assume the community of justice is solely composed of contemporary human beings. There are, in this sense, two gaps in the literature of justice when applied to socio-ecological issues: one concerning inter-generational justice (temporal aspect) and the other inter-species justice (subjective aspect) (see Figure 2.2).

One of the possible reasons for the existence of these theoretical gaps may be the fact that, traditionally, legal relationships, including those of justice, are regarded as relations maintained purely between contemporary human beings. The responsibilities of current generations for future generations and for the rest of the natural world are usually considered a matter of ethics and morality, but not of justice. However, as emphasized by Bosselmann (2016), there is no notion of justice without underlying ethical assumptions.

Indeed, Bosselmann (2016) defends the issue of non-human entities being subject to rights or justice as a matter for philosophers to decide, not for jurists. From a legal perspective, all types of entities may have rights and be subject to justice; what will define which entities will be part of this community will be the ethical perspective adopted. For Bosselmann, mere extensionism of the liberal and individualist perspective only perpetuates individual and anthropocentric freedom, which is hostile to ecological responsibilities. It is, therefore, necessary to include the ecological dimension at the core of the notion of justice.

As seen in this chapter, the ethics informing the notion of distributive justice, widely used by the 'ecology of the poor' school, is the liberal anthropocentric ethic. Although critical theories of liberal–distributive bias of justice

FIGURE 2.2 Dimensions of the 'who' of justice for socio-ecological demands

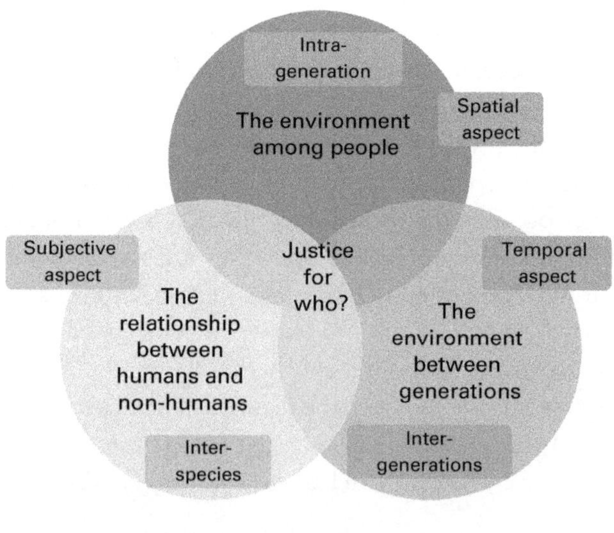

have advanced to overcome the mechanistic paradigm, most of them remain within the anthropocentric boundaries – including Fraser's theory. Currently, however, many authors have sought to develop justice theories that expand their scope of application and overcome the anthropocentric bias of justice. Schlosberg (2009) analyses some of these approaches, from which those dealing specifically with the introduction of inter-generational justice and inter-species justice for socio-ecological issues will be highlighted here.

Undoubtedly, the temporal aspect of inter-generational justice has been less controversial and more accepted, finding grounds in popular notions such as the 'sustainable development' propagated by the United Nations in the Brundtland Report. This approach recognizes the responsibilities of present generations for future generations of humans, expanding the time-frame of the notion of justice. Inter-generational justice requires a principle of equal opportunities between the present and future generations.

The central argument of this approach is that present generations cannot leave less than they have received for future generations. Schlosberg (2009) advocates that this approach, while innovative, remains firmly based on the liberal–distributive idea, focusing on the distribution of environmental goods to future generations of humans. Hence, this approach does not recognize Nature itself as a member of the community of justice, but only as a support to the human system. The recognition and appreciation of this support system are central to this approach, but the extension of the community of justice is still centred on human beings, still based on anthropocentrism, even if its timeframe is extended (Schlosberg, 2009).

Although it is an advantageous model of justice in environmental terms, inter-generational justice does not overcome the liberal anthropocentric paradigm. It is clear then there is a need for ethical debate about traditional legal theories of justice. The two forms predominantly used by theorists for the development of the notion of justice (through the rationality of the monological–liberal/ontological–liberal approach or the public debate of the critical dialogical/discursive approach) have not provided this ethical debate. They all remain in the modern tradition that accepts anthropocentric rationality as the only valid one for ethics, preventing their theories of justice from overcoming this paradigm (Bosselmann, 2016).

Bosselmann (2016) points out that even if the anthropocentric paradigm is challenged in ethical terms by the deep ecology school, ecocentrism cannot be imposed on liberal democracy. Nor will it emerge naturally from a discursive democracy, since both types of democracy have their roots in anthropocentrism. For the author, neither the monological nor the dialogical discourse of justice have so far been able to restrain the significant modifications human beings have made to the conditions of human and non-human life. This inability is due above all to the anthropocentric paradigm and indicates the need to overcome it.

The liberal and anthropocentric perspective of justice tends to use definitions of liberalism on environmental issues as broad, competing ideals (balance between economy, society, and the environment – weak sustainability), ignoring the casual nexus between liberalism, anthropocentrism, and environmental degradation. Against this approach, Bosselmann (2016) defends an ethical paradigm shift, by adopting ecocentrism to construct a notion of ecological justice. Such a notion, presented as opposed to liberal justice, brings the principle of ecological sustainability regarding the intrinsic value of the natural non-human world, which serves to orientate the reformulation of legal concepts, especially the idea of justice.

Ecological justice, therefore, introduces the non-human natural world into the community of justice by building institutional and constitutional milestones of society. It ensures the consideration of Nature in environmental decision-making processes (Nature as a non-negotiable basis – strong sustainability). Thus, the difference between anthropocentrism and ecocentrism does not represent different levels of environmental protection, but different paradigms (Bosselmann, 2016). This proposal of ecological justice seeks to overcome not only liberalism but also, and especially, anthropocentrism.

In order to achieve the ecological paradigm of justice based on ecocentrism, neither monological rationality nor discursive debate on justice is sufficient. Bosselmann (2016) suggests adopting a 'self-reflective discourse' approach to justice, called by him the 'ecological discourse of justice'. Such a discourse, for Bosselmann, allows for openness and transformation within the democratic context. It incorporates ecocentric ethical values into the notion of justice without restricting them to the individualist perspective of traditional theories, whether liberal or critical.

Bosselmann's ecological discourse of justice advocates that the fact the natural world cannot communicate (directly) or be reciprocal of moral favours should not matter to the development of the notion of ecological justice. What matters is the ability of humans to act for themselves and for those who cannot do so (humans and non-humans). Potentially, the whole natural world, human and non-human, subjected to environmental decisions, even if it is not an active participant in moral dialogue, would have their voice heard. The ecological discourse of justice translates the need for recognition and representation of the non-human natural world in socio-ecological issues.

Bosselmann (2016) affirms that there are several ways of realizing such representation, but for this to occur, the notion of ecological justice must be adopted with the internalization of the ecocentric ethical paradigm. Although at this point of the chapter it may already be quite evident, it should be pointed out that, following Bosselmann's line, the terminological differences between 'environmental justice' and 'ecological justice' are not only of linguistic interest. It reflects a truly ethical dilemma between the anthropocentrism of environmental justice and the ecocentrism of ecological justice.

The term 'environment' was shaped by Western (mechanistic) anthropocentric tradition and reduces the relation between human and Nature to a mere environmental relation. The complexities of this environment are captured only by the term 'ecological integrity'. It reflects the existence of natural processes necessary for the maintenance of life support systems on Earth on which humans and non-humans depend. Therefore, Bosselmann (2016) argues that it is not with the environment but with the interactions between the various life forms and with the abiotic elements that justice theories about socio-ecological issues should be concerned.

By focusing on ecological relations, ecological justice recognizes not only the complexities of the natural world but also that humans are part of this world. Perhaps the conceptual differences between the environmental and the ecological terminologies will disappear over time. However, while human societies do not overcome the mechanistic paradigm by adopting a more holistic and systemic worldview, accepting the community of humans and the natural world as parts of this 'environment', it is appropriate to maintain the distinctions between the approaches of environmental justice (anthropocentric and liberal) and ecological justice (ecocentric) (Bosselmann, 2016).

For all these reasons, this book will avoid the use of 'environmental justice' in order to make clear the model of justice proposed here seeks to surpass anthropocentrism and liberalism. On the other hand, the term 'ecological justice' is not best suited for the objectives pursued here either. Usually, the term ecological justice is used in the literature to indicate the inclusion of Nature into the community of justice. However, the model proposed here intends to go beyond that inclusion and, from an ecocentric ethics stance, reinforce the complexity of socio-ecological issues. Therefore, the term ***socio-ecological justice*** seems to better translate this complexity, recognizing that social and ecological problems can be interconnected without being confused, and that humans are part of Nature, but are not reduced to it. The term 'socio-ecological justice' will be adopted as being more appropriate for the migration of the mechanistic paradigm to a more complex paradigm; for better representing the union of social and ecological demands in the same theoretical framework of justice while respecting their differences (unity in diversity, unity without uniformity).

Along the same lines, Schlosberg (2009) highlights that there are many theoretical developments for the construction of a concept of justice for socio-ecological issues. The author seeks to develop his theory going beyond the distributive paradigm, encompassing innovative ideas of contemporary justice theories and political demands of critical social movements. His central focus is to bring other notions, such as recognition and capabilities, to a broader theory of justice for socio-ecological issues. The author does not propose a single universal concept of justice. He seeks to examine the potential of engaging diverse discourses of justice and a variety of approaches to understand and extend practices of justice that include the natural non-human world (individual animals and Nature as a whole) into their concerns.

Stressing the disconnection between most of the theories of environmental justice and ecological justice, Schlosberg (2009) makes clear his intention to expand the discourse of these theories to expose a common language of justice found in both perspectives. The author intends to provide a shared toolbox that can be used to address both environmental justice issues and ecological justice issues. Schlosberg seeks to include the demands for intra-generational, inter-generational and inter-species justice in the same unified, though not uniform, theoretical framework.

In this sense, Schlosberg highlights the essential need to include considerations of intra-generational justice in the debate on inter-generational justice. For him, there is no justification for committing injustices to some contemporaries in the name of justice for future generations. Concerns for future generations must always be accompanied by the provision of justice to the present. Regarding inter-species justice, Schlosberg (2009) affirms that there are different strategies for considering the natural non-human world as a subject of justice.

Inspired by this pluralistic approach of different theories of justice used by Schlosberg, some basic assumptions to fill the notion of justice in real-life cases involving environmental decisions will be developed here. These assumptions will then be applied in Chapter 6 of this book, following the line of *a posteriori* (or pragmatic) justice proposed in this chapter. To achieve this, Fraser's three-dimensional theory and Nussbaum's theory of capabilities, with the ecological direction offered by Bosselmann and Schlosberg, will be used to define a theoretical framework of justice that is broader and critical of the liberal and anthropocentric model.

At first, the environmental justice approach may seem adequate to deal with the current socio-ecological crisis, especially when related to cross-border waste transfer (since it works with the intra-generational justice axis). However, a closer analysis of the issue shows the inadequacy of this approach. Most contemporary socio-ecological problems disrespect not only state borders, but also temporal boundaries and those between humans and non-humans. The same applies to the waste crisis, given that the impacts and degradation generated by it affect human beings, the non-human natural world and their current and future systemic interrelationships.

For all the reasons identified, this book advocates that waste is not only a problem of justice but is also a problem of socio-ecological justice. The challenge that arises is which theoretical framework can be used to propose the minimum assumptions necessary for the realization of *a posteriori* justice in real-life cases of environmental decisions, such as those involving cross-border waste transfer. It is on this point that the following section will now focus.

SOCIO-ECOLOGICAL JUSTICE

Fraser's theory of justice works with two levels of analysis: the political and the meta-political. The first one deals with what Fraser calls first-order injustices that occur within the scope of ordinary–politics. Such injustices take place whenever a subject or a community is prevented from participating in public decisions in parity with other subjects of justice. This can happen, according to Fraser, through three different problems (interconnected or independent): maldistribution (economic dimension); non-recognition (cultural dimension); and misrepresentation (political dimension).

Fraser thus developed her three-dimensional theory of justice with the unifying principle of parity participation. Such a theory is adequate for overcoming the liberal–distributive paradigm and is applicable to the theory of environmental justice (ecology of the poor). That is because the issues typically addressed by environmental justice literature essentially involve intra-generational justice. However, Fraser's proposal comes up against a major barrier if applied to issues of inter-generational justice and inter-species justice. Her unifying principle (participatory parity) cannot be applied to future generations and, especially, to Nature without being seriously questioned.

The vast majority of the authors who work with and defend the adoption of an ecocentric ethical perspective are careful to clarify that the inclusion of the non-human natural world in the scope of justice does not require equity between humans and non-humans. In other words, it does not mean the natural non-human world will have a moral equivalence to or the same rights as humans (Bosselmann, 2016; Schlosberg, 2009). Although due recognition and due representation of Nature can be defended under an ecocentric paradigm, participatory parity as the normative core for the configuration of injustices already finds limitations in the ethical debate, since such principle represents an eminently democratic and human value.

The notion of limiting the basic capabilities necessary for full functioning, used by Nussbaum as the referential normative core for the configuration of injustices, shows a better potential for the development of a theory of socio-ecological justice. It escapes from the reference of eminently human values and can be applied to both humans and the non-human natural world. Nussbaum herself extends the capabilities approach of justice to individual non-human animals, although not addressing species (communities), Nature as a whole and their systemic interrelationships.

For this reason, the capabilities approach could be criticized for being associated with moral atomism and liberal individualism, having limited use and impeding a systemic and communitarian approach. It could be seen as an attempt to extend individualism, perpetuating some of the underlying assumptions of the mechanistic paradigm, even though it defies some of its frontiers. Nussbaum's list of basic capabilities for the establishment of an existential minimum for the guarantee of human and non-human dignity is handy for revealing the similarities between humans and non-human animals. Still, it fails to recognize other elements and ecosystemic relations, to which the idea of dignity would not apply directly.

Even with different approaches, the theories of Fraser and Nussbaum have as a point of contact the fact that both recognize injustices occur with the absence of specific economic, socio-cultural and political conditions. Fraser works specifically with these conditions as dimensions of her theory of justice. Nussbaum, on the other hand, makes clear that such conditions are part of her notion of 'combined capabilities'. After highlighting some differences and proximities, potentialities and limitations of both approaches, the possibility of interaction and dialogue between them for the proposition of a broader and critical theoretical framework of socio-ecological justice can be envisioned.

The proposal here is to use the critical theoretical structure developed by Fraser, with the essential substitution of its normative core for the notion presented by the capabilities approach. This is because the normative core of the capabilities approach demonstrates excellent potential for the construction of a more comprehensive and ecologically inclusive notion of justice.

The idea is that injustice will happen whenever basic capabilities are limited impeding the full functioning of an individual, community or system.

These limitations may occur through one or more dimensions of justice, such as maldistribution, non-recognition, and/or misrepresentation. This nuclear substitution seeks to overcome the intra-generational anthropocentric character of Fraser's theory (Figure 2.3).

Instead of being restricted to the promotion of dignity, the establishment of an existential minimum by the development of basic capabilities necessary for full functioning can promote not only dignity but also other essential elements for the continuity and balance of life. For this purpose, it is fundamental to recognize the intrinsic value of humans and the non-human natural world as an ethical presupposition for the inclusion in the justice community not only of human and non-human individuals but also of communities and Nature as a whole, with their inter-systemic relationships. This change seeks to overcome the individualistic perspective and the limited application to humans and some non-human animals of Nussbaum's theory.

According to Bosselmann (2016), the adoption of the ecocentric paradigm implies the recognition of the intrinsic value of the non-human natural world. Without such recognition, environmental decisions are based only on the argument of potential risks to human health. The adoption of inter-species justice can thus improve the results of environmental decision-making processes. The author emphasizes, however, that the mere inclusion of the inter-species element in traditional theories of justice is insufficient. That is because the barriers to just treatment of non-human life are deeply embedded in the modern (mechanistic) paradigm and can only be removed by questioning such a paradigm.

FIGURE 2.3 The 'what' of justice in the socio-ecological justice proposal

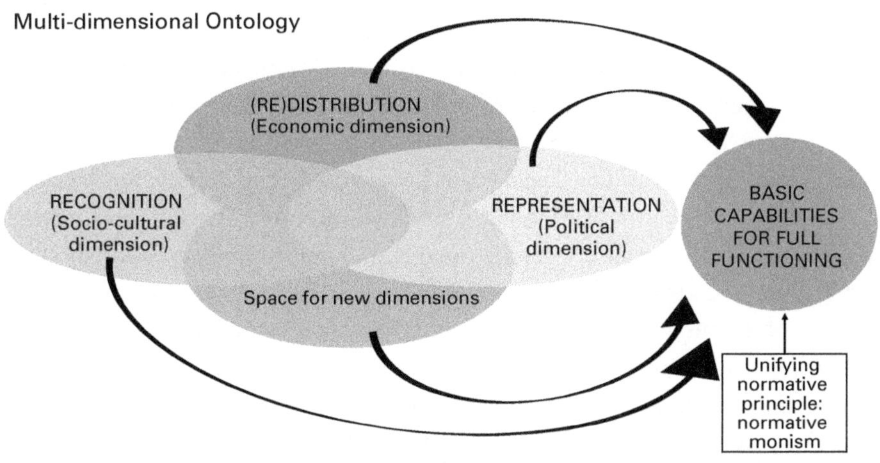

It should be pointed out, furthermore, that the recognition of the intrinsic value proposed here follows the line of Fraser's theory: it is not related to the individual psychological aspect, but the status. In other words, it is an objective recognition that both humans and the non-human natural world have an intrinsic value within (and outside) the socio-economic system of which they are part. As emphasized by Schlosberg (2009), the status approach does not aim to place a value on individual or group identity, or recognition of the problematic psychological situation of victims of injustices, but in overcoming their subordination.

This general theoretical framework for defining the 'what' of socio-ecological justice proposed here at the level of ordinary–politics leads to the second point of analysis, concerning the 'who' of justice. Regarding the meta-political level, the unifying principle Fraser uses for her pragmatic definition of the 'who' of justice is that of 'all-subjected'. She focuses on the extension of the territorial space beyond the state to frame specific issues of justice in a transnational way. Doing so, Fraser expands the spatial aspect of the definition of the justice community among contemporary humans, provided they are all subjected to a particular governance structure.

In this respect Fraser's theory can also be used as the basis for the demands typically made by the environmental justice movement since it expands the community of subjects that can demand justice in transnational environmental issues. However, after analysing her unifying principle of 'all-subjected', the consideration of temporal and subjective aspects for the inclusion of inter-generational and inter-species justice emerges as an argumentative possibility. On environmental issues, all the rules established by environmental governance structures not only affect, but subject Nature, as well as future generations of humans and non-humans. Still, such governance structures do not consider Nature and future generations as members of the community of justice – which, in terms of Fraser's notion of justice, characterizes a meta-political injustice of misframing. With this, a redefinition of the 'who' of justice would also be made to encompass intra-generational, inter-generational and inter-species justices.

The entry of the non-human natural world into the community of justice means that it will be a recipient of justice, but not an agent, following the same logic mentioned above that such inclusion does not mean equity between humans and non-humans, nor that they will have equal rights. According to Schlosberg (2009), although this may seem a paradigmatic proposition, it already exists and is applied, for example, to humans who cannot be agents of justice because of the impossibility of expressing their own conceptions of goods, such as children, for example.

From the combination of these distinct elements emerges the theoretical framework proposed here for the *a posteriori* realization of the 'what' and the 'who' of socio-ecological justice. In short, this framework elects the development of basic capabilities for the full functioning of individuals and communities of humans and non-humans, as well as Nature as a whole with its ecosystemic interrelationships, as its unifying normative core for the notion of socio-ecological justice. Also, it understands that at least equitable distribution, due recognition, and representation are necessary conditions for socio-ecological justice to materialize.

Concerning the recognition dimension, justice between human beings, whether contemporary or not (intra-generational and inter-generational) on socio-ecological issues, should serve to overcome any subordination that is limiting their basic capabilities for full functioning, promoting the necessary and minimal dignity of life. For the relations between humans and the non-human natural world (inter-species), the recognition proposed here (of intrinsic value) should serve this same goal. However, unlike for humans, the recognition of the intrinsic value of Nature will promote the ecological integrity of the natural world as a whole and its ecosystemic interrelationships.

Specifically concerning non-human animals endowed with certain qualities such as sentience and self-consciousness, for example, it is possible to defend the promotion of their dignity of life as individuals, without appealing to the language of psychological individuality. This promotion of dignity, however, must avoid the anthropomorphization of these subjects of justice, also considering their ecosystemic functions, ie their roles for the maintenance of ecological integrity.

Following the same line of Schlosberg (2009), the recognition of Nature using Fraser's model based on status allows the non-human natural world to be considered in demands for justice. It also leads to a human obligation not to interfere in the realization of ecosystemic means and potentialities. The due recognition of present and future generations of humans and non-humans will also provide the necessary means for proper environmental distribution and political representation for all subjects of justice.

Recognition of intrinsic value for developing basic capabilities for full functioning should promote different values when applied to humans (dignity of life), non-human animals (dignity of life and ecological integrity) and Nature and their inter-systemic relationships (ecological integrity). Thus, it has the potential to overcome the risk of a rigorous monism that ecocentric ethics can lead to. This proposal allows, in specific cases, the relations between humans and non-humans to be recognized (their connections), avoiding a general and abstract hierarchy. At the same time, it permits their differences to be identified (their individuality), avoiding confusion between them.

Those are, therefore, the minimum elements for the general assumptions of the notion of socio-ecological justice for its materialization in real-life cases of decisions on socio-ecological issues (see Figure 2.4).

Although the concept of the dignity of life for humans and non-human animals is a well-developed theme in the legal literature, it is believed that its realization should be achieved through dialogue and democracy, reflecting the local realities of all subjected to the governance structure in question. The term 'dignity of life' should not be understood from a Western universalist perspective but in the context of the perceptions (directly or by representation) of those submitted to the justice or injustice situations discussed. In this sense, other notions could supplant or even replace the term 'dignity of life', such as the notion of 'good living' (*buen vivir* in Spanish, or *sumak kawsay* in Quechua), if the situations under discussion are subjecting Andean communities, for example.

The concept of ecological integrity, on the other hand, is relatively new to Law and specifically to theories of justice. According to Bosselmann (2016), debates about sustainability and justice have been separated from each other, and their lost connection is precisely the recognition of ecological integrity. For the author, caring for present and future human generations will always be insufficient if ecological integrity – ie natural processes that sustain life – is at risk.

In other words, under the distributive view of justice, Bosselmann (2016) aptly argues that, despite the impossibility of determining the needs of future generations, it is true that present generations must at least pass on the integrity of the planetary ecosystem. That is, present generations of humans must preserve and transmit to future generations minimum ecological conditions necessary for the continuation of life on the planet. The notion of sustainability will only migrate from its 'weak' to its 'strong' version if concern for the non-human world is included in the notion of justice, distinguishing a mere social justice from ecological justice. The notion of socio-ecological justice thus encompasses intra-generational, inter-generational and inter-species justices.

Schlosberg (2009) identifies ecological integrity as a particular characteristic shared by humans and non-humans. He states that the recognition of both humans and the non-human natural world can be achieved by understanding the importance of ecological integrity, which leads to the development of the potential to flourish, to autonomy, resilience or respect for autopoiesis.

FIGURE 2.4 The 'what' and the 'who' of justice in the socio-ecological justice proposal

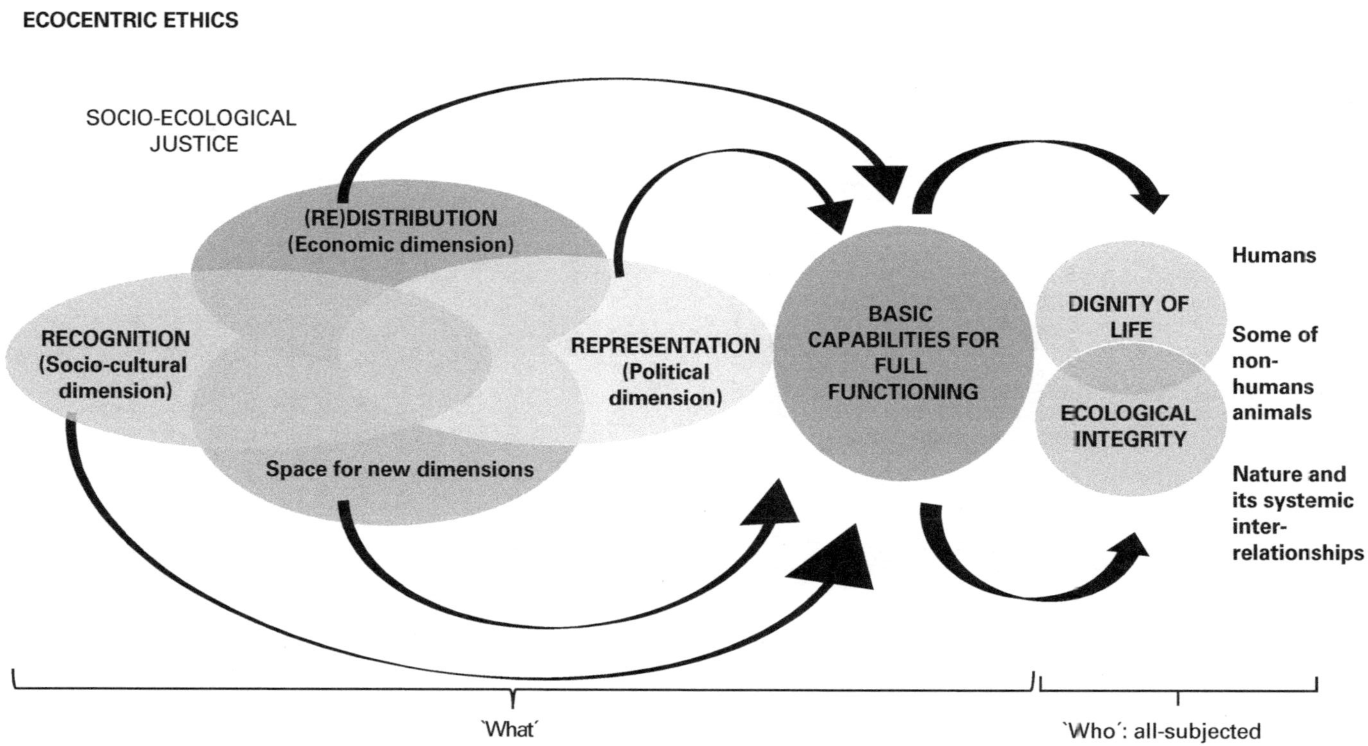

For the author, the debate on the minimum conditions necessary for a good life or full functioning is not restricted only to human or non-human animals but can be applied to other life forms and wider ecosystems.

> This theme will be returned to in Chapter 3, but **autopoiesis** can be described as the quality (of an entity or system) of self-direction, self-regulation or self-correction.

Like humans, Nature must have the opportunity to survive with its ecological integrity intact in the environment of diversity and autonomy characteristic of the biosphere. Although there are other elements in common between humans and the non-human natural world, the incorporation of the notion of ecological integrity in the justice debate helps to include in the justice community not only individuals (human and non-human), but habitats and full ecosystems with their essential relationships. The recognition of the need to develop (potential) basic capabilities of individuals and ecological communities to maintain their integrity and flourish is therefore fundamental (Schlosberg, 2009).

Final considerations

This chapter has confirmed how the environmental movement, representative of organized socio-environmental conflicts, is a plural, multi-faceted and heterogeneous movement. The streams highlighted here, with their respective theoretical developments (schools of thought), have positive elements and strategies as well as characteristics that can be questioned.

The cult of wilderness stream, promoted mainly by the deep ecology school, succeeds in challenging the dichotomized and fragmentary logic of the mechanistic paradigm, proposing an ecocentric approach in opposition to anthropocentrism. However, this is also where its greatest fragility can be found. This school reintroduced human beings into Nature, without emphasizing what differentiates them. It generated a monism, a unity without identifying the differences, not thinking about the complexity of the relations between humans and the natural world.

The gospel of eco-efficiency stream, based on the perspective of the shallow ecology school, stands out in its differentiation between technical and biological cycles; in the search for a rupture of the linear logic of productive systems; and working for the reduction (even if relative) of the extraction of new resources, as well as of the generation of waste. This stream has also been successful in gaining a more significant political commitment to the greening of economic and development models. Still, its anthropocentric and economistic perspective is criticized for granting an instrumental and utilitarian value to Nature, not recognizing its intrinsic value or other form of valuation. In addition, the technological faith it propagates is insufficient to cope with the socio-ecological crises, since it does not take into account the biophysical limits of the planet, especially those related to the entropy law.

It was also noted that both deep ecology and shallow ecology schools do not present a compelling critique of the development model based on the idea of infinite growth, nor about the disproportionality the most vulnerable countries and communities bear concerning the ecological (and waste) crisis. The strategies traditionally developed by these schools ultimately transfer the negative consequences of such crises to less favoured contemporaries (spatial transfer) and/or to future generations (temporal transfer), both human and non-human (subjective transfer).

Finally, it was seen how the environmental justice stream and its respective ecology of the poor school triumphed in the inclusion of contemporary social problems generated by environmental issues to the environmental movement debates. It demonstrates how ecological crises affect those with different levels of power (economic, political, among others) in distinct forms and levels. On the other hand, the ecology of the poor school has many limitations, such as: uncertainties related to the definition and scope of its notion of justice; its primarily intra-generational and anthropocentric focus; the common adoption of a liberal perspective with emphasis on the distributive aspect; as well as a focus on more visible environmental problems.

Although under different approaches, the problem of waste has been addressed both by gospel of eco-efficiency and by environmental justice. Different discourses have been presented to solve the problem of unequal ecological exchange (which includes the cross-border waste transfer), such as: the search for the internalization of negative externalities; the recognition of the existence of an ecological debt or an environmental liability by the ecologically benefited countries for this exchange; and the demands for justice.

It remains clear that waste is, besides being an ecological problem, a problem of justice. The discourse of justice was chosen in this book as a strategy to help to overcome the waste crisis. That is due to the potential of this discourse to avoid the monetary (chrematistic) language of economic strategies and, at the same time, because it can bring together ecological and social demands on waste. In this sense, this chapter sought to construct a theoretical framework for *a posteriori* definition of justice for socio-ecological issues. In other words, a theory for a pragmatic definition of justice that rejects the universal as a starting point.

From the dialogue and engagement between different ecological discourses and between different perspectives of justice, this chapter aimed to define general assumptions that could be applied to different socio-ecological demands in a unified, though not uniform way. It adopted pluralism as a way of avoiding the exclusion of equally valid arguments and strategies. In this sense, the theoretical limits of the 'what' and the 'who' of justice were developed to fulfil the notion of 'socio-ecological justice'.

The option for the term 'socio-ecological justice' rather than 'environmental justice' arose from the conclusion that, in addition to the plural, inclusive and heterogeneous consideration of the 'what' of justice, the process of overcoming theoretically the liberal–distributive justice model should also seek to surpass anthropocentrism and individualism typical of the Western modern mechanistic paradigm.

The term 'ecological justice' was not selected for the theoretical model of justice proposed here to avoid uniformity (or undifferentiation) between social and ecological demands. To reflect the complexity of socio-ecological issues, such as those related to waste, the recognition of Nature as a subject of justice and humans as part of Nature made by ecological justice is crucial. However, it is also essential to understand that even being interconnected, Nature goes beyond humankind and humankind transcends Nature while being completely dependent on it to survive.

From these reflections, the notion of socio-ecological justice, with the adoption of both dialogical and ecocentric perspectives, has emerged as a possible way to deal with intra-generational, inter-generational and inter-species justice problems, treating them not as exclusionary justices, but as related problems that have been deepened by the mechanistic paradigm.

The construction of a theoretical framework of socio-ecological justice that is broad, plural and inclusive was then presented. The conclusion is that the 'what' of this justice must be multi-dimensional, composed at least of the dimensions of:

- distribution;
- recognition; and
- representation…

... all in search of the development of basic capabilities for the full functioning of individuals and communities, the normative core of the 'what' of justice. From the recognition of the intrinsic value of both human beings and the non-human natural world (ecocentric ethics), it was concluded that such a core should be sought for the promotion of the dignity of life and ecological integrity.

As a result of the adoption of this ecocentric perspective, it was understood that the 'who' of socio-ecological justice should also be broad and inclusive, using the normative principle of 'all-subjected', dealing with socio-ecological impacts in order to consider:

- Nature itself (aspect of subjective expansion);
- contemporary human beings (aspect of spatial expansion) and;
- future generations of humans and non-humans (aspect of temporal expansion).

These should be, *a priori*, the minimum assumptions for the development of a concept of justice for socio-ecological issues in real-life cases, ie *a posteriori*.

Inspired by different authors, this chapter sought to develop a pluralistic approach, allowing the union of different concerns and movements, avoiding uniformity and universalization. It is believed that adopting this perspective of 'socio-ecological justice', not only the demands worked by the ecology of the poor school must be respected, but also those worked by other schools. There is no doubt about the validity of the ecocentric perspective defended by deep ecology, as well as of the search for a greater eco-efficiency in the socio-economic processes demanded by shallow ecology, and for social justice advocated by ecologism of the poor. Those developments of each school offering greatest potential should thus be considered as being able to cope adequately with real socio-ecological complexities.

03

Models to overcome the waste crisis

Introduction

The analyses carried out using the biophysical (or social metabolism) perspective, as well as the socio-ecological justice approach, evidenced the complexity that surrounds the subject of waste. As noted in Chapter 1, the waste crisis, as part of the ecological crises, is multi-causal, and the reasons for its emergence arising from different sources. The thread linking these sources, leading to the catastrophic socio-ecological consequences currently being experienced, starts from deeper levels of analysis. It begins with the way knowledge is produced, going through global geopolitical relationships and the establishment of economic and legal models that reproduce and reaffirm this paradigm, here called 'mechanistic'.

Before focusing on the waste issue as a legal problem, or developing possible strategies to ensure that the legislation regulating waste and its cross-border transfer complies with the assumptions of socio-ecological justice, it is important to pay attention to these deeper levels of analysis, which will ultimately lead to the surpassing of the mechanistic paradigm. The models presented here are theoretical proposals that, although far from being implemented, should be viewed as ideals to be pursued for the broad questioning of the mechanistic paradigm as a first step, and, if possible, after questioning it, surpassing its models.

The models presented in this chapter should not be treated as the only answers to the problems pointed out in Chapter 1. Nor should they be regarded as finished and definitive models, but rather as living proposals, open to dialogue and evolution according to real social and ecological conditions. Thus, the central objective of this chapter is the search for possible epistemological, geopolitical, economic and legal models for overcoming the mechanistic paradigm and, as a consequence, the ecological crises embracing the waste crisis.

Epistemological and geopolitical paradigm shifts

Complex thinking

As seen in Chapter 1, throughout our history, humankind has built reference frameworks for organizing the activities of daily existence, known as worldviews or paradigms. In this sense, Modernity brought a new paradigm, established by a set of scientific discoveries that changed the concepts on which science and the whole worldview were based. Today, it is referred to by some authors as the 'mechanistic paradigm'.

This paradigm profoundly altered the epistemological bases of knowledge, affecting the knowledge and actions of Western Modernity. The way in which diverse fields produced knowledge and Western social and geopolitical actions arose, began to be guided by characteristics typical of this paradigm: objectivity, quantification, reductionism, determinism, certainty, universality, fragmentation, isolation, abstraction, linear causality, dichotomization, and neutrality are just a few of them.

Although this paradigm still prevails in epistemological, geopolitical, social, economic, and legal terms, a second scientific revolution took place from the middle of the 19th century onwards. The discovery of the second law of thermodynamics, the entropy law, inaugurated this revolution (Morin, 2008). In general terms, the two laws of thermodynamics can be summarized as follows: the total energy content of the universe is constant, and the total entropy is continually increasing. That means that it is impossible to create or destroy energy. Energy can only be transformed from one form to another (Rifkin *et al*, 1980).

While the first law (of conservation) states that energy is constant in the universe, the second law (of entropy) states that each time energy is transformed from one state to another, some of that energy is lost, becoming unavailable. Entropy, therefore, is the measure of the amount of energy unavailable for use, ie the increase in entropy (unavailable energy) means a decrease in available energy. The term 'entropy' was coined by the German physicist Rudolf Clausius in 1868, when he discovered that in a closed system, the difference in energy levels always tends to disappear, reaching a state of equilibrium at the end (Rifkin *et al*, 1980).

This state of equilibrium achieved in closed systems is a state where entropy has reached its maximum, and there is no more available energy. Clausius concluded that, as a result of the second law of thermodynamics, in the world (which is a closed system), entropy always moves towards its highest level. From this conclusion, Rifkin *et al* (1980) went on to state that, on planet Earth, material entropy is continually increasing, and ultimately will reach its maximum. That is because Earth is a closed system in relation to the universe, exchanging only energy but not matter with its environment.

Another way of looking at the entropy law is through the elements of order and disorder. This law states that all energy in isolated systems (systems that do not exchange energy or matter with the environment) always moves from a state of order to a state of disorder. Thus, a state of minimum entropy is a state of maximum order. In contrast, a state of maximum entropy is a state of maximum disorder (Rifkin *et al*, 1980).

Both approaches to the entropy law highlight a fundamental difference between the mechanistic paradigm and the entropy paradigm: the way time is portrayed. While in the former, time is reversible (or disregarded), for the entropy paradigm it matters and is irreversible, ie it just goes in one direction: forward. Time, in the paradigm of entropy, always goes on because energy itself is always in motion from an available state to unavailability, from concentrated to diffuse, from ordered to disordered. Thus, when the final state of equilibrium is reached, there will be no more energy available, resulting in the collapse of the system (Rifkin *et al*, 1980).

The entropy paradigm, therefore, challenged the original perfection of the mechanistic paradigm, whose universe is illustrated in a state of perfect order. The notion of the irreversibility of the entropic process holds that time cannot be stopped or reversed. Only the speed of time can be accelerated or decelerated as the entropic processes (of matter and energy use) occur. In this perspective, although human actions cannot stop the entropic process of the world, they can accelerate or diminish it (Rifkin *et al*, 1980).

This perspective envisaged the world as a closed mechanical system, fighting against the forces of corrosion and decay and moving inexorably towards its end or collapse. Although it represented a crucial paradigmatic change, the entropy law did not replace the view of the world as a machine, even though that machine is in a state of degradation and decay (Morin, 2008).

It happens that, along with the loss of available energy, another process seems to occur over time, initially revealed through the discoveries of Charles Darwin. Darwin brought a new perspective to the worldview by suggesting that life on Earth began in a simple form and has transformed into more complex forms over time. In this sense, while the mechanistic world is temporally static, and the entropic world is in the process of decay, Darwin's world is becoming more complex, ie evolving (Morin, 2008).

Darwin, thus, presented a third scientific perspective on the world: neither perfect machine, nor decaying machine, but an explosion of life, reproducing and modifying itself. In this process, the time element took an active and creative role and, after the world-clock and the decadent world of previous perspectives, has generated a creative world. The crucial difference in the development of these different worldviews lies in the relation between order and disorder (Morin, 2008).

As seen in Chapter 1, the mechanistic paradigm only accepts the order as truth. According to this perspective, the universe is perfectly ordered, and everything that is not in order will be against the laws of Nature. In turn, the entropy paradigm, by including the notion of irreversible time, understands the world as irreversibly moving from order to disorder; nothing can stop this process, only accelerate it or slow it down. Thus, order and disorder are placed in opposition in both paradigms, with a dichotomized and hierarchical view.

The biological perspective (evolution) contrasts with the physical perspective (entropy). However, both trends (of development and decay) can be identified in living systems, indicating that this is an apparent contradiction. It was with Ilya Prigogine's theory of dissipative structures that the connection between the two perspectives occurred. Reaffirming the validity of the second law of thermodynamics (entropy law), Prigogine interconnected the main characteristics of living systems, reconceptualizing fundamental notions that allowed the co-existence of structure and change, order and disorder, and stillness and movement in the same system (Capra, 1996).

Thus, Darwin's creative world, coupled with other theoretical developments in physics and biology, has brought many new ideas. One of them was a new articulation between order and disorder. Such articulation is presented as part of the core elements of Edgar Morin's work, who proposes the need for a new worldview, a new paradigm that he calls complex. Through the development of the dialogical principle of the paradigm of complexity, Morin created a tetragram (a system consisting of four concepts – order, disorder, interactions, organization) by which he demonstrates that the interactions between order and disorder are at the centre of the organization of any system.

Different theories give the foundations to Morin's theory of complexity, such as chaos theory, dynamical systems theory, and theory of dissipative structures.

The paradigm of complexity shows that order and disorder can be conceived in dialogical terms. At the same time as acting as enemies, in some cases they collaborate and produce organization and complexity, ie they create generative processes of new forms of organization. For Morin (2008), an organization without disorder creates a homogeneous and sterile system, where no change and innovation are possible. On the other hand, complete disorder without any order hinders the organization of a system. It is only with the complex interaction between order and disorder that a system remains organized and, at the same time, open to change.

Although the discovery of the second law of thermodynamics represented a real paradigmatic revolution, it has shown a partial reality to the world, because the entropy law applies only to closed and isolated systems. The entropy law has revealed the starkly sad reality of a planet moving towards total disorder. A planet where entropic levels are rising even faster due to socio-economic processes, increasingly creating unavailable energy and matter in the form of pollution and waste. Morin (2008) emphasizes, however, that the ecosystems contained in the ecosphere (the life on Earth) are open (as are all other subsystems contained therein, such as the socio-economic), and in these systems, there are also processes that create order, generative processes.

Thus, the notion of 'open system', which originated in thermodynamics, eventually created a bridge between this knowledge and the biological sciences. It emerges in opposition to the physical notion that dichotomizes balance/imbalance, encompassing both. Living systems, in their metabolic processes, have the capacity to maintain their structures (balance) even though their constituent parts are in the process of alteration (imbalance). That is only possible because of the ability of these open systems to close themselves to the external world to maintain their structures while opening up to the flows of matter and energy essential to their functioning (Morin, 2008).

It seems that the entropy law does not apply to living systems. That is because, together with (and as a consequence of) the process of entropic disorganization occurring in a living system, the phenomenon of reorganization of this system occurs called 'negentropic process'. Entropy thus contributes to the organization by generating degradation in living systems which, in turn, cannot complexify themselves (evolve) without the disorder. This logic highlighted by Morin (2008) is fundamental for the development of another key principle of complex thinking: the principle of 'self-eco-organization'.

Living systems are, according to Morin, systems of self-organization that stand out and differentiate themselves from the environment through their autonomy and individuality. However, along with this process of distinction, living systems connect to the environment by increasing their openness and performing the exchanges that accompany the process of complexity. More than self-organizing systems, living systems are dynamic systems of self-eco-organization. It means that, while living systems maintain their individuality, they depend on their relationships with the environment (Morin, 2008).

Therefore, the principle of self-eco-organization comes from two typical characteristics of living systems. The first is the organizational closure of the system, a characteristic of self-organization, self-generation or autopoiesis, through which internal metabolic processes performed by the network of the parts of the living system continuously regenerate it. In this way, living systems undergo continuous internal structural changes while preserving their patterns of network organization. The second characteristic is the opening of this system to the flow of energy and matter, that is, to exchange with the environment, since living systems are closed in terms of their organization (they are autopoietic networks) but are open from a material and energetic point of view (Capra, 1996; 2002).

Thus, living systems, at the same time as being autonomous, need the flow of matter and energy that comes from the environment (food) and that is eliminated to the environment (waste). The theory of dissipative structures made the connection between these two functions (closure and openness of the living system), which also brought the vital concept of 'bifurcation or instability point'. Such a point is met when the internal disorder state of the system reaches a high level (entropic process), leading to the emergence of a new organization that re-establishes order (negentropic process). This theory helped to understand the spontaneous emergence of order, as well as the complexification of living systems (Capra, 1996; 2002). In the words of Capra, a living structure is:

> [...] materially and energetically open, using a constant flow of matter and energy to produce, repair, and perpetuate itself; and that it operates far from equilibrium, where new structures and new forms of order may spontaneously emerge, thus leading to development and evolution.
>
> SOURCE Capra, 2002: 31

The environment, therefore, is inside and outside these systems and is also modified by their interactions. Reality is, in this sense, much more in the connection (relation) between these open systems and the environment than in their distinctions. From this come the principles of organizational recursion and retroactivity, which emphasize that products and effects are both causes and producers of what produces them. That is, ideas of recursion and retroaction break with linear causality (cause–effect) and with other dichotomies, such as subject–object, because everything that is product returns to the producer in a circular loop (Morin, 2008).

In society, for example, individuals produce the society which, in turn, produces the individuals. Subject and object act reciprocally on each other and become inseparable. Both subject and object are interrelated consequences of the relationship between self-organizing systems and the ecosystem. Directly connected to these principles is another key principle of complex thinking, known as the holographic principle (or the principle of 'hologram'). Such a principle states that not only is the part within the whole, but also the whole is within the part. This idea seeks to overcome both the reductionism that focuses only on the parts and the reductionism that focuses only on the whole (Morin, 2008).

The use of the systemic–organizational principle by the complexity paradigm transforms 'objects' into 'systems', meaning that they have an organization. This organization is what keeps the system as a whole composed of a group of different connected elements. According to this principle, the essential properties of a living system are properties of the whole which arise from the interactions and relations between the parts. Thus, the systemic vision brings a new way of thinking in terms of connectivity, relationships, and context (Capra, 1996).

With the adoption of the systemic principle, the paradigm of complexity brings the understanding that the whole is more than the sum of its parts since it contains qualities and properties that only emerge with the whole, through the interaction between the different elements (parts) contained therein. These 'emerging' qualities can, in turn, return to the parts. However, the whole is also less than the sum of its parts, since the organization of the whole imposes constraints and inhibitions on the parts that constitute it, so they do not have total freedom as part of the whole (Morin, 2008).

An essential element for a new paradigm of knowledge, emphasized by Capra (1996), derives from the understanding of the characteristics of living systems: it is the pattern of organization present in all life forms, whose most important property is that of a network pattern. According to Capra, the components of every living system are arranged as a network. The first and most obvious property of the network is non-linearity – the network relationships are non-linear and cyclic (they work in feedback loops). Because of this property, living systems are capable of self-regulation and self-organization. Thus, the pattern of life is a network pattern capable of self-organization.

These are just some of the ideas and principles developed by the complexity paradigm, which in general terms recognizes the innovative nature of biophysical processes, emphasizing the collective, cooperative and organizational aspects of Nature with the adoption of a synthetic and holistic perspective. Morin's theory seeks, through transdisciplinarity, a new way of organizing thought, a new epistemology that rejects the reductionism of the mechanistic paradigm.

The paradigm of complexity brings to the surface an epistemology that recognizes ambiguity, uncertainty, contradiction, and disorder as opportunities for the development of creativity and new forms of organization, rather than sources of anxiety. This new way of thinking and organizing thought seeks, through some conceptual instruments (or principles), like the ones mentioned above, to overcome the mechanistic (disjunctive/reductionist/one-dimensional) paradigm by a distinction–conjunction paradigm (Morin, 2008) that will allow the distinction without disjunction and connection without identification.

That means that the paradigm of complexity does not entirely exclude previous paradigms, but integrates their truths, taking into account their limitations. It seeks the contextualization, the complexity, the connections or interrelationships, and the multi-dimensionality of the phenomena. Such a paradigm recognizes the subject and its location as part of the research process itself. So, the complex thinking refutes objectivity and universality, presenting itself as a contextualized thought, situated in a particular place and time, which allows the limits of its validity to be verified (Morin, 2008; Morin *et al*, 1999).

Morin thus presents the notion of an 'open epistemology', which, instead of controlling all knowledge with sovereign power, rejecting all opposing theories and giving the monopoly of investigation and truth to only one of them, will be a place of uncertainties and dialogue. Whenever a theory is unable to integrate observations that have become important in the real world, it will enter into a state of disorder, and dialogue with other theories will bring about a new organization that will integrate old and new theories (Morin, 2008).

This epistemological system will present the characteristic of self-eco-organization: at the same time as being autonomous, it will be open and self-reflexive in its eco- and meta-systemic relations. As a direct result of these characteristics, the epistemology of complexity proposes a unity of science containing diversity, which Morin (2008) calls *Scienza Nuova*.

He proposes this new science as a critique of specialized and compartmentalized knowledge as a form of abstraction, which rejects its connections and intercommunications with the environment and, consequently, breaks the systemicity and multi-dimensionality of phenomena (Morin *et al*, 1999). Physics, biology, and anthropology would compose this transdisciplinary science, ceasing to be closed entities (disciplines), without losing their identities (Morin, 2008).

With this proposal, Morin (2008) seeks to reintegrate the human being into Nature while distinguishing him from it – not reducing the human being to Nature. This epistemology of complexity will, according to Morin, be appropriate for the organization of knowledge in order to capture human complexity without claiming to be a single, complete, and endowed with certainty answer. From this perspective, the Earth is a complex physical–biological–anthropological totality. Life emerged from the Earth and the human being from life. So, at the same time as a natural entity, the human being is a supernatural entity, which has its origin in living and physical Nature, but emerges from it by culture, thought and consciousness (Morin *et al*, 1999).

Ecology of knowledges

As Morin points out, his model of open epistemology, endowed with transdisciplinarity, is a model that, in addition to being complex, is contextualized, ie situated in time and space. Such a model has indeed arisen from the North, although it does not present itself as universal, closed to the dialogue and holder of the monopoly of knowledge. Following this logic, Boaventura de Sousa Santos (2010a) proposes that, in addition to overcoming the visible abyssal line that separates science from other non-scientific knowledge of the North, it is also crucial that this new epistemological model surpasses the invisible abyssal line between North and South, extending the dialogue for the inclusion of other types of knowledge from the South.

The recognition of the existence of abyssal lines (notably the invisible one that separates North from South) is essential for such inclusion. Without this recognition, any proposed epistemology critical of the mechanistic model will continue to reproduce the abyssal lines, however anti-abyssal it may be. Thinking in post-abyssal terms means thinking from the perspective of the 'other side' (from the South) precisely because it represents the domain of the unthinkable in Western Modernity. Thus, Boaventura presents the epistemological model of the 'ecology of knowledges' (Santos, 2010a).

The term 'ecology' was chosen by Santos to represent the recognition of the plurality of heterogeneous knowledge existing in the world (including the sciences of the North). Besides, it translates the sustainable and dynamic interactions that take place between those types of knowledge, without, however, compromising their autonomy. In the ecology of knowledges, the different types of knowledge are intersected as well as the different types of ignorance. It recognizes that there is no unity of knowledge or of ignorance; they are all multiple and interconnected. Thus, for the model of the ecology of knowledges, knowledge is, in fact, inter-knowledge (Santos, 2010a).

Santos (2010a) states that the ecology of knowledges enables us to take a broader view of what we know, as well as what we do not know. It also warns us that what we do not know is our ignorance, not ignorance in general. The epistemological vigilance required by the ecology of knowledges transforms post-abyssal thinking into a profound exercise of self-reflexivity.

It is important to emphasize that the ecology of knowledges, as a post-abyssal epistemological model, seeks the credibility of non-scientific knowledge without seeking to discredit scientific knowledge. According to Santos (2010a), this model implies, in fact, the counter-hegemonic use of scientific knowledge, exploring the internal plurality of sciences while promoting interaction and interdependence between scientific knowledge and non-scientific knowledge.

Connecting this model with some of the notions of complex thinking, we can see that the ecology of knowledges sees the different types of knowledge as self-eco-organizing systems: at the same time as being autonomous, they interact with the environment, with their context and with other types of knowledge, creating a dialogical network organization. That is clear when Santos (2010a) stresses that one of the basic premises of his model is that every knowledge has internal and external limits. Internal limits concern the interactions a type of knowledge allows to happen in the real world and the external ones concern the recognition of alternative interventions by other forms of knowledge.

According to Santos, the hegemonic knowledge only recognizes the internal limits (the limitation of the disciplines to intervene in the real world). Thus, the counter-hegemonic use of modern sciences is only possible with the recognition of its external limits. However, it will only make sense in the context of an ecology of knowledges, with the end of the invisible abyssal line between the knowledge of the North and the knowledge of the South (Santos, 2010a).

In this sense, the visibility of the internal limits of sciences (disciplines) is only the first stage for the development of an alternative proposal of epistemology (Nunes, 2010). Then the concept of epistemology itself (as a place for the establishment of epistemic sovereignty that allows us to define what is and is not valid knowledge) should be questioned (external limits).

Once these two stages are completed, the possibility of a new epistemology emerges. The epistemology of the South proposed by Santos is a general epistemology of the impossibility of a general epistemology. It is a model of the impossibility of attributing epistemic sovereignty to only one form of knowledge that would have the power to define the existence and validity of all other forms of knowledge. This epistemology, moreover, proposes the dialogue and the active involvement with the critical epistemologies of the North (like that of complexity), having as a horizon the decolonization of the epistemological reflection (Nunes, 2010).

From this, another element of the ecology of knowledges arises in affinity with the open epistemological model of complexity: the prudence of the recognition of incompleteness. Like Morin, Santos argues that no single form of knowledge can account for all possible interventions in the world; all types of knowledge are, in different ways, incomplete (Santos, 2010a). Thus, at the epistemological level, beyond the end of the subject–object dichotomy, the ecology of knowledges seeks to overcome the dichotomy between scientific and non-scientific knowledge, placing them in a position of interaction rather than opposition.

To incompleteness, the ecology of knowledges adds the notions of inter-scalarity and inter-temporality (to know and act on different scales and over different durations), given that different knowledge practices take place at different spatial scales and according to different durations and rhythms (Santos, 2010a).

Facing this proposal arises the questioning about how different knowledge can dialogue with each other. To solve this challenge, Santos (2010a) presents two strategies: the inter-cultural translation and the contextualized hierarchies. The first one deals with the issue of the incommensurability between the different types of knowledge, considered by the mechanistic epistemology as a factor of exclusion of all the knowledge that does not speak the same language of the Western sciences.

Inter-cultural translation (of languages and cultures) presents itself as a possibility for the identification of mutual concerns, complementary approaches, and insurmountable contradictions between the different types of knowledge. It highlights the potential of the ecology of knowledges model to cross the abyssal lines (Santos, 2010a).

For Nunes (2010), the epistemology of the South promoted by the ecology of knowledges model appears not from a reflection centred on the sciences of the North. On the contrary, it emerges from practices, experiences, and knowledge capable of defining the limits and conditions under which specific knowledge can be translated or appropriate in new circumstances without the pretension of being constituted as universal. Inter-cultural translation, therefore, does not seek the recognition of other types of knowledge in terms of the dominant knowledge (scientific validation of non-scientific knowledge) for dialogue to take place, but in terms of what connects and what separates such types of knowledge.

Understanding that every practice is guided by and produces one or more types of knowledge, the second strategy presented by Santos to establish dialogue between the different types of knowledge (contextualized hierarchies) comes from the assumption of the need to create a hierarchy among the types related to a given practice. However, instead of an *a priori*, unique, universal and abstract hierarchy, the model of the ecology of knowledges proposes hierarchies dependent on context, defined according to the concrete results intended or reached by the different types of knowledge (Santos, 2010a). The hierarchies of types of knowledge are not defined from the epistemic sovereignty of a type of knowledge. It occurs inseparably from the situated practices producing knowledge (Nunes, 2010).

The pragmatic idea of evaluating the interventions and real relationships in society and Nature that the different types of knowledge provide to decide then which knowledge should be privileged in a specific case provides the foundation for this model. In addition, whenever more than one knowledge can be used in a real intervention, the choice of the knowledge to be privileged will be made according to the precautionary principle. For Santos, the use of such a principle means that the chosen knowledge should be the one that guarantees the greatest possible participation of social groups involved in the design, execution, control, and enjoyment of the intervention (Santos, 2010a).

Thus, the ecology of knowledges model emerges as a pragmatic epistemology (it does not conceive knowledge in the abstract) that starts, according to Nunes (2010) from two postulates:

- **i** the recognition of the validity of all types of knowledge, so none of them are disqualified before its pertinence and validity are verified under situated conditions; and
- **ii** the refusal of relativism, so that no form of knowledge is privileged as more adequate and valid without its consequences or effects being evaluated under situated conditions.

The operations of validation of knowledge (the creation of a hierarchy among them) are thus derived from the situated consideration of the relation between them, configuring the ecology of knowledges (Nunes, 2010).

Post-abyssal geopolitics

Just as the mechanistic epistemological model can be directly related to the global geopolitical dichotomization between North and South, the debate on new epistemological models is essential for overcoming this dichotomization. As seen in Chapter 1, Boaventura de Sousa Santos (2010a) states that global social injustice is closely linked to global cognitive injustice. In this sense, the fight for global justice must also be a fight for cognitive justice, requiring a new thinking, post-abyssal. (Geo)political resistance must, therefore, have as its postulate the epistemological resistance and the logic of appropriation–violence employed in the South can only be faced if the epistemological perspective adopted is situated in the social experience of the South.

Far from being a theoretical abstraction, Santos (2010a) argues that post-abyssal thinking has emerged from social movements that constitute counter-hegemonic globalization. He understands such globalization as a vast set of networks, organizations, and movements fighting against economic, social, political and cultural exclusions generated by the current epistemological, developmental, economic and legal models based on the mechanistic paradigm. From the understanding that these exclusions are products of unequal power relations (coloniality of power), such movements are driven by demands that go far beyond the redistribution of material resources, also encompassing social, cultural, and political matters.

Santos also argues that this counter-hegemonic globalization driven by such movements (which he calls 'subaltern cosmopolitanism') presents as a novelty the profound sense of incompleteness without, however, aiming for completeness. In other words, it recognizes that the understanding of the world exceeds the Western view while assuming the inexhaustible diversity of other, non-Western worldviews (Santos, 2010a).

It is possible to affirm that, like other movements, the environmental movement, with all its different streams, is part of the search for a counter-hegemonic globalization. The consideration of the plurality of this movement and the dialogue between its theoretical bases and strategies showed that many of its elements could be retrieved here for the proposition of a possible theoretical framework of socio-ecological justice. Many of the elements composing this complex notion of socio-ecological justice connect with the epistemological notions defended by the new models emerging from the critique of the mechanistic paradigm.

The socio-ecological justice model proposed favours the diversity of demands, which go beyond the material distributive aspect. It also encompasses the need for recognition and due representation while leaving room for new demands to be added whenever they are innovative and relevant. This definition of the 'what' of socio-ecological justice shows the multidimensional, integrative, and incompleteness aspects of the model. The organization of this model of socio-ecological justice is maintained with the order of the minimum demands already identified (distribution, recognition, and representation) while remaining open to the possibility of new demands (a disorder that will lead to a new order and, consequently, organization). In this way, order–disorder–organization are placed in relation by this model.

The demand for proper representation, together with the use of the normative principle of 'all-subjected', seeks to apply the precautionary principle mentioned by Santos: to guarantee the greatest possible participation of those who are subordinated to the intervention in the real world in the specific case.

The elements of socio-ecological justice proposed here are intended to comply with the fundamental pragmatism and contextualization to overcome the mechanistic paradigm. It also allows simultaneous openness and closeness in the definition of possible demands for socio-ecological justice (the 'what' of justice), as well as the voices that can get actively involved in such demands in specific cases (the 'who' of justice). At the same time that it allows the opening of the 'who' of justice beyond state borders, the model proposed here allows the definition of this 'who' to be closed to those subjected to the acts of the environmental governance structure questioned, avoiding an abstract and universalist generalization. The global and the local, in this perspective, instead of in opposition, are put in relation.

Finally, the unifying normative principle that connects all possible dimensions of socio-ecological justice (distribution, recognition, representation and others – to develop basic capabilities for full functioning to promote ecological integrity as well as the dignity of life) allows connection without identification, at the same time as distinction without disjunction. That is, humans, in this proposal, are placed in relation with non-human animals and with Nature as a whole, without losing what differentiates them. Humans are autonomous at the same time as dependent on their environment.

The notion of 'ecological integrity' emphasizes the importance of the properties that only the whole has, while the notion of the 'dignity of life' emphasizes what only some of the parts have (which, however, is only possible through the guarantee of ecological integrity). In this way, the part is in the whole, and the whole is in the parts, while at the same time it is recognized that the whole is, at the same time, more and less than the sum of its parts. The parts and the whole, therefore, are placed in relation by this model of socio-ecological justice.

These are just some reflections seeking to highlight the potential the environmental movement has to help in the collective construction of a concept and a form of geopolitics that are complex, contextualized, and post-abyssal. A particular element identified in this movement must be highlighted: the relation between local and global as a strategy for overcoming the geopolitical dichotomy between North and South. According to Grosfoguel (2010), social conflicts located at levels below and above the nation-states are strategic spaces for political intervention. These spaces are essential for such interventions to be effective, so the links between local and global of social movements (such as the environmental movement) are of crucial importance.

For Quijano (2010), the only way of overcoming the coloniality of power is through what he calls the 'socialization of power'. This strategy arises in opposition to the classic approaches of statist nationalization (capitalist and socialist) centred in the administration of the state and the hierarchical structures of power. Grosfoguel (2010) clarifies that the strategy of socialization of power in all spheres of social existence favours collective forms of public authority to deal with local and global conflicts.

This concept proposes that all institutions that regulate social life should extend equality and democracy to all spaces of social existence. It is a process of a radical bottom-up democratization, which does not exclude the formation of public institutions to democratize and socialize production, wealth, and resources on a global scale. Such socialization of power would entail the formation of global institutions beyond state borders to guarantee justice and equality in the production, reproduction, and distribution of the world's resources, requiring a kind of self-managed democratic global organization. In short, the socialization of power at the local and global level would imply the creation of a public authority outside state structures (Grosfoguel, 2010).

The socialization of power as a way to overcome the coloniality of power is a radical proposal that, in order to become a reality, demands a decolonial and global power fight. It emerges as a utopian alternative that can only be achieved by overcoming the mechanistic paradigm, ending the dichotomies between Eurocentric and non-Eurocentric fundamentalisms and placing them in dialogue (Grosfoguel, 2010). In this proposal, then, global and local are placed in relation.

The current hegemonic globalization has not been enough to overcome the mechanistic paradigm in epistemological and geopolitical terms. The globalized era brought different perceptions of the world, sometimes contradictory views. At the same time that human beings came to be seen as global citizens, co-existing on this planet with other elements of Nature, the certainty of the differences between the distinct cultures has been evidenced. Together with this infinite human diversity, the finite nature of the planet is now clear. Unity–diversity and finitude–infinity began to co-exist.

Mechanistic thinking is not capable of dealing with these contradictions and ambiguities of the global era. It is precisely for this reason that Morin and Kern (1999) argue that it is insufficient to focus on global aspects, but that it is necessary to take into account local and specific elements in a complex organizing articulation, within the global context. It is necessary to seek the relation of inseparability and inter-retroaction between all phenomena and their context and of every context with the planetary context. Thus, a new thought that ceases to oppose universal and specific, general and singular, is of fundamental importance in order to reach the complex anthropolitical (or geopolitical) formula of a true planetary age, where local thinking/acting and global thinking/acting can co-exist simultaneously.

Chapter 1 exposed the reductionisms and dichotomies generated by the mechanistic paradigm as a first step to overcome the ecological (and waste) crises. The recognition of the existence of new epistemological and geopolitical models critiques of this paradigm, both of the North and the South, seems to be a second step in the search for stronger responses to such crises.

Thus, following the line developed in the first chapter, from these epistemological and geopolitical debates, this book will proceed to more superficial layers of investigation. The next sections will be devoted to a brief study of how these ideas have generated the development of new theoretical models regarding economics, production and consumption patterns and law, all in search of overcoming the current ecological crisis and, consequently, the waste crisis.

Paradigm shifts in economics

Environmental economics

The epistemology and geopolitics resulting from the mechanistic paradigm radically influenced the development of economics, which emerged as a science in Modernity. They also significantly impacted the patterns of production and consumption that predominate to this day. The simplicity and objectivity offered by the mechanistic paradigm also became appealing to those now known as social and human sciences, which began to adopt individualism, materialism, individual freedom, and private property as foundational pillars.

Regarding Economics, the search for material expansion through economic growth has been the primary response given by economic theories for the attainment of progress, prosperity, and, subsequently, development. Those economic goals have become homogenizing notions, reduced to the idea of infinite economic growth measured by gross domestic product (GDP). The mathematical logic of the mechanistic paradigm was thus successfully implemented to socio-economic issues, considering the economic system as a closed and autonomous system.

Interconnected with this process of mechanization of the economy occurred the mechanization (theoretical and empirical) of production and consumption patterns. From a short-term economic point of view, the production of goods became increasingly profitable and efficient as a result of the adoption of a linear and uni-directional model of production. At the same time, consumption began to be increasingly intense, fast, and ephemeral, with consumer goods being more superfluous, disposable, and with shorter life cycles.

Such economic, production and consumption models have been widely questioned. The promise of Modernity for increasing social welfare through progress, prosperity, and development has not materialized. On the contrary, a decrease in the quality of life in general has been identified, as well as the generation and intensification of an ecological crisis, in which there are indications that the planet's carrying capacity will be exhausted even before the natural resources are.

As highlighted in Chapter 2, such consequences have not gone unnoticed. They have been widely questioned by the environmental movement in all its streams, each with its own peculiarities and approaches. In theoretical terms, the wider awareness of the limitations and collapse of the mechanistic paradigm has influenced the development of new epistemological and geopolitical models that reverberate in all areas of knowledge.

The environmentalist stream of the gospel of eco-efficiency has been theoretically represented by, among others, those authors whose work deals with environmental economics, whose central objective is the incorporation of Nature into the market. As its main strategies, environmental economics brings:

- monetary valuation of environmental goods and services;
- the internalization of negative externalities;
- the polluter-pays principle;
- property rights; and more recently
- decoupling through dematerialization and eco-efficiency.

> **Decoupling**, according to Veiga (2010), recommends the reconfiguration of the production process with the increase of eco-efficiency in the supply of goods and services, ie using less matter (dematerialization) and less energy.
> **Dematerialization**, according to Alier (2003), means that the growth rate of energy and materials consumption is lower than the growth rate of GNP.

Environmental economics and its strategies, also known as part of the strategies of the shallow ecology school, represented an advance in relation to the theoretical developments of classical and neoclassical economics. However, if applied by themselves, such strategies are, for various reasons, insufficient to deal with the complex problems generated by current dominant development and economic models.

Central criticisms of this proposal are the assertions that the model offered by environmental economics:

- does not challenge the mechanistic paradigm and its models, only spatially and temporally shifts the consequences of the ecological (and waste) crisis;
- ignores the fact that the socio-economic system is not a closed system and therefore is unable to grow indefinitely due to biophysical limits imposed by ecosystems;
- does not take into account that the ecological crisis affects different communities and countries in an unequal way; and
- does not break with the anthropocentric perspective, adopting a reformist and 'economicist' character.

Many of these criticisms are based on another economic theory, which has its origin identified in the 1970s with the work of the Romanian mathematician and economist Nicholas Georgescu-Roegen, considered the founder of ecological economics. Unlike the authors of environmental economics, Georgescu-Roegen established a fundamental critique of the mechanistic paradigm, linking the socio-economic process to the laws of thermodynamics, revolutionizing economic thinking (Leff, 2010).

Ecological economics

With the adoption of a systemic and transdisciplinary perspective, with references to ecology, the laws of thermodynamics and the flows of matter and energy, Georgescu-Roegen was the first author to defend the view that the economy is not an autonomous, hermetically closed and circular system (Leff, 2010; Montibeller-Filho, 2008). He observed that any production process always implies the transformation of Nature through flows of matter and energy between the ecosystem and the socio-economic system. As a result, matter and energy end up in a state of irreversible degradation (or disorder or maximum entropy) in the environment (Georgescu-Roegen, 2011).

In general terms, ecological economics seeks to analyse and quantify the processes of transformation of Nature occurring in economic activities using the laws of thermodynamics, especially the entropy law. It argues that, from the understanding of the irreversibility of entropic degradation, strategies could be developed for the reduction and deceleration of this process. According to Montibeller-Filho (2008), among these strategies are:

- the use of less exhaustible natural resources;
- the use of more energy-efficient materials;
- the reduction of waste generation.

The proposal of ecological economics represents, undoubtedly, a paradigm shift in this area of knowledge, reflecting the same process that has occurred in epistemological terms. In addition to the adoption of a transdisciplinary approach, one of the significant contributions of this perspective was the understanding of the impossibility of infinite economic growth by considering the economy as a subsystem contained in the ecosphere, whose resources and carrying capacity are finite. According to Leff (2006), this perspective helped to clarify the increasing flow of degraded energy and matter that leads to the ecological unsustainability of current economic processes, as well as the erosion of biodiversity and the exacerbation of socio-environmental conflicts.

For this economic school, ecological sustainability can only be achieved by minimizing the flows of energy and matter passing through this subsystem, requiring the untying between qualitative social advances and quantitative increases in production and consumption (Veiga, 2010). This systemic view of ecological economics also leads to the understanding that the economy as a subsystem, besides being positioned in the ecosphere, also has relationships of direct interdependence with a legal (mechanistic) structure of property rights over environmental resources and services. The distribution of power and resources of this structure is made unequally according to existing social and geopolitical hierarchies and to different intersections, such as race, gender, social status, among others (Alier, 2003).

From the foundations built by Georgescu-Roegen, many theoretical streams were developed within the framework of ecological economics, such as the proposals of Tim Jackson ('prosperity without growth'), Serge Latouche ('degrowth'), and Herman Daly ('steady state'). All these streams make an essential distinction between growth and development, opposing the former. Daly, for example (1993a) argues that growth refers to the increase of service by the increment of the stock of goods and the 'throughout', ie a quantitative expansion in the scale of the economy. Development, on the other hand, refers to the increase of the service while the stock and the throughout are kept constant, implying qualitative changes in the economic and social structure, without necessarily representing a growth.

The term **throughout** is used by ecological economists to define the entropic physical flow of matter and energy from natural sources which passes through the human economy and returns to the sinks of Nature. Thus, this term has the same meaning as the notion of 'metabolic flux'. For more information, see the works of Herman Daly.

In other words, the steady state theory argues that the economy cannot grow beyond the limits that allow the conservation of a basic stock of natural resources, the renewability of natural capital and the absorptive capacity of the waste produced (Leff, 2010). Daly (1993b) argues that the economy must develop without growing (steady state), in the same way that the planet Earth, of which the economy is a subsystem, develops without growing.

It is important to emphasize that ecological economics also addresses many of the issues that are of concern to environmental economics, such as the claims to attribute monetary values to environmental services and losses. However, the major contributions of ecological economics, which differ from environmental economics, are related to its systemic and transdisciplinary approach (Alier, 2003).

In addition to the clear position against the idea of infinite economic growth, this perspective seeks to develop physical and social indicators and references of (un)sustainability with the examination of the economy in terms of social metabolism, going beyond chrematistic values based on the creation of wealth. It also pursues modelling the interactions between the economy and the environment with the use of management tools such as integrated environmental assessment and multi-criteria assessments for environmental decision-making, proposing new environmental policy instruments (Alier, 2003).

Environmental economics seeks the inclusion of Nature in the economic system, proposing the reordering of this to an economic consideration of Nature. In turn, ecological economics sees the economy within the ecosystem, suggesting the reordering of the economy within the ecological framework. It proposes ecological criteria, conditions, and rules to be respected by the economic system, which, from then on, should be subject to the conditions of the general metabolism of Nature (Leff, 2001).

Ecological economics thus intends to adjust economic mechanisms to the laws of thermodynamics and ecological cycles, considering the interrelationships between the economic system and other systems to establish a systemic, transdisciplinary and sustainable economic paradigm (Leff, 2010). Although the boundaries of this economic perspective remain open to the emerging complexities of present times, Leff (2001) argues that ecological economics has not yet severed the umbilical cord that attaches it to the neoclassical economics of natural resources in its view of Nature as a cost or as a limit (not as a potential).

While recognizing the importance of conserving the resource base and ecological balances, ecological economics seems to have underestimated the productive potential of Nature, notably complex, productive, and biodiverse ecosystems (Leff, 2001). Thus, Leff (2010) transcends the critique of the mechanistic economy raised by ecological economics, stating that, despite generating a paradigm shift in the economic field, Georgescu-Roegen did not propose a true bioeconomy, with new measures of values and alternative modes of production.

In other words, Leff (2010) states that, although Georgescu-Roegen considers life to be characterized by the negentropic process (the opposite of the entropic process), he argues that the entropic process of the whole ecosphere (including ecosystems) cannot be avoided. In this sense, in the field of economics, as well as epistemologically, the entropy paradigm translates the world as a decaying machine, focusing on the second law of thermodynamics and seeing it as a death drive intrinsic to life and economic activity.

Productive ecology

As seen at the beginning of this chapter, Morin's complexity paradigm proposes a step forward in relation to the entropy paradigm with the adoption of a view of a creative world. In this world, life and death, order and disorder, negentropy and entropy interrelate and complement each other, generating new and more complex forms of organization. Following the same lines, Leff seeks to break with the theoretical fatalism of ecological economics by proposing an economic paradigm that he calls 'ecotechnological' or 'productive ecology'. In his theory, he highlights the fact that the entropy law does not apply to dissipative, open systems that are far away from a state of equilibrium, such as living organisms and culture (Leff, 2010, 2001).

Adopting this approach, Leff (2010) argues that instead of accelerating the irreversible process of entropic degradation, this economic model would emerge as a potential for the generation of sustainable production through the increase of the negentropic productivity derived from photosynthesis, biological processes, and ecological organization. Leff's ecotechnological productivity paradigm views sustainability as an entropic–negentropic balance of technological and ecological processes, relating material and energy flows to the ecological productivity of goods and services.

> Leff (2010) argues that **photosynthesis** is the most significant negentropic process for the construction of an economy based on the production potential of Nature. Photosynthesis is a physicochemical process carried out by living beings with chlorophyll that produces oxygen, water, and glucose from sunlight, water, and carbon dioxide. It is considered as the primary process of energy transformation in the biosphere, providing organic substances and oxygen to other living beings, as well as absorption of the excess of carbon dioxide, and thus essential for the maintenance of life on Earth. Therefore, this negentropic process assists in the process of reversing the entropic process in living systems, such as the biosphere.

The ecotechnological paradigm requires the differentiation between the distinct processes (levels and scales) of the transformation of matter and energy occurring in Nature and the human economy – ecological, technological, and economic processes. That is, the negentropic ecological processes of formation of matter must be distinguished from other processes of self-organization (like metabolism), as well as from the techno-industrial processes that generate high entropy in the form of heat, contamination, and waste. It contributes to a better understanding of how ecological systems function with a negentropic productive order, whereas technological systems tend to maximize entropic degradation, and what strategies can be used to modify the current production logic (Leff, 2010).

The theoretical and operative bases of this new economic paradigm must highlight the potentialities of a negentropic production based on the ecological organization of the biosphere, as well as on the limits presented by the irreversible entropic degradation generated by the metabolic and technological processes. Thus, Leff's 'productive ecology' seeks to incorporate both the entropic conditions of all mass and energy transformation processes and the negentropic process of biomass formation and absorption of gases and waste to create a sustainable production model capable of satisfying the culturally differentiated human needs (Leff, 2010).

To put into practice this new economic paradigm, it is necessary, initially, to evaluate the current balance between the negentropic production of Nature and the entropic degradation generated by the economic process. A prospective analysis of the necessary changes in knowledge, institutions, and production practices must be done to guide the transition to a co-evolutionary economy. In this new economy, negentropic ecological productivity can stabilize the entropic degradation of the processes of economic and technological transformation. In this sense, the elaboration of public policies with strategies oriented to sustainable productivity and technologies capable of reducing the entropic degradation of production processes is essential (Leff, 2010).

Leff emphasizes that the ecotechnological paradigm will only materialize through cultural values and practices. He believes that ecosystems are complex self-organizing systems affected by the global economy and a diversity of practices of appropriation of Nature based on scientific knowledge and other cultural types of knowledge. Therefore, Leff's proposal accepts the uncertainty with which these ecosystems are modified and explores possible ways to evolve to a sustainable and just productivity from the distinct worldviews and cultural valuations of Nature (Leff, 2010).

According to Leff, creativity and purpose are placed as conditions of real life in the symbolic order and the different types of knowledge can guide, modulate and act on the ecosystemic processes to bring them to states of higher ecotechnological productivity, with the maximum use of exergy and negentropic production. With this, the ecological risks of these processes will be converted into socially intervening processes, controlled and reoriented by knowledge, wisdom, and social action. In addition to thermodynamics, this paradigm of production finds its basis in cultural significations, the participatory management of resources, and the social appropriation of Nature. Decision-making in this paradigm will integrate ecosystem complexity with the environmental complexity that emerges from culture and knowledge in the participatory management of Nature (Leff, 2010).

> In thermodynamic studies, **exergy** is part of the energy available to produce useful work (organized form of energy). Unlike energy, exergy cannot be conserved, since irreversible processes destroy it due to the increase of entropy.

Leff's economic proposal, based on experiences from the South, sets out to develop an economic system based on ecological productivity and human creativity. It would be a new paradigm of production, based on ecological potential and ethnic–cultural plurality, with the decentralization of the economy and the diversification of the types of development, mobilizing the society to re-appropriate its natural patrimony and self-managing its production processes (Leff, 2001).

For Leff, the reconstruction of the economy is essential, and with it must arise a new theory of production that takes into account the ecological and social conditions of different regions. Different cultural values and interests should mediate this process, that, among others, would develop technological patterns to allow ecological regeneration and waste recycling (Leff, 2001). Leff's proposal also approaches the epistemological concepts presented by Santos. It reveals itself as an economic theory of the South that, while conscious of the limits imposed by Nature, opens itself to infinite human diversity as a necessary stage where creative answers are found in the dialogue between different types of knowledge.

Following the same logic of the strategies for overcoming the coloniality of power, Leff affirms that, faced with the reality of increasingly centralized, segmented and unequal societies, submitted to economic globalization and ecological normativity, the proposal to build a negentropic economy based on local and regional economic networks opens new possibilities for a just and sustainable development. Such a model opposes the homogenizing market economy, which erodes the sources of ecological production and cultural diversity (Leff, 2010). Leff's proposal seeks solutions to overcoming the crisis of mechanistic models with the strengthening of local, regional, and transnational communities, in a very similar way to the ideas of socialization of power.

In summary, Leff (2010) argues that only the activation of the desire for life has the potential to transform economic theory to generate a theory of production that brings together the potentials of Nature and the senses of culture. His proposal seeks to initiate a movement of paradigmatic transformation so that the economy is at the service of human and non-human life, cultural diversity, and ecological sustainability.

It is interesting to note that Leff's economic theory fits the assumptions of socio-ecological justice (discussed in Chapter 2 of this book). Productive ecology goes beyond the traditional limits of the economy, expressly including other cultural, social, and political issues. It therefore corresponds to the multi-dimensional approach of possible demands for socio-ecological justice (distribution, recognition, representation, and others). The main goals of striking a balance between entropic and negentropic processes and meeting culturally differentiated human needs harmonize with the main goals of socio-ecological justice to develop basic capabilities for the full functioning of Nature, current and future generations, for the promotion of ecological integrity and the culturally defined dignity of life.

The propositions of ecological economics and productive ecology are in line with the epistemological developments critical of the mechanistic paradigm and present greater coherence in terms of socio-ecological justice. Even so, in practice, the proposals and strategies of environmental economics, when they exist, predominate in public policies of most states and also at the international level. According to Leff (2010), economics has proved to be one of the most resistant fields to the internalization of ecological sustainability conditions in its theoretical structures and instruments. It insists on the use of mechanistic principles, which have long been surpassed by physics itself.

For reasons linked to mechanistic epistemology and the continuing domination of the coloniality of state and non-state power, overcoming the current globalized economic model is very challenging in itself. Besides, it is possible that the absence of more concrete and intelligible strategies within the limits of mechanistic thinking for a transition from the current economic model to fairer and more sustainable models explain the limited scope that propositions such as ecological economics and productive ecology achieve in practical terms. Other economic models have shown greater success influencing public policies in the international context. A good example is the proposal of circular economy, which more recently has gained wide space in the political debate, notably in Europe and Asia.

Circular economy

In general terms, Circular Economy has been defined as an economy that is restorative and regenerative by design and aims at keeping products, components, and materials at their highest utility and value at all times. This proposal distinguishes the technical (industrial) cycle from the biological one (ecological). That is because the resources that are part of the biological cycle (organic matter, for example) can be regenerated entirely (as by composting). However, the resources currently produced in the technical cycle (plastic, for example), can only be recovered and restored (as by recycling).

The proposal of circular economy has historical and theoretical origins in different theories, such as 'cradle to cradle', 'industrial ecology', 'biomimicry', among others, and aims to close the resource cycle. To achieve this goal, this economic model proposes to:

- **i** reuse and recycle resources;
- **ii** improve energy efficiency;
- **iii** use renewable sources of matter and energy; and
- **iv** manufacture products from plant-based materials that biodegrade as fertilizers at the end of their lifespan, replicating the concept of waste as food.

SOURCE EMF, 2015

Circular economy, which initially had a narrower approach, limited to the development of strategies for material reuse and recovery, has been expanded with the addition of new strategies. Based on this, the Ellen MacArthur Foundation, in a report published in 2015, called *Growth Within: A circular economy vision for a competitive Europe*, established three principles on which circular economy should be based:

1 The first is the preservation and improvement of natural capital, controlling the finite stocks of natural resources, and balancing flows of renewable resources.
2 The second is the optimization of the yield of natural resources through the circulation of products, components and materials in use with the highest utility rate at all times, both in technical and biological cycles.
3 The third principle is the promotion of system efficiency with the disclosure and exclusion of negative externalities.

SOURCE EMF, 2015

The report states that the adoption of these three principles, together with changes in the industrial–production system to turn the waste from one system into the input to other production systems and to maximize the full utility of the materials and products in use, can be helpful in building a broader notion of circular economy. Together with these two strategies, the report argues that the addition of new metrics to the current GDP of the existing economic system would also contribute to this goal, seeking to reintegrate the economic system into the planet's system. Regarding the proposal to add new metrics to GDP, it seeks, in summary, to measure and improve the internalization of negative externalities, the eco-efficiency of the production system and the pricing of natural resources (EMF, 2015).

Finally, another highlight of the report is a series of six business actions that can translate the three principles mentioned above, called ReSOLVE (regenerate, share, optimize, loop, virtualize and exchange):

1 Regenerate: the first action refers to the regeneration of the biosphere with the use of renewable energies and materials and the reintroduction of recovered biological resources to the ecosystems.
2 Share: the second action advocates sharing, reusing, and extending product lifetime.
3 Optimize: the third action seeks to optimize the production system, from extraction and logistics to production, use, and collection of materials at the end of life.
4 Loop: the fourth action defends the closure of cycles for components and materials, prioritizing internal cycles, ie those that do not require the entry of new materials into the production system.
5 Virtualize: virtualization is the fifth action, in the sense of maximum dematerialization of goods and services.

6 Exchange: finally, the substitution of non-renewable materials with the use of new technologies and new materials and products is the sixth and last action (EMF, 2015).

The Ellen MacArthur Foundation argues, therefore, that a wholly circular economy must:

1 completely end waste production;
2 build resilience through diversity;
3 move towards the use of energy produced exclusively by renewable sources;
4 adopt systemic thinking; and
5 work with the notion of cascades, ie to return products and materials to the production process at the end of each life cycle so that they can have other life cycles with the least possible throughout, having energy recovery as the last stage.

SOURCE EMF, 2015

Currently, several publications are working not only with theoretical models but also with ways of implementing the circular economy at the micro, meso, and macro levels. At the micro level (application in individual companies and consumers), there is literature on clean production, green consumption, green public procurement, recycling, reuse and composting of products, as well as public policies focused on these topics. At the meso level (referring to the implementation of eco-industrial parks), the literature focuses on eco-industrial systems, networks, and districts of industrial symbiosis, waste trade markets and public policies focused on these topics (Ghisellini *et al*, 2016).

Finally, at the macro level (on the implementation of circular economy in cities, regions, states, and countries), the literature addresses regional eco-industrial productions and networks, ecocities, urban symbiosis, collaborative consumption, zero-waste programmes, innovative urban solid waste management systems, and public policies to safeguard these issues. It seems that the implementation of circular economy strategies in China, for example, has been taking place with a top-down approach, through government decisions that seek to transition the country's economy to this model. In Europe, this process is taking place in the opposite direction, through the bottom-up approach, with extensive mobilization of non-governmental organizations and civil society (Ghisellini *et al*, 2016).

Although the circular economy proposition is quickly winning adepts and gaining ground in terms of public policies and business actions, it has also been criticized. The narrower approaches of the concept, which focus only on efficiency and material recovery, have clear limitations, such as:

- they do not truly challenge business models based on material productivity, only creating an alternative source of inputs for feeding of the current economic model;
- they generate an overall increase in production and consumption; and
- they cause the loss of the added value contained in products when their integrity is maintained (instead of sending them to recovery processes).

SOURCE EMF, 2015

Even the most expanded models of circular economy have been criticized. Although they provide innovative notions and strategies, the idea of circular economy is based mainly on strategies of decoupling, the commodification of natural resources, technological optimism and infinite economic growth, rather than challenging the global increase of production and consumption, especially of the latter. It is interesting to note that the circular economy model touches some of the points raised by ecological economics, mainly by the recognition of the limits of the planet and the need to reintegrate the economy to the ecosphere, which is a significant development. However, it still adopts the typical strategies of environmental economics as central, not representing an alternative discourse to economic growth, but rather an alternative growth discourse (Charonis, 2012).

In virtually every circular economy policy it is possible to find expressions such as 'sustainable growth', 'smart growth', and 'socially inclusive growth'.

Directly linked to these observations are other criticisms raised by various authors. The first one, usually concealed in the discourse of circular economy, is about the possibility of complete closure of the economic system proposed by this model. Moreno *et al* (2014) emphasize that there is enough research today to demonstrate that although it is imperative to maintain a circular flow, the production cycle cannot be closed entirely since many of the materials currently used cannot be reused or recovered after extraction and use. New raw materials will invariably be required to maintain continuity of production.

Rammelt and Crisp (2014) argue that circular economy overestimates the potential for closing ever-growing production systems, ignoring or underestimating the required biophysical basis for such growth. The thermodynamic considerations, mainly of the entropy law, do not receive due attention from this economic approach. It does not take into clear account that the energy after degradation cannot be reintroduced into the production cycle, as well as part of the matter that, after being used, will be irreversibly lost.

These authors argue that, on one hand, within a growing economy, circular economy strategies have the potential to slow throughout growth, prolonging (but not preventing) the time the economic system will take to reach the biophysical limits of the planet. On the other hand, the profits generated by the economy of resources and energy that circular economy can provide will, within the current global economic model, be reinvested for economic growth, accelerating pollution and the depletion of global resources (Rammelt and Crisp, 2014).

This analysis leads to another critical element, not usually discussed in the circular economy debate: the Jevons paradox. This paradox, now known as rebound or 'boomerang effect', occurs when the improvement of eco-efficiency (of matter and energy) in the production system, instead of reducing the consumption of natural resources, generates the opposite reaction – its increase. In general terms, economists clarify that this happens for two main reasons: consumption patterns (consumerism) and population growth (Charonis, 2012).

It seems that the circular economy strategies, at their present stage of development, are not yet capable of preventing the continuous pressure on natural resources. That is because, according to the rebound effect, increases in production and consumption would offset the positive impact of these strategies. Therefore, circular economy has been considered by some authors as unrealistic and contradictory (Charonis, 2012). Following that line, Tim Jackson published the *Prosperity without Growth* report, which was published as a book in its expanded and updated version in 2011.

In his analysis, Tim Jackson proved that, in practice, the strategies of decoupling and dematerialization are not, to date, generating a decrease in the consumption of natural resources and the production of waste. On the contrary: he has shown that they have encouraged even more consumption, now with the 'green label' (Jackson, 2011). Thus, circular economy strategies can help reduce the intensity of natural resource use and waste generation in relative terms. However, increases in consumption and population mean that global resource consumption and waste generation will continue to increase, proving that a more radical and systemic change in the economy is needed (Preston, 2012).

Alier (2003) adds that, apart from decoupling strategies, the geographic displacement of energy and matter sources, as well as of areas for waste disposal and storage of carbon dioxide emissions, are the main reasons for the relative dematerialization of the rich economies. As reported in Chapter 1, the increase in environmental quality in more developed countries is, in many cases, due to the transfer of contamination to other countries, especially to the South.

The argument that there are win–win solutions (to the environment and economic growth) is, to date, far from being true. The alleged dematerialization of Northern countries has not occurred in absolute terms and has generated more local and global geopolitical conflicts over contamination and access to natural resources (Alier, 2003). Thus, it is true that cross-border waste transfer, although not included in the accounting of the economic process of decoupling, has a great responsibility for the relative dematerialization of the Northern countries.

The circular economy proposition demonstrates significant advances over environmental economics, presenting potentially good ideas for overcoming the current economic model. In many respects, the circular economy strategies touch on the ideas of ecological economics, such as:

- the reduction of the use of non-renewable resources;
- the increase of material and energy efficiency of production processes; and
- the minimization of the flows of matter and energy that go through the economic system.

The entropic perspective of ecological economics emerges as an essential element for consideration so that the ecological (and waste) crisis can be overcome. It seems indispensable to establish a macro-economy that, in addition to recognizing the biophysical limits impeding infinite economic expansion, breaks with the social logic of consumerism. These points deserve both greater consideration in the literature and practical actions in the circular economy.

Besides, circular economy brings important elements that approach the idea of a negentropic production, such as:

- the understanding of the difference between technical and biological cycles; and
- the proposal of migration to production patterns increasingly based on biological cycles, from the choice of the materials used to the option of energy sources.

However, the social conditions and cultural meanings that make up ecotechnological production are not directly considered by the circular economy model. It leaves such elements out, as well as the discussion about the focus on local economies that can form transnational economic networks.

Unfortunately, the formulations of ecological economics and, particularly, productive ecology have so far not gained the necessary support to be implemented empirically in global terms. However, such proposals represent paradigmatic changes at the theoretical level and thus offer the potential to spread the seeds of change to overcome the mechanistic paradigm. Proof of this is the approximation of the circular economy with the ecological economics formulations and, to a certain extent, the ecotechnological production of productive ecology.

Beyond the simple contrast between the different economic theories seeking to overcome the mechanistic paradigm, it seems that the dialogue between them proves more productive in exposing their limitations and potentials. All economic streams studied here bring valid arguments and can be used to create a just and ecological economic model. Therefore, the circular economy strategies should undoubtedly be used to overcome the dominant mechanistic economic model, mainly due to their potential for the actual real-life implementation of this new model. However, the circular economy proposal still needs further theoretical development to take into account important issues raised by more critical economic streams, such as those of ecological economics and productive ecology.

The issues of carrying capacity (and its flexibilization with the transfer of social and ecological costs), international trade, the locality and creation of economic networks, and the inclusion of other demands highlighted by the notion of socio-ecological justice must be an object of greater debate in circular economy. Thus, circular economy, in its current form, can be considered as an excellent model of transition for a total (and future) paradigmatic change of the economy. It represents, at the moment, a significant step forward for overcoming the mechanistic model.

In order to achieve a more radical paradigm shift, a greater understanding of ecological and environmental complexity is needed, using both for economic decision-making that will encompass the interests of Nature, current and future generations. The entropic process must be taken into account, as well as the ecological capacities of organization and production (negentropic process) and ethnic–cultural diversity and plurality with their different types of knowledge. In this way, it is possible to achieve an economic model that is concerned with what unites, as well as with what distinguishes the different regions of the planet, working with the finite limits of the planet, the potentialities of Nature and the infinite human creativity and diversity. The demands for socio-ecological justice can, in this sense, collaborate to move towards this new model.

Paradigm shifts in law

Environmental law

As discussed in Chapter 1, Law has undergone a similar process to economics and other fields of knowledge. It has been decisively influenced by the epistemology and geopolitics of the Western modern mechanistic paradigm, which radically transformed law, and inaugurated a new legal paradigm. This paradigm led to the fragmentation of the legal order, which became an aggregate of distinct parts (both of the legally regulated subjects and the structure of knowledge divided into disciplines) governed by the laws of individual human reason.

Mechanistic Law, dominant up to the present day, has abandoned the notion of 'commons', typical of the Middle Ages, to regard individual private property as the most important legal concept, governed and protected by the mechanism of state sovereignty. Western Modernity thus inaugurated a form of law whose atomized order, focused on individuality instead of collectivity, has individual private property and state sovereignty as organizational principles.

This new Law, known as Positive Law, was developed to be a pure system, ontologically separated and differentiated from other areas, such as politics, justice, and morality. With the adoption of this approach, Law started to be seen as a closed and autonomous system concerning its context. A top-down approach of codification of law started to prevail, with the separation of law (object) from the subject, a mere technical, impartial, neutral and objective 'operator' of a mechanical chain of order transmission.

Thus, the mechanistic paradigm allows no room for the community to create its own interpretation of the Law, or significant local variations, transforming Law from a (local) culture to a (global) technology. Those characteristics have generated (and continue to generate) the disempowerment of communities that, with no access to the creation, discussion or implementation of the Law, have no incentives to seek solutions to transform their own reality. Although the separation between Law and communities has generated positive effects, in general, it has been lethal for both communities and Nature. That is because Law has enabled the maximum exploitation of the property as a tool to guarantee the effectiveness of the market (subservience to the economy).

North and South treat this exploratory potential of Law in different ways. The coloniality of power inherent in the mechanistic paradigm has made possible the co-existence of the Law (of the North) and the Non-Law (of the South). In the first one, the logic of regulation–emancipation prevails, while in the second, that of appropriation–violence. The mechanistic paradigm generated the denial not only of Nature but also of part of humanity, submitted to the exploratory logic of Non-Law. Mechanistic Law is thus dominated by (at the same time as being an instrument for the development and continuation of) the epistemological, geopolitical and economic models that allow social and ecologically destructive practices that have caused and intensified the ecological (and waste) crisis.

However, just as in economics, critical epistemological and geopolitical models of the mechanistic paradigm have reverberated in the theoretical debate of Law. Initially, this influence happened through the production of legislation regulating isolated elements (water, air, fishing, among others). However, since the 1970s, especially after the Stockholm Declaration, the outlines of a new legal discipline began to arise: Environmental Law (Pope, 2017).

The international debate promoted by the environmental movement assisted in the elaboration of environmental legal principles with the development of protocols and declarations emanating from various international forums and conferences on the subject. Those principles constitute the current international and national environmental legal basis (Montibeller-Filho, 2008; Benjamin, 2010). This movement led to the opening of Law to environmental issues with the creation of environmental law. Like environmental economics (with the economization of Nature), that new legal field has been seeking the internalization of Nature in Law, ie the juridicization or legal regulation, of Nature.

In general, environmental law can be considered as an independent legal field that has developed more intensely in the last 40 years, creating its own concepts, principles, legislation, jurisprudence, regulatory bodies and public policies at the national, supranational and international levels. The central objective of this legal discipline can be summarized as that of environmental protection, which can be affected, impacted, or threatened by human activities (Pope, 2017).

The development of environmental law has occurred in distinct ways in different countries, especially in states that adopt the common law system versus those based on civil law. Whereas in the former environmental law has been built primarily through jurisprudence, in countries using the civil law system it has been developed primarily through the creation of legislative frameworks and, in some cases, even via constitutional reform (Pope, 2017).

Despite the undeniable advances made by environmental law, after almost five decades of its existence, it is possible to identify many of its limitations and insufficiencies facing the growing complexity of the socio-ecological crisis. Ost (1995) argues that since its birth this new legal field has developed with contradictions. For instance, the state, through its sovereignty (internationally) and its governmental bodies (internally), is one of the main promoters of environmental law while being one of the key actors responsible for the ecological crisis. That is, the state, while having the duty of environmental protection, pursues, at the same time, objectives such as infinite economic growth.

Ost also highlights the contradictions generated by the juxtaposition of environmental norms established in different sectors and regimes. Even when attempts are made to develop legal codes, the overlap of standards leads to a lack of consistency of solutions. In this way, environmental law has, so far, limited itself to providing compensation (almost always late and insufficient) for socio-ecological destruction that nothing seems able to stop. The attempt to safeguard the environment in a society whose dominant paradigm is still 'developmentalism' (development reduced to the idea of infinite growth) is thus in itself a contradiction (Ost, 1995).

Extending the focus from environmental law to the legal system as a whole, the contradiction deepens. That is because the isolation of the few standards designed to protect Nature (or at least some of its elements) becomes clear in a system which favours the appropriation, transformation, and destruction of Nature massively. Thus, environmental law has proved to be ineffective, and its performance is only marginal. Among other reasons, it contradicts more 'powerful' standards that regulate the various human activities that are destructive of Nature (industry, agriculture, transport, real estate, energy, among others) (Ost, 1995).

These contradictions are also evident in the scope of international environmental law, where the distance between the intentions declared in international treaties and conventions addressing environmental issues, and the reality of their application is still enormous. According to Ost (1995), as soon as states sign new international environmental commitments, they rush back to continue with their previous behaviour. Attempts to regulate environmental protection at the international level appear to contradict the global economic and geopolitical forces, and, from this battle, Nature and the most vulnerable communities are often the losers.

Environmental law, although it has contributed significantly to the opening up of the Law to environmental issues, is very limited in its effectiveness at both the national and international level. There are many issues to be addressed, and Environmental Law, as it has been developed so far, has not challenged the mechanistic paradigm. On the contrary, this field of Law has reproduced mechanistic disciplinary fragmentation, representing only external constraints of relatively small impact when confronted with other elements which are given higher priority by the hegemonic legal regime.

Ost (1995) argues that the environment should be an essential component of all public policies and not merely an external limit with low impact in such policies. The safeguarding of Nature should be placed as an essential objective of Law as a whole. A comprehensive legal regime with protective virtues should, thus, be developed to achieve more significant progress towards overcoming the ecological crisis. To date, this lack of capacity for inter-penetration of environmental issues into Law as a whole has been a reflection, among others, of the inability of current societies to combine Nature with development.

Besides the insufficiency of environmental law as a separate and fragmented discipline, the very complexity inherent in socio-ecological issues aggravates the challenge. Bugge (2013) points out that many elements challenge Law when it seeks to deal with the ecological crisis. Among them, he highlights that:

a the characteristics of self-organization and complexity mark Nature, so humans have limited influence on the reactions of Nature;

b many environmental problems are invisible and marked by uncertainties;

c Nature cannot act legally, not having a voice in decision-making processes;

d many environmental problems are interconnected with social and economic issues, both in their causes and in their effects;

e many environmental problems are transboundary and transtemporal;

f complexity and plurality of values are almost always present in environmental decision-making processes.

SOURCE Bugge, 2013

All those characteristics have challenged environmental law, whose limitations caused by the failure to surpass the mechanistic paradigm are increasingly demonstrating its insufficiency as a theoretical model to overcome the environmental (and waste) crisis.

According to Bosselmann (2013), Law has legitimized and legalized excessive growth and ecological destruction. Both domestic and international Law have generally been ignorant of ecological realities because of Western modern beliefs and values, ie the mechanistic values. On the other hand, the dramatic changes in the global context, arising from corporatism, politics and social activism, indicate the vital role that Law can play in this context – either reinforcing and guaranteeing the *status quo* or being a facilitator of change. Thus, the action for the transition to a new paradigm by Law will only be possible with the overcoming of this ignorance regarding the ecological preconditions.

Capra and Mattei (2015) argue that a real paradigm shift will be possible with the understanding that the mechanistic perspective, which currently dominates human laws, is only a perspective. Therefore, as has been happening in other fields of knowledge, such a perspective can be overcome by migrating to a more systemic view of the nature of Law. The current dominant collective perception of Law as an objective or pre-existing structure by which the behaviour of individuals can be classified as legal or illegal is only a cultural and ideological construction of Western Modernity.

Ecological law

In Law, this paradigm shift has not yet occurred and is increasingly necessary and urgent. The main problems of our times are systemic, interconnected, and interdependent, and the global crisis experienced is an 'ecological crisis' in the broadest sense of the term. Capra and Mattei (2015) advocate the need for a profound legal paradigm shift to achieve a new 'Ecology of Law'.

This new legal order would assume a view of social reality as being composed of social networks and communities rather than an aggregate of individuals. Law, from this ecological perspective, is not seen as an objective structure, but as a living network that is created by and emerges from the self-organization of actively engaged citizens and communities.

If, in a strictly scientific sense, ecology refers to the science of relations between members of an ecological community and its environment, the ecology of Law (or Ecological Law) refers to the legal order that is coherent and honours the basic principles of ecology. It implies a process in which legal institutions, from extractive machinery based on the mechanistic approach of private property and state authority, are transformed into institutions based on ecological communities. In copying long-term ecological survival strategies, ecological law would seek an economy designed to nourish and preserve Nature in the interests of current and future generations (Capra and Mattei, 2015).

In addition to the process of 'juridicization' of Nature (the internalization of Nature in Law), a true change of legal paradigm requires the 'ecologization' of Law. Law, from this perspective, comes to be seen as a subsystem contained in the ecosphere and in a relationship with other systems: complex and contextualized. The challenge, however, is great. Instead of institutions with sharp contours, stable criteria and untouchable boundaries based on the mechanistic paradigm, Law will have to deal with complexity, wholeness, processuality, irreversibility, uncertainty, non-linearity, evolutionary conditions and other characteristics of the new paradigm of complexity (Ost, 1995).

In a broader (or metaphorical) understanding, ecology refers to a pattern of relationships that defines the context for a particular phenomenon. In this sense, ecological law proposes a legal order that does not treat Law as an independent social area, separated from politics, economy, justice, morality, among others. The concept of contextualized and complex Law does not separate the domain of facts ('what Law is') from the domain of values ('what Law should be') (Capra and Mattei, 2015).

The adoption of a systemic perspective of Law will seek to overcome the vision of this field of knowledge as a closed system of professionally established rules, based on exclusion and concentration of power patterns. In the mechanistic paradigm, as scientists do with scientific knowledge, jurists use power strategies based on the domination of legal knowledge. Ecological law, on the other hand, is seen as a common and thus must be part of the collective effort to redesign society, providing generative legal institutions to replace the current extractivist ones (Capra and Mattei, 2015).

It does not mean that there is no room for legal professionalism in Ecological Law. It means that jurists, in such a system, can translate the different knowledge into the policies and interpretation of Law, sharing and benefiting from the collective intelligence in their legal action. The resumption of collective ownership over the legal system is a crucial part of the strategy for attuning human laws to Nature and communities. In order for Law to be alive and generative again, it must be seen as a present structure, in constant movement and evolution, as an expression of social and ethical behaviour (Capra and Mattei, 2015).

Similarly, Ost (1995) defends the critical role of the jurist in ecological law. Always aware of the diversity of interests at stake in cases of conflict, the jurist must ensure that opportunities for public debate are multiplied, allowing contradictory opinions, combating practices that limit the transparency of legal processes and engaging in the facilitation of broad participation, the use of adversarial principle and the right to defence of individuals and communities. Socio-ecological issues will be, above all, democratic issues and the jurist will play a fundamental role as guarantor of this principle.

Besides, the ecologization of Law will require the overturning of the anthropocentric perspective of the mechanistic paradigm. According to Ost (1995), the paradigm of complexity leads to the resolving of the dichotomy between subject and object, not rejecting or absolutizing them, but placing them in relation. The new paradigm opens them to each other without confusing them, focusing, above all, on the relationship that constitutes them, on the link that unites them. Ost calls this network of relationships 'medium', surpassing the vision of Nature as an object or as a subject.

Law in the paradigm of complexity must bring the understanding that the relations between human beings and Nature are of reciprocal implications and interaction. The human being should be seen as a product and condition of Nature, just as Nature must be seen as a product and condition of the existence of human beings. Human beings, although representing highly complex living systems, cannot survive without Nature, whereas the opposite is not true. However, even being dependent on Nature for our survival, the human being transcends Nature, giving it meaning (Ost, 1995).

Although environmental law has helped to begin this journey, the complete ecologization of Law will be both necessary and welcome, as Law will be better able to provide legal solutions suitable for socio-ecological protection, whose problems are global, complex and dynamic by Nature. Ost (1995) emphasizes that the ecologization of Law does not mean the exclusion of the juridicization process of Nature with the establishment of human values and guarantees that it must promote.

Capra and Mattei (2015) point out that many differences distinguish ecosystems from human communities, such as lack of language and sense of justice in the former. Although human communities need to reconnect and learn from the laws of Nature, making them a central part of their laws, human values such as justice cannot be learned or copied from Nature. Thus, reproducing the laws of Nature is not enough for the construction of Ecological Law. What is needed is to learn from ecosystems how to live sustainably, transforming and adapting the diversity of human values consistently and coherently with the fundamental values that sustain life on the planet.

The complex view of the nature of Law, coupled with the recognition of the basic principles of ecology (which are generative rather than extractive), would allow the emergence of new legal forms to support the survival of the planet. For Capra and Mattei (2015), this process of transformation of Law requires, as a consequence of ecological thinking, that individual private ownership is moved from the centre of the legal system, in favour of the commons. The core structure of Law needs to be rethought, to migrate from a professionalized, pre-existing and objective structure, separate from the behaviour it regulates and tries to determine, to reflect the basic principles of ecology and complex thinking. That is, this new Law must be seen as a common in itself, as a constant process of long-term collective action, in which communities, sharing common purposes and cultures, institutionalize their collective desire for the maintenance of stability during social reproduction (Capra and Mattei, 2015).

Ost (1995) argues that the process of the ecologization of Law can take place by performing three essential steps. The first one (epistemological) seeks to overcome the mechanistic paradigm with the adoption of complex thinking. The second stage is ethics, from which derives the responsibility of bequeathing a viable planet to future generations as a chain of transmission. Finally, the third step would be the legal one, in which Ost proposes to adopt the notion of 'patrimony' (heritage or legacy) as a qualification and legal regime for this mixture of Nature and culture that is the 'medium' (the relation between human being and Nature).

Capra and Mattei (2015) also propose essential stages for the ecologization of Law that, while expressed differently, dialogues directly with the strategy presented by Ost. For Capra and Mattei the first step would be to create an ecologically literate society with the adoption of complex and systemic thinking (a similar proposal to Ost's epistemological stage). Ecological literacy (or eco-literacy) would generate awareness of the power of influence that society has over human laws via our aggregate action. It would also lead to an understanding of how Nature, through the ecological principles of organization, has sustained the web of life. It would help to understand how:

- in living ecosystems, the waste of one species is food for others;
- matters continually circulate through the web of life;
- the energy that drives ecological cycles comes initially from the sun;
- diversity ensures resilience;
- no single organism can exist in isolation; and
- life does not exist on the planet by combating, but by constructing networks.

SOURCE Capra and Mattei, 2015

Being ecologically literate should become a crucial skill for politicians, business leaders, and professionals in general (especially lawyers), making it the most important part of education at all levels. Knowledge about natural processes and patterns of organization is no longer restricted to environmental academics and philosophers, and it is clear that understanding the basic principles of ecology and to live according to them is imperative for the survival of humans on the planet (Capra and Mattei, 2015).

Once a certain degree of ecological literacy has been achieved, urgent changes in Law and Economics must be made. Like the laws of Nature, human laws must be understood as manifestations of a relational order in which the individual is not alone but connected and sharing power with other living inhabitants of the planet (human and non-human), who are also entitled to equal access to global commons (Capra and Mattei, 2015). Capra and Mattei call ecodesign or ecological design the next step for the process of the ecologization of Law.

Ecodesign is the radical redesign of social technologies and institutions in order to close the current gap between the design of human social life and sustainable natural systems. From the ecological point of view, design is the moulding of energy and matter flows to meet human needs. The principles of ecodesign must reflect the principles of organization that Nature has developed to sustain the web of life over billions of years, nurturing a sense of community rather than the separation between humans and the rest of Nature (Capra and Mattei, 2015). Ecodesign ultimately proposes the reformulation of different socio-economic systems by the adoption of ecological responsibility (a proposal analogous to Ost's ethical stage).

A connection can be made with the notion of socio-ecological justice proposed here, which presents the potential for the ecological redesign of the different social institutions in their processes of operation and decision-making. It is understood that ensuring at least just distribution, full recognition, and due representation for the development of basic capabilities for the full functioning of individuals and communities can promote ecological integrity and dignity of life. This notion of socio-ecological justice can therefore help to close the gap between the design of human social life and sustainable natural systems, imposing responsibility on social institutions for the present and future generations of humans and non-humans, as well as for Nature as a whole.

In Law, ecodesign would seek to transform the current mechanistic legal order into ecological law, into an ecolegal order. That represents the third stage of the ecologization process of Law presented by Capra and Mattei. In order to achieve this, a narrative that values and emphasizes ecological behaviour must first be constructed, focusing on the change of quantity to quality, and on the wisdom derived from Nature that well-being is not linked to consuming more than necessary. Translated into social categories, this means that freedom is a right to be enjoyed in relationships, having no connection to the accumulation of goods (Capra and Mattei, 2015).

The construction of ecological law also requires that new legal principles, ecologically oriented and opposed to extractive logic, are developed so that they can translate the ecological vision into institutional theory and practice. Such principles would have the potential of guiding communities in their decision to follow the Law in the present terms or to exercise their right of resistance in the interests of their own survival (and of future generations), avoiding individualistic and destructive socio-ecological behaviour (Capra and Mattei, 2015).

In this sense, the notion of socio-ecological justice proposed here could appear as one of these legal principles of ecological law for the guidance of decision-making processes involving socio-ecological issues. It would be a way of legal regulating and reintroducing the notion of justice to Law, addressing socio-ecological issues in a complex and contextualized way and taking into account the interests of current and future generations, of humans and non-humans, with the attendance of fair distribution, full recognition, and due representation. Judicial decisions guided by the principle of socio-ecological justice would have as their main objective the guarantee of the development of the basic capabilities for the full functioning of individuals and communities, promoting ecological integrity and dignity of life culturally defined.

> Chapters 1 and 2 showed how the mechanistic paradigm generated the separation of the notion of justice from Law, so that, in an ecological law, this reintroduction makes sense.

Thus, migration to an ecological law, founded conceptually in an ecological literacy and whose design is not abstract and created with a top-down approach, but, instead, comes from the specific needs of all members of the communities and of the different experiences of existing commons (bottom-up approach), emerges as a possibility for overcoming the ecological crisis. The ecological community will be the central axis of this ecolegal order and will foster the diffusion of power (socialization of power), participatory democracy, inclusion, justice and ideals of sharing (Capra and Mattei, 2015).

Finally, Capra and Mattei (2015) argue that this strategy for the migration of the mechanistic law to ecological law requires the pursuit of at least three objectives considered essential for such a goal to become politically relevant, namely:

- **i** disconnect Law from power and violence;
- **ii** transform communities into sovereign entities; and
- **iii** make property generative (rather than extractive).

These objectives have the potential to assist in the realization of the process of the ecologization of Law.

As for the first strategic objective, the authors argue that it is necessary to restore some harmony between human laws and the laws of Nature, bringing Law back to social networks and ecological communities and rejecting the individualism induced for economic reasons. It is necessary to see the ideology behind the current dominant legal system, which is abstract and mechanistic, owned by the state and kept away from the communities and their members by the professionalizing culture of corporate jurists (Capra and Mattei, 2015).

According to Capra and Mattei, being aware of the nature of Law and the ideology of its dominant discourses will allow politically viable solutions to be produced by communities, who will see Law as a constantly evolving common good capable of reflecting local conditions and fundamental needs. This holistic, ecological, and systemic view sees Law as a process of constant negotiation of the construction of cultural connections, making the communities be and feel part of it. It overcomes the perception of people as individualized abstractions and sees them functioning in complex networks of social and ecological relationships based on quality (Capra and Mattei, 2015).

Mechanistic law, grounded in the dichotomy of individualistic market competition and state hierarchy, is based on the concentration of power. It has thus proved incapable of presenting adequate solutions to complex post-modern problems, legitimizing, ultimately, the use of institutionalized violence for the maintenance of social unity. In order for such a perspective to be overcome, new human laws based on ecological principles must emerge from a bottom-up, context-specific, community-based process (Capra and Mattei, 2015).

That is a paradigm shift that leaves the false dilemma between individual versus state and between public law and private law. It breaks with the violent logic of domination and consumption inherent in both the 'state hierarchy' and 'market competition' and recognizes that both (false) alternatives are metaphors of exclusion and concentration of power in the quest for economic growth (Capra and Mattei, 2015).

In this same sense, Ost (1995) argues that the alternative between private appropriation or public management of Nature is false and must be rejected. In reality, the private is maintained by the public, which in turn depends on the private, representing a tight interlocking of interests. Thus, both perspectives on which mechanistic law is based – market and state – are only seemingly antagonistic and share the same concept of Nature as a simple object at disposal for unbridled exploitation. For Ost, only the adoption of another logic, a logic of real public participation, is capable of leading to a re-appropriation by individuals and communities of responsibility for their environment.

Bringing Law back to the communities, disconnecting it from power and violence, means that, rather than being alienated from the laws that govern them, communities would be the legislators and enforcers themselves, spreading power among their members, practising inclusion in an internalized and self-enforcing legal system based on the communion of values and goals. Overcoming the artificial isolation between public and private spheres, such communities would thrive in social production and reproduction (Capra and Mattei, 2015).

The interpretation of Law would be a non-professional exercise of sharing its collective meaning. Far from the concentration of power and violence, it would be a path, such as language, culture, and art, through which the collective communicates and decides about itself. There is no reason to believe that, with the future of humanity at risk, a network of generative legal orders cannot be organized, that political and practical relevance cannot be given to this new systemic understanding translated into a new legal vision that allows humankind to survive and thrive (Capra and Mattei, 2015).

As a direct consequence of the removal of Law from power and violence, the second strategic objective proposed by Capra and Mattei (2015) refers to the transformation of communities into sovereign entities. According to the authors, Modernity, with the scientific revolution and the intense urbanization process, allowed the domination of the future and Nature, disconnecting the majority of the (urban) population from Nature and its natural cycles. The authors point out as a fundamental element for the construction of an ecological legal order the transference of the sovereignty of the state and the individual to the living, present, and future communities.

Having living communities as sovereign entities of ecological law would make the commons the centre of the ecolegal order. Thus, both private property and the state would be conditioned to the protection of the commons and generative development. Common spaces could become private properties, provided they are generative (non-extractive) and with the purpose of founding life. Otherwise, the community of that territory, as a sovereign entity, would have the power to revoke and/or transfer such property rights (Capra and Mattei, 2015).

According to Ost (1995), the collective property system has ensured for centuries a satisfactory use of many common resources. Hence, contrary to the myth of the 'tragedy of commons', common property does not necessarily lead to the ruin of all. The success of the commons model will depend on the symbolic regulation of the communities of users of these goods, sharing the culture, traditions, and values that engender the complementary use of common goods with moderation.

Such community regulation of the commons should not be reduced to purely economic exploitation, nor to totalitarianism or bureaucratic arbitrariness. It should appeal to other forms of rationality that go beyond the current logic of maximum exploitation for individual profit. The adoption of this new perspective would also place Nature halfway between the conception of the 'Nature museum', which forbids any possibility of the use of natural elements by human beings (deep ecology), and the concept of commodification of Nature (shallow ecology), which is inappropriate for the protection of ecological integrity and promotes overexploitation of Nature (Ost, 1995).

Capra and Mattei (2015) provide an example of a practical instrument for the relocation of the commons at the centre of the legal order, with living communities as sovereign: the use of 'sunset' laws in private properties. Such laws establish timescales and situations in which property rights must become commons again, allowing the community to transfer them to other members for further generative purposes.

The same conditionality, although more controversial, should be applied to the state. For the authors, from an ecological perspective, state sovereignty should only be maintained if it is capable of protecting the commons of its territory against the extractive use of private properties. If the state protects private property more intensely than the commons (which occurs in mechanistic law), it would lose its legitimacy due to its inability to protect the commons, the most direct expression of the sovereignty of the community of a specific territory. In this case, the community would assume the collective duty of resisting extractivism of private property. Finally, the authors argue that pursuing political representation would not be a contradiction, since ecological communities would develop through negotiated consensus rather than majority rule – which often represents, in fact, only the interests of a small minority (Capra and Mattei, 2015).

Capra and Mattei (2015) present the transformation of property into a generative one as the last strategic objective to achieve the ecologization of Law, also interconnected with the first two objectives. They have borrowed the concept of 'generative ownership' from the journalist Marjorie Kelly, who advocates it as the one that serves the needs of life, tends to be socially just and ecologically sustainable, built into the fabric of organizational structures, generating welfare and real prosperity of life. This model contrasts with the conventional corporate ownership model, called by Kelly 'extractive ownership', whose main characteristic is maximum extraction and financial exploitation.

Generative ownership is part of a process of 'commoning', in which the resources necessary to satisfy common needs, be they material or spiritual, are understood as common goods. Such resources should, then, be governed according to the principle of solidarity to meet the needs of all members of the community and for the whole community to thrive. For Capra and

Mattei (2015), commoning practices (adapted to current needs) should become the new standard of rules to be implemented if the modern arrangement (based on the dichotomy 'private property–sovereignty') fails to be generative and act in the public interest.

According to Capra and Mattei (2015), most commoning practices are local (such as service exchanges within communities), but some needs share a crucial global dimension – such as the need for knowledge, which is incapable of dealing with the current complex demands provincially. In the current political system, the relevance of decision-making is global, given that the strongest political actors are global corporations. Therefore, it is fundamental that the process of commoning starts from the local and extends to the global, with different communities assuming their global political role through networks and sharing of experiences (transnational movement). That would, according to Capra and Mattei, be the foundation of an international legal order based on the commons legally organized and independent.

Once again, it is possible to link this analysis to the notion of socio-ecological justice defended here, which presents the potential for the development of the commoning process. The demands for socio-ecological justice have as their unifying normative principle the development of basic capabilities for the full functioning of individuals and communities (satisfaction of their needs) for the flourishing of all. Such demands will be raised whenever existing national and international institutions fail to prove themselves as generative and to act in the public interest (not promoting ecological integrity and dignity of life). The notion of socio-ecological justice, with its extended definition of the 'who' of justice, allows the demands raised to be from local to transnational.

The idea of generative ownership is capable of generating new political relations, local and global. It produces, through collaborative networks, an alternative economy that, emerging, will progressively replace the extractive logic of the current dominant economic model. Generative ecological law will support this new economy, this collaborative network from local to global (Capra and Mattei, 2015).

This point of Capra and Mattei's proposal also resembles Ost's (1995) approach, which advocates overcoming the models of commodification and public management with the adoption of the notion of 'patrimony' (or heritage) as a legal status of the environment. For Ost patrimony, as a continuation of the concept of the commons, can channel modes of production and consumption into ways that preserve the capacity of regeneration of natural resources and, more generally, local and global cycles, processes and balances that ensure the reproduction of living beings.

A complex, hybrid concept of mixed, transtemporal and translocal content, the legal regime of patrimony (heritage), according to Ost, can translate the ethical concern of taking responsibility for future generations (and, we add, for present generations and Nature as a whole). It imposes, in particular, a moderation in the use of Nature in order to ensure socio-ecological justice among all these subjects. Whether it will be to a legal model of generative property or heritage, the exploratory property model should certainly be replaced urgently if the current ecological crisis is to be overcome.

The main characteristics of Capra and Mattei's ecological law proposal are presented in Table 3.1.

TABLE 3.1 Comparison between mechanistic and ecological paradigms in science and law

PARADIGM SHIFTS IN SCIENCE AND LAW	
SCIENCE	LAW
The Mechanistic Paradigm ('World as Machine')	
Physical reality is an aggregate of separate building blocks.	Social reality is an aggregate of discrete individuals.
Scientific knowledge is used to dominate and control nature.	Law is used to protect extractive ownership as an individual right.
Scientific truth (the 'laws of nature') can be arrived at through reasoning.	Natural law is based on human reason.
Scientific descriptions are objective, independent of the human observer.	Law is an objective framework separate from a human interpreter.
The Systemic, Ecological Paradigm ('World as Network')	
Physical reality is a network of inseparable relationships.	Social reality is composed of social networks and communities.
Scientific knowledge ('ecological literacy') is to be used to learn from and cooperate with nature.	The new ecological legal order is to be used by eco-literate citizens to protect and generate commons.
Scientific knowledge is always approximate; it emerges from a process of establishing consensus in the scientific community.	Law emerges from actively engaged citizens in self-organizing communities.
Scientific descriptions depend on the human observer and on the process of acquiring knowledge.	Law is what is deemed to be law by civic and legal communities; it depends on human interpretations of social reality.

SOURCE Reprinted with permission of the publisher. From *The Ecology of Law: Toward a legal system in tune with nature and community*, copyright© 2015 by Fritjof Capra and Ugo Mattei, Berrett-Koehler Publishers, Inc, San Francisco, CA. All rights reserved. www.bkconnection.com

That Law contemplates ecological issues is an agenda developed by environmental law. However, it has encountered substantial obstacles in achieving its objectives, especially in times of crisis when there is economic and political instability. The mechanistic structure of modern law limits the environmental law that has been developed to date (even using a multidisciplinary approach). Isolated as an autonomous discipline, it has little influence on the old and robust individualist and extractivist bases of mechanistic law. It is reasonable to think that the solution to a problem will never be found using the same kind of thinking that created the problem. Environmental law, which, although with essential progressive contributions, has encountered obstacles in the very structure of Law for its full and effective realization, is at a crossroads.

Capra and Mattei's proposal goes beyond the creation of a new field of Law to address ecological issues. The ecologization of Law suggests a transdisciplinary transformation of the very foundations of Law. With the support of knowledge acquired from other sources of knowledge, such as ecology, biophysics, and traditional knowledge, together with the adoption of a complex and systemic approach, this new model would reconnect the laws of humans and the laws of Nature in the construction of a Law consistent with the values of socio-ecological justice. Such logic would apply across all disciplines regulated by Law. It would be the basic pattern of relationship, the thread that connects each of those interconnected and interdependent parts, composing the legal network. Ecological Law is also seen as a subsystem, interconnected with and interdependent of other subsystems, such as the economy and politics, all inserted in the ecosystem.

As Capra and Mattei (2015) admit, the current concept of Law has indeed allowed, through environmental law, the exceptional introduction of environmental regulations to limit property rights and free enterprise. However, ecological law goes beyond and, by placing the commons in the centre of its tutelage, turns the exception into the rule.

Final considerations

This chapter presented new epistemological, geopolitical, economic, and legal models that aim to overcome the mechanistic paradigm. It was seen how the laws of thermodynamics inaugurated a new paradigm, termed the entropy paradigm. In the worldview of this new paradigm, the world seen as a perfect machine in the mechanistic paradigm, whose time would not interfere with its functioning, became a machine in decadence, irreversibly moving towards complete disorder.

Incorporating the lessons, as well as the limitations of both the mechanistic and the entropy paradigms, a proposed new worldview has more recently emerged: the complexity paradigm. This new paradigm emphasizes the innovative character of biophysical processes, as well as the collective, cooperative, and organizational aspects of Nature, with the adoption of a synthetic and holistic perspective. It presents some principles as strategies for overcoming the reductionism and fragmentation of mechanicism, such as the dialogical, self-eco-organization, organizational recursion and retroactivity, holographic and systemic-organizational principles.

In epistemological terms, the complexity paradigm presents as a proposal an 'open epistemology', for the construction of knowledge in a simultaneously complex and contextualized way. This epistemology can promote distinction without disjunction, as well as conjunction without homogenization. It seeks to end the visible abyssal lines between the different fields of knowledge, unifying them in the same theoretical framework, without abolishing their autonomies (unity without uniformity), evidencing the inner limits of science.

Following the same line of reasoning, there is the proposal of the 'Epistemologies of the South', which, through the strategy of the ecology of knowledges, intends to overcome the invisible abyssal lines of the dominant epistemological model. It highlights the external limits of epistemology and provides dialogue and engagement among different types of knowledge (scientific and non-scientific) through inter-cultural translation and the creation of contextualized hierarchies. The two proposals of epistemology, one from the North and the other from the South, thus seek the inauguration of post-abyssal thinking.

Complex and contextualized post-abyssal thinking has the potential to generate geopolitical impacts to overcome the logic of appropriation–violence applied to the global South. Proof of this are the social movements of counter-hegemonic globalization, such as the environmental movement, which has brought different demands and uses different discourses that go beyond the economicist, universalist and reductionist logic of the mechanistic paradigm. These movements, in organizing themselves in transnational networks, deal with both the local and the global to overcome the North–South dichotomy hitherto reproduced by the model of international organization founded on nation-states and the market.

In order to overcome the coloniality of power, the socialization of power is necessary, with the creation of collective forms of public authority capable of creating a radical democracy with a bottom-up approach. The socialization of power does not seek to exclude global public institutions, but rather ensures that such institutions are open to a democratically self-managed global organization. The flag of the global, in this way, must be articulated with a complex organization that contextualizes it in order to allow the realization of 'local thinking/acting' at the same time as 'global thinking/acting'.

In Economics, evidence shows how environmental economics has been internalizing Nature in the market, with different strategies to overcome the ecological crisis. Its propositions, however, remain within the mechanistic paradigm. In turn, ecological economics, under the entropy paradigm, represents a real paradigm shift in this area of knowledge. This economic school includes the economy in the ecosystem and thus clearly stands against infinite economic growth, proven to be biophysically impossible by the entropy law.

Just as on an epistemological level, ecological economics does not take into account that, together with the process of entropic degradation, there are productive and organizational negentropic processes in the planet. Starting from this logic and in line with the paradigm of complexity, the proposal of 'productive ecology' seeks an entropic-negentropic balance between technological and ecological processes, with strategies based on ecological productivity and human creativity. This theoretical economic model proposes the participative management of Nature, with the creation of local and regional economic networks.

Although the ecological economics and productive ecology proposals represent paradigm shifts, it has been found that they have not yet provoked significant political mobilization, perhaps due to the absence of strategies more intelligible to the mechanistic models for their practical implementation. In this sense, circular economy has been highlighted in its political impact, with clearer strategies, such as the search for the closure of the materials cycle, dematerialization, and eco-efficiency.

It was concluded that circular economy, although not as radical as ecological economics or productive ecology, presents many advances in relation to environmental economics. If it incorporates more specific strategies that consider:

- the biophysical (entropic) limits of the planet;
- the necessity of overcoming consumerism;
- the importance of including social conditions and cultural meanings in their formulations; as well as

- the redirection to decentralized and interconnected local and regional economic networks...

... circular economy has the potential to be an excellent model of transition to an economy based on the paradigm of complexity.

As for Law, it has been demonstrated how environmental law has been essential for Nature to be the object of some consideration by this area of knowledge. However, the internalization of Nature into this field through the proposition of its legal regulation has been done without challenging the mechanistic paradigm, proving to be insufficient and contradictory for several reasons. The proposal of ecological law emerges as a possibility of paradigmatic change to migrate from mechanicism to complexity with the simultaneous realization of the processes of the ecologization of Law and the legal regulation of Nature. Such a model seeks the rescue of Law as a living network that emerges from the self-organization of society, seen as social and community networks.

Ecological Law is thus seen as a common in itself and proposes the substitution of individual private property for the commons as a central element of the legal network, in addition to overcoming the market–state dichotomy, electing the community as a sovereign entity. In general, the strategies for the ecologization of Law must necessarily go through:

i an epistemological change, which may occur through an ecological literacy capable of diffusing the paradigm of complexity;

ii the ethical redesign of social institutions so they can establish a new type of responsibility, covering present and future generations of humans and non-humans; and finally

iii the establishment of a new legal status for Nature, able to cover all its complexity and inter-systemic relations, basing its legal regulation on this new paradigm of complexity, in constant dialogue with the ecologization of Law.

At the international level, the ecologization of Law would reproduce the process of commoning from local experiences until reaching the global through transnational movements. Thus, the proposal of ecological law converges with the ideas of socialization of power and the relationship between complexity and contextualization, as a way of overcoming the North–South dichotomy maintained by the coloniality of power.

Given some epistemological, economic and legal models for overcoming the mechanistic paradigm and, ultimately, the socio-ecological crisis currently experienced, it is worth emphasizing once again that such theoretical models do not represent single, finished and definitive answers. That does not mean that they are invalid; quite the contrary. The complexity, incompleteness, and uncertainty inherent in such models are in line with an uncertain and evolving reality. Although far from being fully realized, such models serve as inspiration for the creation of new paths to be traced by humanity in the hope of a more just and sustainable future.

Based on this understanding, Part Two of this book will study the current waste legislation (Chapter 4), and cross-border transfer as a key operation for the classification of waste management as global (Chapter 5). This investigation seeks to ascertain why waste has become a legal problem. The study of the selected normative instruments will also indicate if the current legal regulation of cross-border waste transfer adopts a socio-ecological perspective or an economistic one and, therefore, a reductionist perspective of the theme.

The sixth and final chapter will then address the specific theme of global governance and waste management. Drawing on the models presented in this chapter, as well as on the other theoretical developments carried out in the first two chapters, notably on social metabolism and socio-ecological justice, strategies will be sought to overcome the problems pointed out in Chapters 4 and 5. Such strategies will, ultimately, represent possible ways to assist in the process of overcoming the waste crisis.

PART TWO

A new waste management system for a sustainable and just future

04

Waste as a legal problem

Introduction

It became clear in Chapter 1 of this book that the waste crisis has many causes and involves factors which are not only biophysical/metabolic, but also social, economic, geopolitical and even epistemological. The result is the current overproduction of waste surpassing the limits that can be withstood by the world's ecosystems, generating irreversible ecological and social impacts. However, such limits are masked and manipulated by the transboundary (and internal) transfer of socio-ecological costs, a typical example of the externalization of socio-ecological costs, unsustainability and ecological injustice in global terms.

As seen in Chapter 2, the increasingly damning evidence of such problems has sparked socio-environmental conflicts and those related to waste, all addressed by the environmental movement in its various streams. Although also touched on by other forms of discourse, waste has been treated as a problem of justice. As highlighted in Chapter 3, in theoretical terms new epistemological, geopolitical, economic and legal models have been developed and proposed as possible paths for overcoming the ecological crises and the waste crisis, understood as an ecological and justice problem.

Indeed, in legal terms, the waste crisis has undoubtedly involved a search for political–regulatory solutions at the national and international levels. However, the complexity inherent in the issue of waste has rendered the legal handling of the subject particularly challenging. In this light, this chapter chiefly aims to identify why waste has also become a legal problem. It will investigate if the way in which waste is currently regulated by the legislation selected here contributes to bringing Law closer to its ontological bases by means of the adoption of a more holistic and systemic perspective on the subject.

The need to establish a legal framework for waste and its management (its juridicization) has become increasingly pressing and relevant – both in terms of ecological protection and the mediation of related social conflicts. How this juridicization is done seems, however, to constitute an essential element of the socio-ecological effectiveness of the legislation. With this in mind, both this and the following chapter place a particular focus on the legal regulation of waste across various levels of legislation, whether national (Brazilian), regional (European Union) or international. Special attention will be paid to cross-border waste transfer under five important regulatory instruments in order to describe the subject and perform a comparative study.

Although there are many other regulatory instruments applied to waste seeking to control its transboundary movement (cross-border transfer), a selection has been made of those with the most significant impact at international level, as well as those relevant to the methodological scope adopted in this book. The outlining of the five regulatory instruments in this chapter and the following one do not, however, exclude the need to analyse other instruments for the fulfilment of the regulations, such as other European directives and other Brazilian regulations, notably issued by IBAMA and CONAMA. Such regulatory instruments incorporate an enormous set of definitions, objectives and control systems related to waste, waste management operations and the subjects responsible for the waste, among others.

The focus shall be placed on two international regulations, two regulations of the European Union, and one national regulation, as follows:

1. Basel Convention (referred to here as the Convention);
2. Decision C(2001)107/Final by the Organization of Economic Cooperation and Development (referred to here as the OECD Decision);
3. EU Directive no. 2008/98/EC (referred to here as the Waste Framework Directive);
4. EU Regulation no. 1013/2006/EC (referred to here as the Regulation); and
5. Brazilian Law no. 12.305/2010 (referred to here as the Brazilian National Solid Waste Policy – NSWP).

THE BASEL CONVENTION

The Basel Convention regulates the transboundary movements of hazardous wastes and 'other waste substances' and their disposal among the parties to the Convention, and emerged at the end of the 1980s in response to various environmental accidents caused by the international transfer of waste. After lengthy negotiations between Northern and Southern Parties, which had different interests on the subject, the Convention finally came into effect in 1992.

OECD DECISION C(2001)107/FINAL

The OECD Decision, on the other hand, arose from a series of decisions made by members of the OECD in the 1980s. Such Decisions are regarded as international agreements effectively binding all of the OECD's Member States. The signing of the Basel Convention by all of the OECD's members sparked a process by which these decisions were reviewed in order to align them with the Convention, resulting in Decision C(2001)107/Final, adopted in 2002 and remaining in force today.

EU DIRECTIVE NO. 2008/98/EC AND EU REGULATION NO. 1013/2006/EC

Many cases involving the cross-border waste transfer to Brazil and other countries in the South originated and still originate in EU Member States. For this reason, and due to its status as a party to the Basel Convention and its partnership with the OECD, the EU also regulated the subject at Community level. The European Regulation should not, however, be studied in isolation without an understanding of the Waste Framework Directive, a key legislative document containing general definitions, fundamental principles and basic obligations for different subjects all related to the theme of waste. Both instruments are binding for all of the Member States; the Regulation with direct application and the Waste Framework Directive by means of its transposition into national legislation.

BRAZILIAN LAW NO. 12.305/2010

Finally, for the first time in the country, Brazilian Law no. 12.305 established a National Solid Waste Policy (NSWP), with principles, objectives, instruments, responsibilities and control systems for environmental management of waste. As Brazil is the destination of many cross-border waste transfers, as well as being party to the Basel Convention and an OECD partner, the country executed its right to prohibit the import of hazardous waste, regulating all other types of waste importation by means of IBAMA and CONAMA standards inspired by the aforementioned international instruments.

The rationale to be applied in studying these regulatory instruments in this chapter and Chapter 5 will range from the general to the more specific, entailing different assessment orders. This chapter will investigate the general aspects of the extent to which the subject of waste has been incorporated into the regulatory instruments selected. This analysis will be used as a basis for assessing how well cross-border waste transfer is legally regulated in order to create an effective global waste management system from a socio-ecological perspective. The rationale of moving from the general to the more specific will result in an assessment ranging from the most local to the most global between the chosen law, ie, from national, to regional to international legislation. As will be seen, local instruments establish general norms on waste within their spheres of application, while international instruments focus on the specific operation of waste transfer.

Furthermore, most local regulations have higher levels of detail and local relevance in comparison to global instruments, which tend to be more abstract. For these reasons, this chapter shall first study the juridicization of waste by means of the regulatory instruments selected, emphasizing the concept of waste and its legal classifications. It will then move on to investigating the juridicization of waste management, with its concepts, subjects, legal controls and hierarchies between the different management options. A specific assessment of the operation of cross-border waste transfer, as mentioned, shall be embarked upon in more detail in Chapter 5.

Waste: legal concepts and classifications

Legal concepts of waste

Although the concept of waste may initially seem obvious and intuitive, it has sparked many doubts and controversies between doctrine, jurisprudence, legislators and economic operators (Aragão, 2006). The notion of waste is, merely in itself, replete with ambiguities. What is waste to some may not be waste to others. What may be seen as waste now, may not be in the future. As will be revealed in this section, the very uncertainties inherent in the notion of waste are entirely reflected in the difficulty of establishing a legal concept that is clear and unquestionable.

A clear and precise definition of waste is in fact a critical element in the application and implementation of all legislation applied to the subject, including that related to its transboundary movement. In practice, the difficulties emerging from the interpretation of the subject are rather complex. Between a material which is universally deemed to constitute waste and another universally deemed not to constitute waste, there is a grey area of materials which may be considered waste or not, mainly in the area of used goods.

Valuable resources could be preserved and disastrous environmental harm avoided if a definition were to be consolidated and used consistently across the various levels of legislation. A clearer and more precise definition of what constitutes waste could prevent materials from being discarded when they could be reclaimed, and also preventing materials in circulation without any form of environmental protection from receiving due attention in being framed as waste. It is for this very reason that doctrine, jurisprudence and legislators have made concerted efforts to clarify the concept.

According to Aragão (2006), waste is a by-product of almost all social processes of production and consumption; it is an undesirable result, emerging in an involuntary – and sometimes even forced – fashion. Waste is involuntarily produced due to the negligence of the holder generating it, ie without a clear, direct intention to produce it. Forced waste, on the other hand, is produced when the holder not only has no intention of generating it, but it happens even against their will. In the case of consumers, such a situation occurs when there are no more sustainable alternatives to products essential to meeting basic needs. In terms of producers, the forced generation of waste occurs when the latest technological development does not allow production processes to be more effective in preventing such undesired results (Aragão, 2006).

One example of this situation is the sale of fruit and vegetables packaged in plastic by almost all supermarket chains in Northern countries, leaving consumers with no option but to produce such waste upon purchasing these foods. Another example is producers' use of machinery that, despite representing the most cutting-edge technology on the market, generates large quantities of waste in producing certain consumer goods.

In clarifying several criteria used to differentiate between waste and non-waste materials, Aragão (2003) affirms that, from a static perspective, waste may be identified as by-products of a production process or consumption and due to the fact they can potentially be abandoned, reflecting the lack of interest on these materials by their holder. The solid state does, however, lend certain waste the character of belonging to someone (someone will always own or discard this solid waste) justifying the particularities of its legal regime (responsibility of the producer, but also of the holder; special duties of care in transport; rigorous proportionality in the application of the 'polluter pays principle', among others).

BRAZIL

In Brazil, the National Environmental Policy (Law no. 6.938/81) is recognized as an important national legislative framework for environmental protection. It was the first law in the country that widely and systematically regulates the subject, establishing principles, directives, instruments and concepts related to the environment. Next came the promulgation of the 1988 Federal Constitution, which, in corroborating Law no. 6.938/81, constitutionalized environmental protection, dedicating a specific chapter to the environment and with the topic also running through the text as a whole. For this reason, it is considered to be an Ecological Constitution. Although they provide the essential bases for the regulation of any matter related to environment, such legal instruments do not, however, deal explicitly with waste.

In fact, it was only as of 2010, with the promulgation of Law n. 12.305 implementing the National Solid Waste Policy (NSWP), that waste was specifically addressed by Brazil's federal legislation. Since then, Art. 3, section XVI of this law has considered waste in the Brazilian jurisdiction as all:

> [...] materials, substances, objects or goods discarded as a result of human activities in society, whose final destination has been done, is intended to be done or is required to be done, in a solid or semi-solid state, as well as gases contained in receptacles and liquids whose specific qualities render their release into the public sewage network or water bodies inviable, or that require technical or economically inviable solutions given the best technology available.
>
> SOURCE Brazil, 2010

The Brazilian NSWP definition is rather detailed and may be divided into three different elements: origin, destination and state. The first element is the 'origin' of the materials, substances, objects or goods that may be considered by the Law as waste. Section XVI specifies that only those resulting from human activities shall be classified as waste, therefore excluding all those produced by other living beings and abiotic elements of Nature. For the NSWP, therefore, waste subject to the law will always be that resulting from anthropogenic activities.

Once discarded, the procedure to the final destination is a key element in the Brazilian NSWP's consideration of a material, substance, object or good as waste. This is referred to here as the 'destination' element. In order to be able to completely shape the meaning of such an element in the regulation's definition of waste, 'final destination' must also be defined. According to section VII of Article 3, an environmentally-sound final destination is a:

> [...] waste destination that includes reuse, recycling, composting, energy recovery or other destinations accepted by the Sisnama, SNVS and Sausa

> competent bodies, including the final disposal, meeting specific operational standards in order to prevent harm or risk to public health and safety and to minimize adverse environmental impacts.
>
> SOURCE Brazil, 2010

The Brazilian NSWP considers waste not just materials, substances, objects or goods to be disposed of, ie destined for final disposal in landfills – when they are therefore considered as refuse, without the possibility of treating and recovering the materials of which they are composed (section XV). The definition of waste in section XVI, along with section VII from article 3, do in fact clarify that materials, substances, objects and goods that may be recovered also fall within the framework as waste according to the Brazilian legal system.

The destination element also comes into play in the classification of the discarding of waste according to three different moments:

- i when it has already occurred;
- ii when the intention to do so has been declared; and
- iii when it is obligatory.

Just like the Brazilian NSWP (which was inspired by the current European Waste Framework Directive), the European legislation also takes these three moments into account. For Aragão (2009), the first case would constitute a 'historic' situation, as the material's character as waste was demonstrated through the fact that the holder got rid of it. The second case would be a 'future' situation, as the characterization of the material as waste rests on the holder's intention to get rid of it. Finally, the last case would be a 'legal' situation, given that the material's nature as waste is presumed by law.

The third component that may be extracted from the Brazilian NSWP definition of waste is relative to its 'state', bearing in mind that the law considers that all materials, substances, objects and goods in a solid or semi-solid state could be waste. Materials, substances, objects and goods in a gaseous state may however also be classed as waste, when contained in receptacles, as may liquids, when they cannot be added to the public sewage network or to bodies of water due to the threat posed to public health and the environment (Table 4.1).

There have been no further efforts in Brazilian jurisprudence, and only very few attempts in the wider literature, to define the concept of waste in the national context, which may perhaps point to the quality of the definition provided by the Brazilian NSWP. On the other hand, the absence of such a

TABLE 4.1 Elements of the concept of waste according to Brazil's NSWP

	Origin	Destination	State
Elements of the NSWP concept of waste	Resulting from human activities	Discarded followed by its final destination (includes disposal and recovery) according to historical, psychological or legal circumstances	Solid, semi-solid, gaseous, (when contained in a receptacle) and liquid (when discarding it in bodies of water or the sewage network is not environmentally appropriate)

debate in Brazil may highlight the subject's relative novelty, given that it was only recently that specific legislation was introduced into the national sphere (in 2010), and that it has only been in recent years that it has been proven to constitute a problem for the environment and public health in Brazil. Such a conclusion becomes even more pertinent in light of the complexity of the concept of waste as evidenced by the longstanding debates in countries in which the extent of the waste crisis became clear years before its emergence in Brazil.

EUROPEAN UNION

Within the European Union, the legal framework governing solid waste was for years that of Community Directive 75/442/EEC. According to this Directive, waste was defined as 'any substance or object in the categories set out in Annex I which the holder discards or intends or is required to discard' (article 1, 'a'). In the second paragraph of the same item, the Directive tasks the European Union Commission with the creation of a list of the types of waste to belong in the categories listed in Annex I of the Directive.

In December 1993, the European Commission thus presented and adopted such a list by means of Decision 94/3/EC, designating it the European Waste Catalogue (EWC) and seeking to encompass all waste, regardless of whether or not it was intended to be disposed of or recovered. Legally, the EWC is a Decision addressed to the EU's Member States and is a non-exhaustive list, which should be (and in fact, is) reassessed and reviewed when necessary. The Decision also highlighted the fact that the materials listed in the EWC only constitute waste when they could be framed in the definition of waste of the Directive 75/442/EEC.

Decision 2000/532/EC of May 2000 substituted and updated the EWC. According to the European Commission, the main objective of the EWC is to establish common reference terms for the entire European Union, improving the efficiency of the various activities entailed in waste management. Despite recognizing the importance of the role the EWC plays in the attaining of objectives, such as facilitating the control of intra-community production and circulation of waste, and principally in the promotion of solid bases for the creation and development of an internal market of recoverable waste materials, Aragão (2009) categorically affirms that such a catalogue did not significantly contribute to the defining of the concept of waste. That is due to the fact that any material listed in the EWC may or may not be characterized as waste, depending on the holder's 'intention' to discard the material or not.

This Directive was revoked and replaced by Directive 2006/12/EC, which established the new legal framework for the treatment of waste in the European Union, still however keeping the same definition of waste of the Directive 75/442/EEC. Since the 1970s, therefore, the essential element defining waste within the European Union has been the holder's intention to 'discard' the material.

Aragão (2009) states that European literature and jurisprudence have striven to clarify the best interpretation of 'discard', and, in wider terms, 'waste'. According to Aragão there are two main interpretations of the meaning of 'discard'. The first is a subjective perspective in which to discard waste means to abandon it – which is always illegal – or to dispose of it, whether legally or illegally. This concept carries a pejorative connotation, applying a value judgement on the holder that abandons the waste material. There is also a neutral-value objective meaning according to which the act of discarding a waste material merely corresponds to altering its destiny, ie to its transfer. The objective concept is obviously wider, encompassing not only the intention of disposal, but also the recovery of waste materials (Aragão, 2009).

The decision to opt for the subjective or objective concept of the notion of 'discard' – which is an essential element of the concept of waste in the European Union – has a direct impact on the applicability of any legal instrument regulating waste materials. If the more restrictive subjective concept is adopted, only materials destined to be disposed of are considered to be waste. Adopting the objective perspective, on the other hand, defines waste as materials destined for either disposal or recovery. The adoption of the wider, objective concept implies the application of the legal regulations governing waste in cases of both disposal and recovery.

In arguing for the adoption of the objective concept of 'discarding' waste, Aragão (2009) emphasizes environmental reasons and those promoting harmony among different legal jurisdictions. The environmental reasons are the obvious environmental impacts caused by waste materials. Furthermore, many recovery and disposal operations are physically identical, mainly constituting intermediary operations such as screening, disassembly and crushing. Excluding waste materials which may be recovered from the concept of waste pushes them outside the current legal regulations guiding the matter, allowing for the waiving of environmental protection and the permission of harm generated by recovery processes.

It is important to highlight that a waste material's potential for recovery should not serve as a parameter for not classifying it as waste. That is because under current economic and social standards, the decision to recover a material depends on elements varying according to location and time, such as technological and scientific progress, the cost of the virgin raw materials to be substituted by the recovered materials, and the cost of the recovery and disposal processes, among others (Aragão, 2009).

With so many ambiguities and doubts surrounding the concept of waste, Directive no. 2006/12/EC wasted an opportunity to clarify the matter, merely repeating the definition of waste in the abolished Directive. Since then, the European Parliament and the Council have demonstrated an understanding of the need for greater clarification and distinctions between the concepts of waste and non-waste, as well as between recovery and disposal. They have also advocated for the importance of reinforcing preventive measures for waste generation and an approach taking into consideration the entire life cycles of products and materials. For the sake of clarity and legibility, such EU bodies therefore defended the need to also revoke and substitute Directive 2006/12/EC.

February 2007 saw the European Commission issue Communication COM(2007)59 to the Council and the European Parliament on the interpretation of the distinction between waste and by-products. In this Communication, the Commission suggested several possible interpretations according to recent decisions on the subject issued by the EU Court of Justice, expressing the difficulties generated by the absence of clarity on the legal concept of waste. It affirmed that:

> The evolving jurisprudence and relative absence of legal clarity has made in some cases the application of the definition of waste on this issue difficult for competent authorities and economic operators alike. There is some evidence

> of differing case-by-case solutions on similar facts by competent authorities in different Member States – this leads to inequalities in the treatment of economic operators and obstacles in the internal market. An excessively wide interpretation of the definition of waste imposes unnecessary costs on the businesses concerned, and can reduce the attractiveness of materials that would otherwise be returned into the economy. An excessively narrow interpretation could lead to environmental damage, and undermine Community waste law and common standards for waste in the EU.
>
> SOURCE EU, 2007: 4-5

The EU Court of Justice has played an important role in the clarifying of the legal concept of waste. According to Aragão (2009), the objective concept of 'discard' has dominated in European jurisprudence, reaffirming the understanding that waste is any substance that the holder intends to discard, regardless of whether it is destined for disposal or recovery. In the European context, therefore, the case law understanding consolidated over the years suggests that even materials that may be recovered (whose holders intend to recover them or that have a positive economic value) are subject to the general rules of the law governing waste, and should always be managed in a way which is compatible with the environment.

It was in light of these difficulties that Directive no. 2008/98/EC was approved as the new European Waste Framework Directive. Such a Directive established environmental protection measures and those protecting human health in order to prevent or reduce the adverse impacts derived from the generation and management of waste, reducing the general impacts of the use of resources and improving the efficiency of this use (article 1). Excluded from the Directive's sphere of application were gaseous effluents, contaminated or non-contaminated soils, radioactive waste, explosives, faecal material, waste water, animal by-products (except those destined for incineration, landfill and energy recovery) and waste resulting from mining and quarrying (article 2).

This new EU legal framework introduced the definition of waste as 'any substance or object the holder discards or intends or is required to discard' (article 3, 1). Unlike the previous directives, Directive 2008/98/EC no longer refers to the EWC to define what constitutes waste, returning the focus merely to the action, intention or obligation to discard of a certain waste material, thereby reinforcing 'to discard' as an essential element determining what constitutes waste within the European Union.

The preamble to Directive no. 2008/98/EC affirms that, in order to prevent confusion on the various aspects of the definition of waste, it tries to clarify when materials or products are considered by-products rather than waste, and when waste is no longer considered as such. Furthermore, for the sake of safety and coherence, the Directive establishes that the Commission may, based on the definition of waste, specify the moment at which materials or objects are transformed into waste, notably in terms of electrical and electronic equipment and vehicles.

Unlike the Brazilian NSWP which uses the terminology of 'final destination' in its concept of waste and specifies its meaning, the European Directive does not include an explicit definition of 'to discard'. A reading of its text does, however, reveal that Directive 2008/98/EC adopts the objective concept, encompassing both materials destined for disposal and recovery.

As a natural result of the extensive debate held at a European legislative and jurisprudential level, the new European Framework Directive thus sought to differentiate waste from by-products in its article 5 according to case law precedents in the EU Court of Justice (see Table 4.2). It clarifies that substances or objects resulting from production processes may only be considered by-products (and therefore not waste) when it is certain that they may be used later on, without the need for any further processing outside the normal industrial practices, or that are produced as an integral part of such a production process. This later use must also satisfy all of the requisites in order not to harm the environment or prove hazardous to public health.

Directive no. 2008/98/EC, article 6, lists the general criteria required for an item of waste to no longer be considered as waste (end-of-waste status) and instead obtain the status of 'product' (or secondary input), as follows:

- **i** it has been subjected to a recovery process;
- **ii** it is normally used for specific purposes;
- **iii** it satisfies the technical requirements, legislation and regulations for the specific purposes for which it was destined;
- **iv** its use does not exert globally adverse effects on the environment or public health; and
- **v** there exists a market or demand for the object or substance.

In a certain sense, the new European Waste Framework Directive gave a more objective definition of waste, clarifying the criteria differentiating waste from non-waste over the course of the text. Furthermore, both materials destined for recovery and for disposal should be subjected to the rules and standards provided by the Directive. Finally, it must be highlighted that all the other Directives and Regulations governing subjects related to waste,

TABLE 4.2 EU criteria for differentiating between waste, by-products and ex-waste

	Waste	By-products	Waste that is no longer waste (becoming products)
EU criteria for differentiating between waste, by-products, and waste that has lost the status of waste	Any substances or objects the holder discards or intends to/is required to discard.	If, although defined as waste: 1) it is certain to be used again without any other processing outside normal industrial practices and without causing environmental damage or harming public health; or 2) it is produced as an integral part of the production process itself.	If, upon definition as waste, it meets the following criteria: i) it has been subjected to a recovery process; ii) it will be commonly used for specific purposes; iii) it meets technical requirements, the legislation and the regulations for the specific purposes it is destined for; iv) its use does not entail impacts which are adverse for the global environment and public health; and v) if there is a market or a demand for these objects or substances.

such as Regulation (EC) no. 1013/2006 on waste transfer, refer to the definition given by the Framework Directive in their use of the concept of waste.

With that said, criticism has still been voiced on the definition of waste provided by the new Framework Directive. Gharfalkar *et al* (2015), for example, believe that such a definition considers products in perfect working order as waste if their holders intend to discard of them. In their view, the 'intention to discard' element makes the definition too wide, including products which are directly reusable in the category of waste, which may potentially excessively encumber the holder and/or other subjects responsible for their management.

INTERNATIONAL LAW

In the international sphere, the Basel Convention defines waste in article 2 as 'substances or objects which are disposed of or are intended to be disposed of or are required to be disposed of by the provisions of national law'. By means of this definition, the Convention seems to restrict its scope of application to waste materials destined for disposal, therefore excluding all transboundary movements of materials destined for recovery. In short, the current definition of waste provided by the Basel Convention ends up excluding waste materials destined for recovery.

In fact, since its signing and adoption in 1989, many of the articles in the Basel Convention have been interpreted in a variety of ways by the different Convention member parties. The many uncertainties generated by basic definitions contained in this legal instrument, including the very definition of what constitutes waste, were recognized by the parties of the Convention as one of the reasons hindering its implementation. By means of Decision BC-10/3, the Conference of the parties, in their 10th encounter (COP10), decided on the need to achieve greater legal clarification on the terms used by the Convention with the aim of improving the instrument's effectiveness (UNEP, 2011a).

> Decision BC-10/3 reflects a project initiated by Indonesia and Switzerland, both party states to the Convention, to develop strategies and search for solutions in order to improve the effectiveness of the Basel Convention. Such a project was approved by Decision 10/3 at COP10 with seven main goals, including: i) ensure the entry into force of the Basel ban (banning the transfer of hazardous waste from OECD member countries to non-member countries); ii) the provision of greater clarity on legal concepts used by the Convention – among them, the very definition of waste; and iii) the adoption of more effective means of tackling the illegal trafficking of hazardous waste.

In that light, the COP10 requested the development of a study on the implementation of the Convention. Given that its enforcement is directly linked to the terminology used, the creation of a Glossary with relevant terms was proposed, such as:

i waste and non-waste;

ii hazardous waste and non-hazardous waste;

iii reuse;

iv direct reuse;

v refurbishment;

vi second-hand goods; and

vii used goods.

SOURCE UNEP, 2011a

On the same occasion, the COP10 requested the development of a report containing a study and possible options for interpreting the terms listed above (UNEP, 2011a).

At its 12th meeting, COP12, Decision BC-12/1 concluded with the need of a more in depth legal clarification of the Convention, by means of a review of Annexes I, III, IV and IX. Upon a summons from the COP, Canada offered to lead and group the opinions of the other countries which were parties to the Convention in the process of reviewing the Annexes mentioned, based on the Glossary draft developed by a group of specialists, which was approved during the meeting for the 13th Conference of the parties in May 2017 (UNEP, 2015a; 2017).

An Expert Working Group on the Review of Annexes of the Convention was then established by COP13 via its Decision BC-13/2. In 2019, this working group had its mandate extended by Decision BC-14/16 during the 14th meeting of the COP to also review other entries (such as those related to plastic waste) and Annexes as well as the implications of those changes (UNEP, 2019).

At the time of writing this book (2020) not a single amendment has been made at the Convention to clarify the terms used, with revisions still being discussed by the parties. The Glossary of terms approved by COP13 is positioned as a document for helping guide the parties in the implementation of the Convention, without jeopardizing the legislations and guides developed on a national level by the parties, except in the sections expressly legally binding. Furthermore, the document publishing the Glossary highlights that it is part of a process reviewing the Convention Annexes and may therefore be subject to modifications during this process (UNEP, 2016).

> All of the Decisions adopted by the COP so far have only been in the sense of recognizing the need to clarify the terminology used by the Convention, with the definitive interpretations of these terms, such as that of 'waste', only emerging with the adoption of an amendment of the Convention and its Annexes, after being signed and ratified by the party states.

Following the thrust of the waste regulation developed by the EU and by other countries such as Brazil, the Basel Convention's Glossary of terms clarifies that the definition of 'waste' provided by the Convention covers three possible situations. The first refers to a substance or object that has already been disposed of. The use of any of the operations of disposal listed in Annex IV qualifies the substance or object as waste. The Glossary states that some of these operations may be used in production processes, therefore with the use of non-waste materials. This means that although materials subjected to disposal operations listed in Annex IV may be considered to constitute waste, other factors must also be taken into account in an analysis to be made on a case-by-case basis (UNEP, 2016).

The second situation refers to substances and objects intended for disposal. The Glossary highlights that this is an important feature of the definition of waste, as it allows for the waste to be subjected to control before being disposed of, facilitating the prevention of possible harm to society and the environment (during storage and transport, for example). The Glossary clarifies that, for the Convention, the intention referred to is not only a subjective and internal desire on the part of the exporter or generator of the waste; it should be inferred from facts and circumstances relative to the material, including results of behaviour by holders that allow for forecasts to be made, as well as more objective features such as the existence of a contract for the disposal of such a substance or object, for example (UNEP, 2016).

The Glossary also underlines that all of the circumstances must be considered on a case-by-case basis in order to verify if there is an intention to dispose of the substance or object. Factors such as origin, destiny, appearance, obsolescence, insufficient functionality, insufficient protection for avoiding damage during transport, loading and unloading may indicate that a material is intended for disposal and therefore be relevant to an analysis of a material's configuration as waste (UNEP, 2016).

Finally, the third situation envisaged in the Convention's definition of waste is of the substances and objects whose disposal is a legal requirement. According to the Glossary, this situation reflects the fact that some substances and objects may be defined as waste in some states, but not in others. As an example, the Glossary specifies that waste derived from production processes which is considered as such in some countries due to the absence of the intention to produce it, may instead be considered to constitute a by-product in another country if it does not fall under the general legal definition of waste (UNEP, 2016).

Continuing with the concept of waste provided by the Basel Convention, the Glossary observes that the current text of the Convention does not make it clear when waste is no longer waste; it therefore lists several possibilities for a material to no longer be considered as waste, falling outside the Convention's scope of application. The first of these possibilities occurs when waste has been prepared for reuse. In order to be certain that used goods are, in fact, being transferred for reuse, the Glossary suggests verifying that the product is in perfect working order by means of a review, repair or cleaning. Although such operations are not currently listed in Annex IV, the Glossary states that the parties may recognize them internally as recovery operations, guaranteeing that the goods are suitable for reuse (UNEP, 2016).

The second possibility occurs when waste has been subjected to recycling, even if only to be transformed into a product which is different to the original, and if the operation has been finalized. Some of the recycling operations are listed in Annex IV. The last possibility listed by the Glossary is when waste is subjected to a recovery operation (UNEP, 2016).

Finally, the Glossary makes an important observation when highlighting that although some types of waste may have a significant economic value in the recovery industry, this value should not be considered in isolation as a suitable criterion for distinguishing between waste and non-waste, although the economic value of materials transferred may be a factor indicating how they should be categorized. If the value of a substance or object moved/transferred is significantly lower than that of equivalent goods, whether new or used, for example, or if a substance or object does not have any economic value, this may indicate that its owner intends to dispose of it, therefore rendering it waste (UNEP, 2016).

Based on clarifications made in the Glossary and approved by the COP13, it may be perceived that, unlike the literal definition of article 2 of the Basel Convention, for the purposes of applying this Convention waste is any material destined not only for disposal, but also for recovery, as long as the transboundary movement is made before the recovery process. Such clarifications presumably render the Basel Convention's definition of waste significantly closer to the concept proposed by the current European Waste Framework Directive. It is, however, important to highlight once more that this interpretation given by the Glossary is not binding as it only serves as a guide for the party states, who can choose whether or not to follow it.

In terms of the OECD Decision currently in force, although it may be observed that the majority of the basic terms used have been harmonized with those provided by the Basel Convention, the OECD also believed it necessary to make small modifications for the sake of clarity. The OECD Decision C(2001)107/Final (Chapter 11, A, 1) defined waste as substances or objects that:

i are disposed of or recovered;

ii are intended to be disposed of or recovered; or

iii are required to be disposed of or recovered according to the provisions of national law.

Just like the European Waste Framework Directive and under the Basel Convention, the OECD concept of waste is also based on the 'destiny' of the material. However, unlike the Basel Convention, which only uses the term 'disposal', the OECD makes an explicit distinction between 'disposal' and 'recovery' (see Table 4.3). The OECD highlights that, although its concept of waste encompasses both materials destined for disposal and recovery, the scope of Decision C(2001)107/Final is restricted to waste destined for recovery, leaving aside that which is destined for disposal to be legally controlled by other instruments, such as the Basel Convention.

The concept of waste therefore may seem deceptively simple at first glance, but a slightly more detailed analysis reveals the ambiguities and

TABLE 4.3 Legal concept of waste according to the legislation selected

	Brazilian NSWP	EU Framework Directive	Basel Convention	OECD decision
Legal concept of waste	All materials, substances, objects or goods discarded as a result of human activities in society, whose final destination occurs, is proposed to occur or is required to occur. It can be in solid or semi-solid state, in gas state contained in receptacles, or in liquid state whose specific qualities render their release into the public sewage network or water bodies unviable, or thus require technical or economically unviable solutions given the best technology available.	Any substances or objects that the holder discards or intends to, or is required to discard.	All substances or objects disposed of, or planned to be disposed of or that must be disposed of according to a legal requirement in the national law.	All substances or objects which i) are disposed of or recovered; ii) are intended to be disposed of or recovered; or iii) are required to be disposed of or recovered under the provisions of national law.

difficulties surrounding the concept and the complexity of the subject. Evidence of this are the various terms found in the regulations used across different levels of legislation to identify which materials may be considered to constitute waste. Terms such as 'discard' and 'get rid of', 'final destination' and 'recovery', 'final disposal' and 'elimination' are used to refer to similar actions, although this is not explicit in all legislative texts. Such ambiguities are reflected not only in the concept of waste, but, as shall be seen in the next section, also in their legal classifications, and principally in terms of the classification between hazardous and non-hazardous waste.

Legal classifications of waste

As has been demonstrated, the decision to categorize a material as waste or non-waste (thus rendering it a product, by-product or a material that is no longer waste) is often difficult and potentially highly controversial. The definition is, however, of fundamental importance to verifying whether a material should be subjected to the legislation governing waste or not. Another important and controversial element of the legal regulation of waste is its legal classification. As each country and trading area establish their own classifications, this complicates the processes of global management. In general terms, the most frequently used classifications in the various legal frameworks are those relative to its origins and those focused on how hazardous it is.

Although the importance of the possible further classifications of waste has been recognized, this book will focus on the classification in terms of its hazardousness, taking into account the methodological scope of the analysis of international regulations on the subject. Just as with the debate on the legal concept of waste, clear and precise classifications on how hazardous a particular form of waste is are also fundamental to the regulations' application, given that hazardous and non-hazardous waste are regulated differently across the various levels of legislation.

BRAZIL

In Brazil, the NSWP established the following two classifications in its text. In terms of origins, article 13, section I, divides waste into:

a household;

b produced in street cleaning;

c urban solids (merging the two previous types);

d produced by retail establishments and service providers;

e produced by basic public sanitation services;

f industrial;

g produced by health services;

h produced by civil construction;

i agrosilvopastoral;

j produced by transport services; and

k produced by mining activities.

In terms of hazardousness, the NSWP classifies waste as: a) hazardous; and b) non-hazardous. According to this law, hazardous waste is:

> that which, due to being inflammable, corrosive, reactive, toxic, pathogenic, carcinogenic, teratogenic and mutagenic constitutes a significant risk to public health or environmental quality, according to the law, regulation or technical standard.
>
> SOURCE Brazil, 2010

Non-hazardous waste, on the other hand, is any type which does not fall under this definition (article 13, section II, items 'a' and 'b').

The NSWP does not however define these above-mentioned hazardous characteristics. To this end, CONAMA published Resolution no. 452 in 2012 in order to regulate the control procedures for importing waste according to the Basel Convention, providing the base definitions for the classification of waste according to how hazardous it is. According to the Resolution, hazardous waste (class I) is all that contained in Annex I of the Basel Convention (categories of hazardous waste) – unless it does not possess any of the characteristics listed in Annex III of the Convention (list of hazardous characteristics) – as well as the waste listed in Annex II and VIII of the Convention (categories of waste that require special consideration and the list of hazardous waste) (article 2, section I).

Although it merely replicated the lists of types of hazardous waste and characteristics of the Basel Convention, the Resolution clarified that such lists may be expanded by means of assessment and deliberation on the part of CONAMA (article 3, single paragraph). Resolution no. 452 also includes the definition of non-hazardous waste, with this divided into inert waste (class IIB) and non-inert waste (class IIA). According to this regulation, non-hazardous inert waste is that which has been subjected to a dynamic and static contact with distilled or deionized water at room temperature, with none of its components solubilized at concentrations higher than the standards for drinking water, except in terms of the appearance, colour, turbidity, hardness and flavour, according to Annex G of ABNT NBR 10.004/2004 (article 2, section III). Non-hazardous non-inert waste, on the other hand, is any that does not fall under the definitions of waste of class I and IIA (hazardous and non-hazardous inert waste) (article 2, section II). Table 4.4 summarizes the various waste classifications.

EUROPEAN UNION

In the sphere of the European Community, the degree of detail and complexity in the legal classification of waste in terms of hazardousness is much greater. Here the classification is mainly made by two instruments: Annex III

TABLE 4.4 Classification of waste in terms of hazardousness in Brazil

Classification according to hazardousness in Brazil	Hazardous waste	Non-hazardous waste
NSWP	That which, due to being inflammable, corrosive, reactive, toxic, pathogenic, carcinogenic, teratogenic and mutagenic constitutes a significant risk to public health or environmental quality, according to the law, regulation or technical standard.	All that which does not fall under the definition of hazardous waste.
Conama Resolution	All that included in Annex I of the Basel Convention (categories of hazardous waste), unless it does not possess any of the characteristics listed in Annex III of the Convention (list of hazardous characteristics), as well as waste listed in Annex II and VIII of the Convention (categories of waste requiring special consideration and list of hazardous waste).	a) Non-hazardous inert waste: all that which has been subjected to dynamic and static contact with distilled or deionized water at room temperature, with none of its components solubilized at concentrations higher than the standards for drinking water, unless in terms of appearance, colour, turbidity, hardness and flavour. b) Non-hazardous non-inert: all that which does not fall under the definitions of waste in classes I and IIA (hazardous and non-hazardous inert).

of the European Waste Framework no. 2008/98/EC and the European Waste Catalogue (Decision no. 2000/532/EC, with later amendments, including the most recent made by Decision no. 2014/955/EU). Other legislative instruments used in the EU also specifically serve to support the classification of hazardous and non-hazardous waste, such as the EU Regulation on Classification, Labelling and Packaging for Hazardous Substances and Preparations (Regulation (EC) no. 1272/2008) and the Regulation on Persistent Organic Pollutants (POPs), Regulation (EC) no. 850/2004.

Apart from providing a definition of waste, Directive no. 2008/98/EC also includes a list of characteristics qualifying waste as hazardous. According to this Directive, hazardous waste is all that which possesses one or more of the 15 characteristics specified as hazardous in Annex III (article 3). All of these characteristics share one feature in common: that of generating hazardous effects on the environment and living beings in general, including humans.

The Directive highlights that hazardous waste must be regulated with rigorous specifications in order to prevent or limit its potential negative effects, whether on the environment or human health. It also expressly prohibits the reclassification of hazardous waste as non-hazardous by means of dilution or mixing to reduce the initial concentration of hazardous substances to values lower than those established in the definition of hazardousness (article 7).

Given the updates made to other regulations of reference and new scientific discoveries, in 2014 the need emerged to update and alter Annex III of the Framework Directive, which was made by Regulation (EU) no. 1357/2014 by the European Commission. The new text of Annex III defines hazardous materials as those which are explosive, oxidizing, inflammable, toxic, carcinogenic, corrosive, infectious or mutagenic, among others. Along with this list, the Annex also lists the definitions for each of the characteristics of hazardous waste, clarifying its peculiarities and differences. Directive no. 2008/98/EC does, however, highlight that the list of hazardous characteristics contained in Annex III must be used in conjunction with the classification system for hazardous waste contained in the EWC in order that the classification is harmonized within the sphere of the EU. This reflects the fact that the EWC is another instrument of reference for the classification of waste within the EU.

As mentioned in the previous section, the EWC was created in the 1990s and recently revised by Decision no. 2014/955/EU, with amendments made to its methodological section in order to align it with the updates to legislation on chemical products. Although it constitutes a non-exhaustive list that is liable to periodic review, the EWC has helped harmonize the classification, facilitating the control of many activities involved in the international management of waste, notably its transfer. The EWC thus provides a list of types of waste divided into 20 chapters, where they are classified according to origin (eg domestic, urban, industrial, hospital). Each of these chapters is divided into subchapters and entries. All 842 entries in the EWC are further classified into 'absolute hazardous entries', 'absolute non-hazardous entries' and 'mirror entries' (EC, 2015b).

The absolute entries of hazardous waste refer to waste that is always classified as hazardous, whether due to its innate constitution or due to being contaminated by substances considered to be hazardous. This category is always highlighted in the list by an asterisk used to indicate it is a hazardous waste. All waste indicated as hazardous on the EWC has at least one of the characteristics of hazardousness mentioned in Annex III of the European Waste Framework (EU, 2014). The absolute entries for non-hazardous waste refer to waste that is always classified as non-hazardous and therefore never indicated with asterisks. Finally, the mirror entries are applied to waste that may be classified either as hazardous or non-hazardous depending on the specific case. It means that these types of waste may be attributed codes for hazardous waste or codes for non-hazardous waste (EU, 2014).

The EWC classifies waste as hazardous by taking into account its origin, composition and, when necessary, the limit values of the concentration of hazardous substances in such waste. This list is binding in terms of the identification of waste classified as hazardous, ie all waste highlighted with an asterisk must necessarily be listed as an absolute entry for hazardous waste and classified as such (article 7, EU, 2008). In terms of waste which may be marked with codes for hazardous or non-hazardous waste, this should be listed as mirror entries and assessed on a case-by-case basis in order to confirm if it features the hazardous characteristics contained in the European Waste Framework Directive (EC, 2015b).

Furthermore, the fact that waste contains a substance classified as hazardous does not necessarily mean that the waste itself is. The classification of waste as hazardous depends on whether or not it contains the characteristics of hazardous waste listed in Annex III of the European Waste Framework Directive, as well as the concentration of hazardous substances in the waste (EU, 2014). In case of doubt, the precautionary principle should be employed, and the waste classed as hazardous. This logic does not apply to waste included in the absolute entries for hazardous waste, in which the concentration of hazardous substances does not influence its classification as hazardous (EC, 2015b).

In classifying waste according to how hazardous it is, many characteristics listed in Annex III of the European Waste Framework Directive refer to criteria provided by Regulation (EC) no. 1272/2008 (see Table 4.5). This Regulation adapts the EU system of classification, labelling and packaging for substances and chemical mixtures to the Globally Harmonized System (GHS), thus determining the criteria for substances to be classified as hazardous (EC, 2015b). Regulation (EU) no. 1272/2008 is not directly applied to waste, as it does not consider it as a substance, mixture or article (under the scope of the standard). It does, however, indirectly influence the instruments governing waste in classifying it according to how hazardous it is. Many of the mirror entries, for example, directly refer to hazardous substances, which must be assessed according to the criteria established by Regulation (EC) no. 1272/2008 (EC, 2015b).

All of this means that the EU system for classifying waste in terms of hazardousness is rather complex and detailed. It is also very important that waste classification is complete and compatible with the EWC in order to allow business operators and competent authorities to decide if the waste is hazardous or not. In light of the difficulties which may emerge from this complex system, the European Commission developed a guide (and thus a non-binding document) to assist the Member States in the process of defining

TABLE 4.5 Classification of waste in terms of hazardousness in the EU

EU classification according to hazardousness	Hazardous waste	Non-hazardous waste
Framework Directive	Any that has 1 or more of the 15 characteristics specified as hazardous in Annex III.	Does not have any of the characteristics specified in Annex III.
EWC	'Absolute hazardous entries': waste indicated to be hazardous. Waste that possesses at least one of the characteristics listed in Annex III; 'Mirror entries': waste that may be hazardous if characteristics listed in Annex III are present.	'Absolute non-hazardous entries': waste that is always classified as non-hazardous, and therefore not highlighted with an asterisk; 'Mirror entries': waste that may be non-hazardous if characteristics listed in Annex III are not present.

and classifying hazardous waste in terms of the application of the EU standards for waste. Based on the legal framework for waste in the EU, this guide unveils this complex web of regulations, establishing a general step-by-step process for identification and classification.

The Guide initially highlights that for a material to be classified as hazardous or non-hazardous waste in the EU, it must be confirmed if the material in question is considered waste according to the European Waste Framework Directive and not excluded from its scope of application. This initial verification is the basic precondition for determining the applicability of the general legislation on waste and the EWC for the material in question (EC, 2015b). After confirming the condition of the waste and the applicability of the waste law, the next step is to identify which entry the waste falls under. If it is classed as an absolute entry, whether hazardous or non-hazardous, the classification is considered to be complete. If the appropriate entry for said waste is a mirror entry, additional analyses must be undertaken to complete the classification (EC, 2015b).

There are two alternatives for mirror entries: mirror entries for hazardous waste and mirror entries for non-hazardous waste. In order to complete the classification and definition for the two options, the waste must be analysed for hazardous components, for whether or not it presents one or more of the characteristics of hazardous waste provided in the Waste Framework Directive and/or for any of the relevant POPs (persistent organic pollutants). It is only then that it may be decided if the waste is hazardous or non-hazardous (EC, 2015b). Within the EU, therefore, knowledge of the composition of waste is of fundamental importance to classifying waste as hazardous. It is important to highlight that, whenever there is uncertainty over the composition of waste and it is impossible to provide further clarifications, such waste will be considered hazardous (EC, 2015b).

When identifying the composition of waste, it must be checked if it contains any substance chemically classed as hazardous, according to Regulation (EC) no. 1272/2008. If the result is positive, two methods are suggested by legislation for classifying mirror-entry waste as hazardous:

i calculation to check if such a hazardous substance exceeds the limits imposed by the legislation (concentration of the hazardous substance); and

ii test to check if the waste in question has the hazardousness characteristics or not (the trial methods outlined under Regulation (EC) no. 440/2000 should be used or other methods which are internationally recognized).

SOURCE EU, 2014

FIGURE 4.1 Road map for classifying hazardous waste in the EU

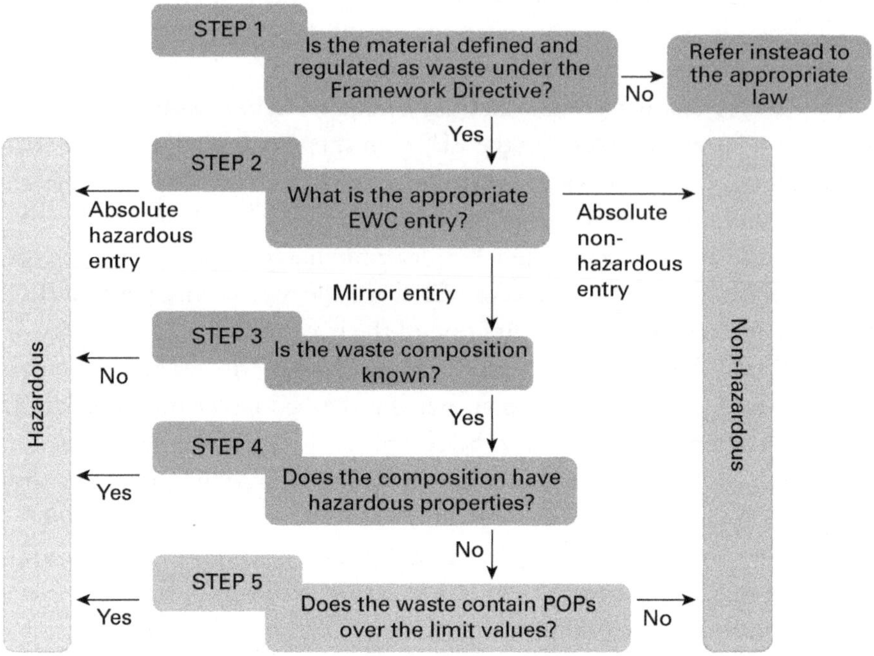

SOURCE Adapted by the author from EC, 2015b

If, during this stage, the waste displays concentrations of hazardous substances above the limit permitted and/or the hazardousness characteristics, it must be classed as hazardous. Otherwise, a final analysis should be carried out in order to check if it contains any of the POPs listed in the Annex to the EWC and if these substances exceed the limit values established by the POPs Regulation. If no POPs are present or if the amount falls under the limit concentration established by law, the mirror-entry waste may ultimately be classed as non-hazardous waste. Otherwise, it should be classified as hazardous (EU, 2014). In short, the path described in Figure 4.1 should be followed in order to classify waste according to its hazardousness in the EU.

INTERNATIONAL LAW

In the international sphere, the Basel Convention defines hazardous waste as that which, when subjected to transboundary movements:

a belongs to any of the categories contained in Annex I of the Convention (except when it does not display any of the characteristics outlined in Annex III); and

b although not covered by paragraph (a), is defined or considered hazardous waste by the national legislation in the country of export, import or transit that is a Party of the Convention (article 1).

Annex I of the Basel Convention lists the categories of waste that must be controlled by the Convention under the terms of article 1 mentioned above. The Convention does, however, make it clear that the fact that a waste material is listed in Annex I does not impede the use of Annex III to demonstrate the possibly non-hazardous nature of the waste. In other words, the Convention leaves open the possibility for proving a type of waste is not hazardous even if it is listed in Annex I, as long as it does not display any of the characteristics listed in Annex III.

According to Annex I, the categories of waste materials whose flow should be controlled are clinical, pharmaceutical, explosive in nature, featuring mineral oils, organic solvents, wood preservative chemicals, among others. The Annex also lists metallic, organic and non-organic materials, which, if part of the composition of waste subject to transboundary movements, shall be considered as hazardous: examples include copper compounds, zinc compounds, arsenic, mercury, non-organic fluoride compounds, acidic solutions, asbestos, organic phosphorous compounds, and organic solvents.

In 1998, an amendment was added to the Basel Convention in order to facilitate the Convention's application, introducing Annex VIII with its extensive list of waste considered to be hazardous by the Convention under the terms of article 1. Annex IX was also added, listing waste materials not covered by article 1 and therefore not considered to constitute hazardous waste. The amendment made did, however, highlight that such materials' inclusion in the lists provided in Annex VIII and IX did not mean that in some cases a demonstration could be made as to their non-hazardous nature (referring to Annex VIII) or hazardous nature (referring to Annex IX), based on Annex III.

In general terms, Annex VIII classifies the following waste as hazardous:

i metallic waste or any which contains metal (A1);

ii waste mainly composed of non-organic elements, but which may contain metal or organic materials (A2);

iii waste mainly composed of organic elements, however which may contain metals and non-organic materials (A3); and

iv waste that may contain both non-organic and organic elements (A4).

Annex IX provides a relatively extensive list of waste, which is not theoretically considered to be hazardous by the Convention. Article 1 of the Convention does not apply to the waste listed on Annex IX, rendering them therefore outside its scope, except if they contain any of the materials listed in Annex I in an amount which makes them display any of the hazardousness characteristics contained in Annex III.

Examples of waste not considered to be hazardous under the Basel Convention are:

a battery waste, except for those made with lead, cadmium or mercury;

b electronic assemblies consisting only of metals or metal alloys;

c electric or electronic assemblies (including printed circuit boards, electronic components and wires) destined for direct reuse (the reuse may include repair, refurbishment or updating of the material, but not large-scale reassembly) and that which is not destined for recycling or final disposal;

d waste from vehicles at the end of their lives that does not contain liquids or any other hazardous components;

e non-dispersible waste derived from mining activities; and

f pneumatic waste, as long as it is not destined for final disposal.

Finally, Annex III lists the characteristics qualifying waste as hazardous, including:

- explosive;
- inflammable liquid;
- inflammable solid;
- oxidizing;
- poisonous;
- infectious;
- corrosive;
- toxic and
- ecotoxic.

Along with each of them, the Annex also provides a definition. It must be highlighted that Annex III, upon listing the characteristics that render a substance or waste hazardous, also observes that the potential for hazardousness posed by certain types of waste is not wholly known and documented. It also declares that no tests are available to quantitatively establish that such hazards do not exist.

Annex III therefore highlights the need for greater investment into research in order to develop means of characterizing potential hazards to human beings and/or the environment derived from such waste. Finally, it also demonstrates that some standardized tests have been undertaken on pure substances and materials and that many countries have developed national tests which may be applied to the materials listed in Annex I in order to facilitate decisions on whether or not they present any of the characteristics listed in Annex III.

Table 4.6 provides a summary of the classification of waste under the Convention.

Finally, the OECD Decision defines hazardous waste as that which belongs to any of the categories contained in Appendix 1 of the instrument, unless it does not possess any of the characteristics contained in Appendix 2 (see Table 4.7 for a summary). The Decision also states that

TABLE 4.6 Classification of waste in terms of hazardousness under the Basel Convention

	Hazardous waste	Non-hazardous waste
Classification in terms of hazardousness under the Basel Convention	a) All that which belongs to any of the categories contained in Annex I of the Convention (Categories of hazardous waste and substances), except if it does not display any of the characteristics listed in Annex III (List of hazardous characteristics); and b) That which, although not covered by paragraph (a), is defined or considered to be hazardous waste by the national legislation of country of export, import or transit which is a party to the Convention. c) Waste listed in Annex VIII.	a) Not framed as hazardous according to article 1. b) Waste listed in Annex IX.

TABLE 4.7 Classification of waste in terms of hazardousness under the OECD Decision

	Hazardous waste	Non-hazardous waste
Classification of hazardous waste according to the OECD decision	All that which belongs to any of the categories contained in Appendix 1 of the instrument (Categories of hazardous waste), unless it does not possess any of the characteristics contained in Appendix 2 (list of hazardous characteristics); and All that which, although not covered by this general definition, is defined or considered to be hazardous under the national legislation in countries of export, import or transit of waste.	All that is not defined as hazardous.

waste not covered by this general definition, but that is defined or considered to be hazardous by national legislation in countries of waste export, import or transit, shall also be classified as hazardous (Chapter II, A.2). Appendix 1 of the OECD Decision lists the categories of waste to be controlled by this instrument, while Appendix 2 provides a list of the characteristics considered to be hazardous. Both Appendices of the Decision are, however, identical to Annex I and Annex III of the Basel Convention, respectively.

As has been demonstrated, although the concept of waste and its classification in terms of hazardousness may initially appear to involve objective and clear elements in the process of the juridicization of waste, the reality is that it is highly complex. Waste emerges as a legal problem in terms of both its concept and its classification. The challenges posed by the subject do not end there; the legal guidance on waste management is also permeated by uncertainties and ambiguities across the various levels of legislation. The next section therefore pays particular attention to the legal regimes on waste management in the regulatory instruments selected for study in this book, emphasizing several concepts, subjects and types of responsibility for management, legal controls for preventing poor management, the duty of best management, the legal waste hierarchy and the legal differences between recovery and disposal operations. Such an analysis will be of fundamental importance to the later assessment of the adequacy of cross-border waste transfer legislation to develop the best global waste management system.

Waste management: legal regimes

Concepts, subjects and legal controls of waste management

KEY CONCEPTS

Apart from key concepts, Brazil's NSWP also lists principles, objectives, instruments and directives on the management and handling of solid waste in the national sphere, as well as on the responsibilities of its producers and public authorities in this process (article 1). One interesting and innovative aspect of the Brazilian Law is the distinction made between the 'management' and 'handling' of solid waste. Law no. 12.305 of 2010 establishes the integrated management of waste as a management model to be promoted and implemented in the country (article 7, section VII). Article 3, section VI defines it as the:

> [...] set of actions aimed at the search for solutions for solid waste, in order to consider the political, economic, environmental, cultural and social dimensions, with social control and under the premise of sustainable development.
>
> SOURCE Brazil, 2010

Section X, on the other hand, defines solid waste handling as the:

> [...] set of actions exercised, whether directly or indirectly, in the stages of collection, transport, transhipment, treatment and environmentally-sound final destination for solid waste as well as final disposal of refuse, according to the municipal plan for the integrated management of solid waste or the handling plan for solid waste, demanded by means of this Law.
>
> SOURCE Brazil, 2010

The notion of management expressed in Brazil's NSWP is wider, and includes handling. Planned by the public authorities, 'management' refers to the general approach and objectives of the management of solid waste in a systemic fashion, encompassing the environmental, social, cultural, economic and technological variables and those related to public health (article 6, section III). 'Handling', for its part, is the responsibility of those who have generated the waste, referring to the specific actions that should be taken across all stages and operations in waste management. The two, are, however, interconnected, with management encompassing handling and handling following the general guidelines of management while materializing it (Leite, 2015).

In the European Union, article 3 of Directive no. 2008/98/EC defines waste management as 'the collection, transport, recovery and disposal of waste, including the supervision of such operations and the after-care of disposal sites, and including actions taken as a dealer or broker' (EU, 2008). The further EU Directives and Regulations on waste do not generally provide a definition of waste management and, if so, refer to the concept in the Framework Directive. There is therefore no distinction made between management and handling in the European Waste Framework Directive. By means of the definition given in this instrument, it is clear that its notion of management is similar to that of handling under the Brazilian NSWP, except for the fact that the latter does not refer to future actions to be taken after the closure of disposal sites in its concept of handling.

Under the Basel Convention, management refers to the 'collection, transport and disposal of hazardous waste and other wastes, including after-care of disposal sites' (article 2.2). For purposes of the Convention, environmentally-sound management of hazardous waste and other wastes means 'taking all practicable steps to ensure that hazardous waste and other wastes are managed in a manner which will protect human health and the environment against the adverse effects which may result from such wastes' (article 2.8). The OECD Decision does not, however, provide any definition of waste management for its scope of application.

This chapter makes a distinction between **poor**, **sound** and **best management**. This distinction has been developed in the literature and cannot be found in the legislation regulating the subject. The legal concept of 'sound' waste management, when it is expressly made by the law, refers to (and is restricted to) the prohibition of poor management. Hence the importance of making the conceptual distinction between poor, sound and best management.

The notion of management under the Basel Convention adopts the same approach as the European regulation, not explicitly including the recovery operation. However, the legal interpretation of the term 'disposal' used in the Convention reflects a shift towards the inclusion of recovery operations, what aligns the concept of waste management under the Convention with that of the European Union (see Table 4.8 for a summary). Although they list 'transport', none of the legal concepts of management highlighted here explicitly establish 'waste transfer' as a possible stage in the management, given that transport cannot be confused with transfer (or movement).

TABLE 4.8 Legal concept of waste management according to the legislation selected

	NSWP	EU	Basel	OECD
Legal concept of waste management	**Management:** set of actions aimed at searching for solutions for solid waste, considering the political, economic, environmental and cultural dimensions, with social control and under the premise of sustainable development. **Handling:** set of actions exercised, directly or indirectly, in the stages of collection, transport, trans-shipment, treatment and environmentally-sound final destination of solid waste as well as environmentally-sound final disposal of refuse.	The collection, transport and recovery and disposal of waste, including the supervision of these operations, the maintenance of disposal sites upon closure and the measures taken by dealers or brokers.	**Management:** the collection, transport and disposal of hazardous waste and other waste, including later care in the sites of disposal for the waste. **Environmentally-sound management:** taking the practical measures to guarantee that hazardous waste and other waste is managed in such a way to protect human health and the environment against the adverse effects which may result from such waste.	No definition.

LEGAL SUBJECTS AND RESPONSIBILITY SYSTEMS

In terms of the legal subjects involved in the management of solid waste, a consideration of the peculiarities and the scope of each of the regulatory instruments analysed reveals such instruments generally outline similar subjects and responsibilities. In Brazil, a principle fundamental to the implementation of the integrated management of solid waste provided by the NSWP is the 'shared responsibility for the life cycle of products' (article 6, section VII), defined in article 3, section XVII as the:

> [...] set of individualized and connected attributes among manufacturers, importers, distributors and dealers, consumers and providers of public urban cleaning services and those for handling solid waste, in order to minimize the volume of solid waste and refuse generated, as well as to reduce the impacts on human health and environmental quality derived from the life cycle of products, according to the terms of this Law.
>
> SOURCE Brazil, 2010

This model of responsibility demonstrates that, although it is a policy oriented towards waste management, the Brazilian NSWP holds a holistic and systemic vision of the problem, encompassing the entire life cycle of products and clarifying the objective of minimizing waste and the impact it generates. The shared responsibility for the life cycle of products is therefore applied before and after the generation of waste, in a preventive fashion, although it is also oriented towards the management and handling of waste to have already been produced. It imposes responsibilities on all those who are part of the chain of production and consumption of goods that result in waste in order to jointly fulfil the objectives of sound management and handling of waste laid out by the law (Leite, 2015).

In this light, the shared responsibility entailed by Brazil's NSWP therefore extends beyond the typical subjects of 'generators, holders and public authorities', attributing responsibilities in an individualized and connected fashion to manufacturers, importers, distributors, dealers, consumers and providers of public urban cleaning and waste handling services, tasking each of these actors with different duties (article 30).

An analysis of the law reveals that the greatest responsibilities in the management of solid waste have been conferred on the subjects linked to the productive sector, seemingly due to a logic reflecting the 'polluter pays principle'. These subjects' preventive responsibility hinges on such a principle, as it does on the principles of eco-efficiency, prevention, precaution and sustainable development. The responsibility of manufacturers, importers, distributors and dealers encompasses the whole life cycle of products, from cradle to cradle, starting from the ecological conception of the products (in order to prevent the production of waste) to the post-consumption collection and environmentally-sound final destination, always with the aim of reintroducing waste materials to the production cycle (Leite, 2015).

The reverse logistics systems mentioned by the Brazilian NSWP must be structured and implemented by manufacturers, importers, distributors and dealers. While dealers and distributors must store and return waste received from consumers to manufacturers and importers, the last two must provide an environmentally-sound final destination for the waste returned (article 33, paragraphs 5 and 6). The provider of public urban cleaning and waste handling services is responsible for establishing selective collection systems and for offering methods and organizing measures with the productive sector for reuse, recycling, reintroducing waste materials into the production cycle, composting and ensuring an environmentally-sound final destination for the waste. With a sector-wide agreement or term of engagement signed by the productive sector, the public authorities will carry out the activities they are tasked with by this instrument, according to the fees involved (article 36).

Finally, the consumer is positioned as a third subject under Brazil's NSWP. Consumers have the responsibility to properly package and make available their waste generated, whether organic, reusable or recyclable, for normal collection or return by means of the reverse logistics system (articles 33 and 35). The Brazilian law therefore adopts a wider-reaching perspective of the concept of waste management, which is reflected in the systemization of a unique regime of responsibility and of subjects encompassing both the preventive and the post-generation aspects of waste management.

Remaining with the Brazilian context, emphasis must be placed on the subjects listed under CONAMA Resolution no. 452 of 2012 and IBAMA Regulatory Instrument no. 12 of 2013. These are the legal instruments regulating the import of waste into the country (cross-border waste transfer to Brazil). Under the Resolution (article 2, section IX), the actor responsible for the final destination of waste is the 'natural or legal person, whether public or private, that performs activities encompassing the environmentally-sound final destination for solid waste' (article 2, section VII), as well as waste importers, defined as 'determiners of the final destination of waste or the third parties hired by them' (Brazil, 2010).

According to the Regulation, only natural or legal persons engaged in the determining of waste's final destination may import it, whether directly or by hiring third parties. Apart from these subjects, the IBAMA Regulatory Instrument also mentions the 'transporter of controlled waste' in section X of article 2, defining this as the 'natural or legal person, whether public or private, that performs activities encompassing the collection or transport of solid waste in any of the phases of the handling of this waste' (Brazil, 2010).

In the EU, the European Waste Framework Directive adopts a more restrictive perspective on the notion of waste management and establishes two distinct types of responsibility:

i the expanded producer responsibility, linked more to the preventive aspect of waste generation, exclusively linked to producers and encompassing the entire life cycle of products; and

ii the responsibility for waste management, applicable for management operations (those for handling, in the Brazilian concept).

Although they utilize different systems of responsibility, the subjects and the responsibilities outlined under Brazilian legislation and in the EU Directive are quite similar, with both adopting a systemic approach when encompassing the entire life cycle of products.

Initially, the Framework Directive provides definitions for some of the subjects mentioned throughout the text. The first of these is the producer of the waste, defined as 'anyone whose activities produce waste (original waste producer) or anyone who carries out pre-processing, mixing or other operations resulting in a change in the nature or composition of this waste' (article 3.5). The holder of waste, on the other hand, is the 'waste producer or the natural or legal person who is in possession of the waste' (article 3.6).

According to the Directive, a dealer is 'any undertaking which acts in the role of principal to purchase and subsequently sell waste, including such dealers who do not take physical possession of the waste' (article 3.7). And finally, a broker is defined by this legislation as 'any undertaking arranging the recovery or disposal of waste on behalf of others, including such brokers who do not take physical possession of the waste' (article 3.8) (EU, 2008).

With these definitions and the extended producer responsibility in mind, Directive no. 2008/98/EC highlights in its preamble that the introduction of this type of responsibility serves as one of the means for aligning the concept and production of goods in ecological terms, considering the efficient use of resources during the entire life cycle, also promoting their repair, reuse, dismantling and recycling. It therefore refers to a responsibility that seeks to promote the non-generation, reduction and recovery of waste materials.

In article 8, the Directive highlights that the Member States may take measures – whether legislative or not – to reinforce the preventive character of waste management, establishing the producer extended responsibility regime to individuals or collectives who are professionals in the development, manufacture, transformation, treatment, sale or import of products. It also highlights that the extended producer responsibility is applied without prejudice to the responsibility for the management of waste as laid out in article 15 and without prejudice to the specific legislation on products and waste streams. The responsibility for waste management under the EU Waste Framework Directive, on the other hand, prescribes that the Member States must take the measures necessary in order to guarantee that the initial producer or other holders are primarily responsible for the treatment of waste. This responsibility may be directly fulfilled or delegated to companies engaged in waste treatment operations (article 15).

Each Member State must specify the conditions of this responsibility and decide when the initial producer is responsible for the entire treatment chain and in which cases this responsibility may be shared or delegated among those intervening in the waste treatment chain (article 15). Therefore, although there is no direct mention in the text, the Waste Framework Directive leaves open the possibility for the Member States to establish a shared responsibility for waste management within their territories.

Finally, the Directive affirms that the Member States must impose sanctions that are effective, proportionate and dissuasive to those managing waste, namely the 'producers, holders, brokers, dealers, transporters and collectors, establishments or undertakings which carry out waste treatment operations and waste management schemes', in case these violate their management responsibilities outlined in the Directive. In such cases, the Member States must secure waste management systems and seek to recover the costs for non-compliance and possible recovery measures that may be necessary, without prejudice to Directive no. 2004/35/EC on environmental responsibility in terms of the prevention and remedying of environmental damage (Preface).

Also within the sphere of the EU is Regulation (EC) no. 1013/2006, governing cross-border waste transfer, just like the Basel Convention and the OECD Decision. All subjects mentioned in this instrument are therefore tasked with responsibilities linked to this type of operation, to this specific stage of waste management when it exists. All three instruments refer to two large groups of subjects, one composed of the states (countries) and the other of natural or legal persons, all involved in waste transfer. Each of these subjects has distinct responsibilities within this type of operation, to be further assessed in Chapter 5.

The subjects mentioned in the EU Regulation are the producer, holder, dealer, broker, collector, consignee, notifier, country of dispatch, country of destination and country of transit. The first four of these subjects are defined here in the same terms as those of the Waste Framework Directive. The collector is anyone who collects waste. The consignee is the person or company, under the jurisdiction of the country of destination, to whom the waste is transferred. The notifier is defined as any natural or legal person who intends to carry out a transfer of waste or intends to have a transfer of waste carried out and to whom the duty to notify is assigned (article 2).

According to article 2, if the waste transfer originates in an EU Member State, the notifier must be defined in terms of the hierarchy established by the Regulation, as follows:

- **i** original producer;
- **ii** the licensed new producer who carries out operations before the transfer;
- **iii** a licensed collector who, by means of various small quantities of the same types of waste collected from a wide variety of sources, brings the waste together for the transfer, which must be initiated from a single notified location;
- **iv** a registered dealer with written authorization from the original producer, new producer or licensed collector to act on their behalf as notifier;
- **v** a registered broker with written authorization from the original producer, new producer or licensed collector to act on behalf of the notifier; or, finally,
- **vi** if all of the aforementioned persons are unknown or insolvent, the holder of the waste.

If the transfer originates in a country which is not an EU Member State, the notifier is the natural or legal person designated by the legislation in the country of destination, or, in the absence of such a designation, the holder of the waste at the time when the export took place (article 2).

Regulation (EC) no. 1013/2006 (article 2) also mentions the countries involved in waste transfer operations as responsible subjects. The list is as follows:

- **i** country of dispatch – that in which the waste transfer initiated or is planned to initiate;

ii country of destination – to which waste is transferred or planned to be transferred, whether for recovery, disposal or loading prior to recovery or disposal in an area not under the national jurisdiction of any country; and

iii country of transit – which may be any country other than those of dispatch and destination, through which the waste transferred passes or is planned to pass.

In the international sphere, the Basel Convention lists subjects similar to those in Regulation (EC) no. 1013/2006 in terms of the states, however the natural or legal persons defined as subjects by the former may be distinguished from the terms of the latter. The subjects of the Convention are, generally speaking, the state of export, state of import, state of transit, exporter, importer, carrier, generator and disposer. The first three subjects listed here are defined under the Convention in a similar fashion as in the EU Regulation (article 2). In terms of the natural or legal persons listed as subjects under the Basel Convention, along with the exporter and importer, which are the persons that, under the jurisdiction of the states of export and import, undertake the business transaction of the transfer, are two further subjects: that which transports the waste during its transfer (carrier), as well as that to which the waste is transferred and which is responsible for its final destination (disposer). Finally, the generator is the subject whose activities produced the waste, or, if unknown, the natural or legal person who holds and/or controls such waste (article 2).

The OECD Decision uses similar subjects to the Basel Convention, with slight differences. In terms of the states, the country of export and country of transit mentioned under the Decision have exactly the same definition as in the Basel Convention. On the other hand, the country of import is defined in a simpler fashion as the Member State to which waste is transferred or is planned to be transferred.

In terms of the persons mentioned as subjects, the Basel Convention defines exporter and importer as those responsible for the acts of import and export, while the OECD Decision defines these as:

i the person who, under the jurisdiction of the country of export, initiates the waste transfer, or who has, at the moment at which the transfer is planned to begin, the possession or other legal forms of control over the waste; and

ii the person to whom, under the jurisdiction of the country of import, the possession or other legal forms of control over the waste has been conceded at the moment at which the waste is received in the country of import.

Unlike under the Basel Convention, the exporters and importers in the OECD Decision are not necessarily responsible for the business transaction of export and import but are the holders of waste at the moment of export and import. Due to this distinction, the OECD Decision lists a further subject, named the 'recognized trader'. This is the person who, under the jurisdiction of a member country and with the appropriate authorization from the countries involved and interested in the waste transfer, acts as the main buyer and subsequent vendor of the waste transferred.

According to the Decision, the recognized dealer may undertake and facilitate the cross-border waste transfers destined for recovery operations. This means that the OECD Decision recognizes that, apart from the waste holders at the beginning and the end of the cross-border waste transfer, there may also be a third subject, specifically involved in the business transaction of export and import. It must be highlighted that the person of the recognized trader may, in some cases, be confused with that of exporter or importer, under the terms of the definitions of this Decision.

Finally, the OECD Decision posits the generator as a subject part of waste management, defining it as any person to generate waste by means of their activities. This definition of generator, unlike the definition provided by the Basel Convention, excludes the holder, which is referred to as the 'exporter' and 'importer'. Furthermore, the OECD Decision does not mention the carrier and the person directly responsible for the recovery of the waste as subjects with specific responsibilities in the management of waste.

Table 4.9 provides a summary of how the various regulatory regimes deal with subjects and responsibilities for waste management.

THE PROHIBITION OF POOR WASTE MANAGEMENT AND LEGAL CONTROL INSTRUMENTS

Given the concepts of waste management and its responsible subjects under the regulatory instruments selected here and according to the stance taken by Aragão (2003), it appears that good waste management from the *post-factum* perspective (referred to as 'handling' in Brazil) generally requires the legal prevision and application of two types of duties issued to these subjects:

a negative duties, derived from the general prohibition of bad (or poor) waste management; and

b positive duties, resulting from the particular imperative to manage waste in the best way possible (the duty of best waste management).

TABLE 4.9 Subjects and responsibilities for waste management under the legislation selected

Subjects and responsibilities in waste management	NSWP	EU	Basel	OECD
Subjects	Productive sector (manufacturers, importers, distributors, dealers), consumers and public authorities; Determiner of waste's final destination, importer and transporter of waste.	Producer, holder, dealer, broker, transporter, collector; Consignee, notifier, country of dispatch, country of destination and country of transit.	State of export, state of import, state of transit, exporter, importer, transporter, generator and disposer.	State of export, state of import, state of transit, exporter, importer, generator, disposer, recognized dealer.
Responsibilities	Shared responsibility for the life cycle of products: applicable before and after the generation of waste in an individualized and chained format linking all subjects (including preventive and *post-factum* aspects).	Producer extended responsibility: oriented at the preventive aspect and at the productive sector. Responsibility for waste management: applicable after the generation of waste, related to management operations. Transfer regulation: see Chapter 5.	See Chapter 5.	See Chapter 5.

According to Aragão, prohibiting poor management is not synonymous with the duty towards best waste management, with the two duties instead complementing each other. That is to say, the actions/operations taken for waste destination may not be poor (from an environmental or sanitation point of view), but they may not be the best either, ie not the best possible management solutions from an environmental point of view (Aragão, 2003). In order to better understand the meaning of prohibiting poor management and the imperative to manage waste in the best way possible, reference must be made to the ecological existential minimum and the non-regression principles (Figure 4.2).

The principle of the 'ecological existential minimum' has emerged as an ecological re-interpretation of the concept of human dignity and imposes the right to receive sufficient materials to maintain a decent ecological baseline for all life forms. It therefore implies the duty to protect all processes, relations and goods indispensable to maintaining the minimum existential conditions from an ecological point of view (Leite, 2015). In this context, Ayala (2011) highlights that such a principle justifies the consideration of minimum environmental protection standards in light of existential risks which, based on this minimum, may be considered intolerable or unacceptable.

An ecological existential minimum thus demarcates the protection of an existential zone which must be maintained and reproduced, upon which there is no possibility for legislative/regulation changes typical of the democratic system. This establishes the link between the principle of ecological existential minimum and the principle of non-regression, protecting this

FIGURE 4.2 Progression from poor management to best management

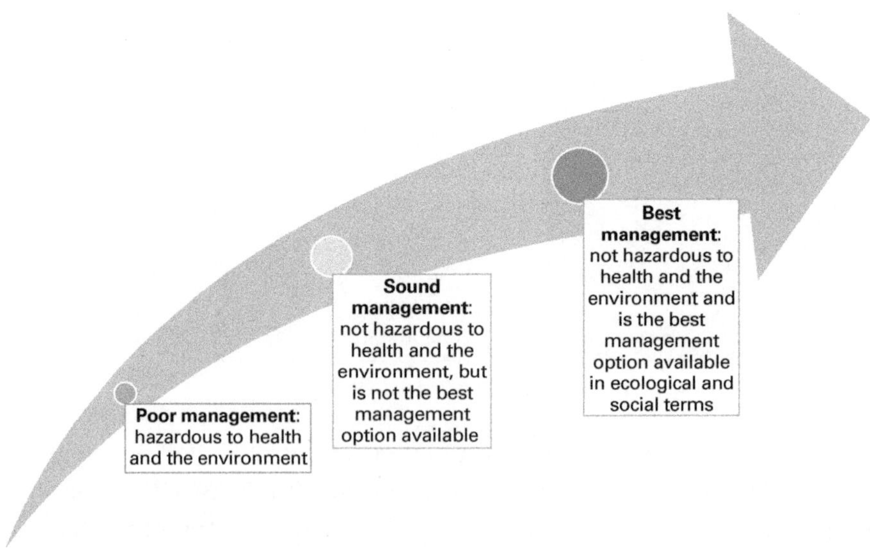

existential minimum against regressive initiatives which may threaten the ecological standards fundamental to existence (Leite, 2015).

If the application of the ecological existential minimum for the protection of natural resources is evident, such a principle is not as clear in the context of the generation and management of waste. However, it is impossible to conceive of an entirely protected existential zone without considering the socio-ecological impacts derived from waste. The principle has implications for the need to prevent waste generation, but also it is possible for minimum standards of environmental protection based on the ecological existential minimum to be established in light of the risks derived from waste management. As non-compliance with these minimum standards would characterize poor waste management, prohibiting this poor management may be deduced from the principle of the ecological existential minimum.

Aragão (2003) defines poor waste management as all and any operation related to waste that may constitute a danger to health or to the environment. Although the prohibition of poor waste management may be deduced from Chapter VI (articles 47, 48 and 49) of Brazil's NSWP, which specifically prohibits acts related to waste management, there is nothing establishing such a prohibition in wider terms. The European Waste Framework Directive, for its part, provides a general clause that consecrates the prohibition of poor waste management, stating in article 13 that:

> Member States shall take the necessary measures to ensure that waste management is carried out without endangering human health, without harming the environment and, in particular: a) without risk to water, air, soil, plants or animals; b) without causing a nuisance through noise or odours; and (c) without adversely affecting the countryside or places of special interest.
>
> SOURCE EU, 2008

The Framework Directive also mentions the possibility of establishing baseline technical standards for the activities of waste treatment, if proven that these will be beneficial to public health and the environment. Such norms must – among other objectives – focus on the main environmental impacts of waste treatment activities, preventing poor waste management prohibited under article 13 (article 27). These baseline standards provided by the Directive are no more than baseline environmental protection standards for waste management derived from the principle of the ecological existential minimum.

Regulation (EC) no. 1013/2006 and the Basel Convention both state that sound management of hazardous waste and other waste requires the adoption of all the necessary measures to guarantee the waste will be managed in such a way to protect human health and the environment against the adverse effects which may result from this type of waste. It may therefore be concluded that both instruments view management which fails to adopt these measures as poor waste management. Such instruments also affirm that, along with guaranteeing that all persons involved in waste management take the measures fundamental to preventing or minimizing the harm generated by hazardous and other waste, each state, member or party must prohibit the export and import of waste upon an analysis that it will not be managed in an environmentally-sound fashion. Finally, there is no mention in the OECD Decision that expressly prohibits poor management in a general sense.

According to Aragão (2003), poor waste management may be prevented by checking the reputation of the waste operator and the person in charge, and more importantly, how environmentally sound the chosen waste management option is. Such checks are essentially performed by means of administrative controls on the economic activities that produce and manage waste. Furthermore, the prevention of poor waste management may occur by means of diffuse controls, such as those resulting from public participation in the decision-making processes on environmental matters.

In terms of the more specific administrative controls legally governing waste management, the requirement to obtain prior authorization for waste management operations must be mentioned, as must the requirement to produce records and updates on its generation and management. Apart from this, the obtaining of an environmental licence for activities that produce and manage waste and the responsibility to provide assessments of the environmental impact of the management facilities are examples of administrative controls in terms of the wider legal environmental management applied to waste (Aragão, 2003).

On this subject, article 8 of Law no. 12.305/10 expressly listed the main instruments in Brazil's NSWP which must be used for implementing waste management policies and controls in the country. The NSWP's specific provisions for governing waste include:

- **a** inventories and the annual system for declaring solid waste (section II);
- **b** the National Information System on Solid Waste Management (section XI); and
- **c** the National Register of Hazardous Waste Operators (section XV).

Examples of administrative controls oriented towards the protection of the environment, as laid out in Article 8 of the NSWP, include (among others):

- **i** environmental, sanitary and agricultural monitoring and inspection (section V);
- **ii** environmental quality standards (section XVII, a);
- **iii** the assessment of environmental impacts (section XVII, d);
- **iv** licensing and the review of activities that are effectively or potentially polluting (section XVII, f);
- **v** the term of engagement and term of adjustment of conduct (or prosecution agreements) (section XVIII).

In the context of the EU, Directive no. 2008/98/EC also mentions several instruments for controlling the management of solid waste, such as:

- permits;
- registrations;
- inspections.

In terms of permits, the Directive states that Member States must demand that all establishments or companies planning to treat waste obtain a licence from the competent authority. According to the Directive, this permit must include content that specifies the types and quantities of waste that may be treated, the safety measures and precautions to be taken and the monitoring and control operations, among other demands (article 23).

The Registration applies to entities that are not subject to permits. In such cases, the competent authority in each Member State must maintain a record of: establishments or undertakings who professionally collect or transport waste; the dealers or brokers; and establishments or undertakings who benefit from exemptions in the permit requirements (article 26).

The Framework Directive also expressly mentions the need for establishments or undertakings engaged in waste collection, transport and treatment operations, as well as brokers, dealers and those who produce hazardous waste, to be subjected to periodic inspections by the competent authorities (article 34). Such subjects should also maintain a chronological record of the quantity, nature and origin of the waste, and, if relevant, the destination, frequency of collection, mode of transport and method of treatment, with this record to be made available to the competent authorities. Such demands may be made to both those managing hazardous and non-hazardous waste at the discretion of each Member State (article 35).

The IBAMA and CONAMA standardizations in Brazil, Regulation (EC) no. 1013/2006 in the EU, the Basel Convention and the OECD Decision also introduce several instruments specifically controlling the cross-border waste transfer, within their respective scopes, which will be further analysed in Chapter 5. Generally speaking, the Basel Convention provides instruments such as:

- authorizations for transporting and disposing of waste (article 4, 7, a);
- packaging and labels for hazardous waste transported (article 4, 7, b);
- monitoring for waste transferred by means of a movement document used from the beginning to the end of the transfer (article 4, 7, c).

The notification and prior written consent (article 6) and the responsibility to provide information between the Member States and the Secretariat of the Basel Convention (article 13) are outlined as instruments essential to the operations governed by the Convention. Regulation (EC) no. 1013/2006 also mentions prior notification and consent as fundamental instruments, as well as contracts and information documents. The main instruments in the OECD Decision are the notification and movement documents. A summary of the administrative controls under the various legislations can be found in Table 4.10.

It is interesting to observe that, although the explicit aim of Brazil's NSWP and the EU Waste Framework Directive is to adopt a systemic approach, seeking to preside over the entire life cycle of products, neither of these legislations mentions an analysis of the life cycle of products as an instrument of control against poor waste management. In truth, the responsibilities related to the preventive aspect of waste generation are not generally met with binding measures in these laws, approaching the subject more in the sense of 'encouraging' the productive sector to adopt more ecological concepts for its products. The laws governing the international transfer of waste, on the other hand, adopt a more restrictive approach on waste management and do not even mention an analysis of the life cycle of products.

As outlined by Aragão (2003), there is also potential for the exercise of a more diffuse form of control against poor solid waste management, with public participation a key instrument to this effect. The Aarhus Convention of 1998 on public participation in decision-making processes in environmental matters expressly mentions such a right in terms of waste management. Participation assured under the terms of the Aarhus Convention should assist

TABLE 4.10 Administrative controls for waste management under the legislation selected

	NSWP	EU	Basel	OECD
Administrative controls for waste management	Inventories; declarations for control systems; environmental, sanitary and agricultural monitoring and inspection; environmental quality standards; assessment of environmental impacts; licensing and review of activities; term of engagement; term of adjustments of conduct, among others.	Permits; registrations; inspections; notification; prior consent; contract; information documents, among others.	Authorization to transport and dispose of waste; document of the movement of the waste transferred; notification; prior written consent; responsibility to provide information, among others.	Notification and movement documents.

the decision-making process authorizing waste management activities, just as it should be carried out in a reasonable period and should be effective.

To this end, Aragão (2003) defends the need for prior, complete and timely information on the activity proposed, on the nature of the decision which may fall upon it and the public authority responsible for the decision, among other elements. Aragão also highlights the fact that possibilities for public participation are broader than those foreseen for the subjection of such activities and projects to an environmental impact assessment. It is thus important that public participation remains separate from such a procedure, adding to and counting as a further source of monitoring of the management of solid waste.

As demonstrated, Brazil's NSWP understands the integrated solid waste management as a set of actions oriented at a search for systemic solutions on solid waste, with social control and under the premise of sustainability. This means that solid waste management in Brazil should also be undertaken with the support of social control, which, according to the law, is expressed in article 3, section VI, in the following terms:

> [...] set of mechanisms and procedures that guarantee society information and participation in the processes of formulating, implementing and assessing public policies related to solid waste.

SOURCE Brazil, 2010

Law no. 12.305/10 also mentions society's right to information and social control as one of its guiding principles (article 6, section X). It thus ensures that plans for solid waste will be amply publicized, as well as promising social control in their formulation, implementation and functionality (article 14, single paragraph) by means of a process of social mobilization and participation, including the holding of public hearings and consultations (article 15, single paragraph).

The European Waste Framework Directive also makes express mention of the importance of public participation in solid waste management processes. It states that the Member States must ensure that, apart from the interested parties and authorities, the general public also has the opportunity to participate in the development of waste management plans and waste prevention programmes, also ensuring easy access to such plans once devised (article 31).

Regulation (EC) no. 1013/2006, on the other hand, states that the Member States are required, under the terms of the Aarhus Convention, to guarantee the provision of information relative to transfer notifications, as long as such information is not confidential under national or EU legislation, published in the appropriate media by the competent authorities (Preface and article 21). There is, however, no provision for public participation and social control mentioned in the international instruments of the Basel Convention and the OECD Decision (see Table 4.11).

Given the legal concepts of waste management, the types of responsibilities, the subjects and the provisions for administrative and diffuse control devised to prevent poor waste management under the legislation studied, the next and final section of this chapter will attempt to understand the juridicization of the positive duty of best management, as well as the legal hierarchies entailed in waste management.

The duty of best management and the legal waste hierarchies

IS SOUND MANAGEMENT ENOUGH? THE NEED TO SEEK THE BEST WASTE MANAGEMENT

In terms of the affirmative duty resulting from the imperative to manage waste in the best means possible – the duty of best waste management – it must once again be emphasized that this does not mean the opposite of the negative duty to prevent poor management. While the prohibition of poor management implies that such activities must not cause serious harm to soil, water, plants, animals and the environment as a whole, the duty of best management requires the use of the best possible solutions (Aragão, 2003).

TABLE 4.11 Diffuse control of waste management under the legislation selected

	NSWP	EU	Basel	OECD
Diffuse control of waste management	Management integrated with social control: set of mechanisms and procedures guaranteeing society information and participation in the processes formulating, implementing and assessing public policies related to solid waste. Ample publicizing of the contents of solid waste plans, as well as social control in their formulating, implementing and functionality, by means of a process of social mobilization and participation, including the holding of public hearings and consultations.	Member States must ensure that the general public has the opportunity to participate in the development of waste management plans and waste prevention programmes, also providing easy access to such plans once devised. Member States are required to guarantee that information on transfer notifications is published in the relevant media and by the competent authorities, as long as such information is not confidential under national or EU law.	No mention.	No mention.

If the ecological existential minimum principle assists in understanding the prohibition of poor waste management, the ecological legal 'principle of non-regression' provides key elements to an understanding of the duty of best management. The legal definition of the ecological principle of non-regression is based not only on a recognition of ecological existential minimum standards but also on the clause of progression and of improvement of the minimum standards already reached. The definition of the non-regression principle is therefore always fulfilled by means of two stages: recognizing the minimum environmental protection standards already achieved and then implementing progressive, permanent and uninterrupted efforts to provide socio-ecological quality of life for all life forms (Leite, 2015).

Directly linked to the ecological legal principle of non-regression is the obligation to use the 'best available technology' (BAT). BAT has emerged as an obligation in terms of results rather than means, ie within the range of technologies available to carry out a particular objective, eg solid waste management, use must be made of the technology which manages to achieve the objectives set by the regulatory standard. Although this is a complex and multi-faceted concept, the BAT concept is generally linked to the idea of environmental and economic efficiency oriented towards the prevention of the transfer of pollution between different environments (Silveira, 2016).

Of the regulatory instruments studied here, only the European Waste Framework Directive and Regulation (EC) no. 1013/2006 mention the use of BAT. The Framework Directive includes the need to use BAT at different points in the text and adopts the concept provided by the since-revoked Directive no. 96/61/EC on integrated prevention and control of pollution. In this sense, BAT in the context of this standard refers to:

> [...] the most effective and advanced stage in the development of activities and their methods of operation which indicate the practical suitability of particular techniques for providing in principle the basis for emission limit values designed to prevent and, where that is not practicable, generally to reduce emissions and the impact on the environment as a whole.
>
> SOURCE EU, 1996

The Directive also clarifies that 'best' refers to the most efficient techniques for achieving a high general level of environmental protection. 'Technique' (substituting the term 'technology') would encompass both the 'technology used and the way in which the installation is designed, built, maintained, operated and decommissioned'. Finally, the term 'available' would involve:

> [...] those (techniques) developed on a scale which allows implementation in the relevant industrial sector, under economically and technically viable conditions, taking into consideration the costs and advantages, whether or not the techniques are used or produced inside the Member State in question, as long as they are reasonably accessible to the operator (article 2).
>
> SOURCE EU, 1996

Although it does not mention the BAT concept in its text, Regulation (EC) no. 1013/2006 refers to the concept used in Directive no. 2006/12/EC by affirming that the Member States must guarantee that waste management facilities apply BAT according to the licence held by them, regardless of whether the situation involves waste transfer for disposal or recovery.

Articles 11 and 12 of the Regulation thus mention the possibility that the authorities responsible for approving the cross-border waste transfer may raise objections to its transfer based on the argument that the facilities to which the waste is destined do not use BAT, whether involving disposal or recovery.

In terms of waste management, it is therefore possible to infer that the duty of best management extends beyond the prohibition of poor management. It recognizes the baseline ecological standards to be respected by the waste management process, establishing the requirement for progressive efforts to be made towards improving the socio-ecological efficiency of these processes with the use of the best technology available. With the aim of ensuring sound waste management, most of the regulatory instruments in force for the juridicization of waste provide specific legal hierarchies for the alternatives available in waste management. If the general objectives of such legislation focus on promoting non-generation, reduction and recovery of waste materials, efforts to fulfil the duty of best waste management should be made to this effect.

WASTE HIERARCHY AND BEST LOCATION AS LEGAL TOOLS TO IDENTIFY THE BEST MANAGEMENT

The waste hierarchy is commonly described as an order of priority of at least three waste management options based on the presumed environmental impacts. The order of priority should contain at least three options in order not to be considered as a simple preference. There must be a real order to be followed, as well as a commitment to adhere to this order, or otherwise it cannot be considered a hierarchy (see Table 4.12 on page 252). The likely origin of the waste hierarchy is recognized in the literature as the result of prioritizing the reduction, recycling and reuse of hazardous waste over its treatment and disposal. Discussed first in the 1980s, the idea became popular and is currently taken as one of the instruments fundamental to achieving sustainable waste management (Ewijk *et al*, 2014).

According to Ewijk *et al* (2014), the original concerns culminating in the creation of a waste hierarchy may be summarized as the desire to deflect insofar as possible the sending of waste materials for disposal in landfill. It would appear that the migration of waste management based on disposal (final disposal) to management based on a waste hierarchy was the direct result of the triple waste crisis linked to costs, contamination and capacities first encountered in industrialized countries in the 1980s.

The waste hierarchy therefore seeks to identify various measures which may be applied to materials before and/or after they become waste.

The majority of the current regulations are strongly influenced by the waste hierarchy, generally recommending a priority order ranging from prevention to disposal as the final option (Gharfalkar *et al*, 2015). This is, however, another point of controversy that generates many difficulties of interpretation and implementation of juridicization of waste.

The waste hierarchy was introduced by the NSWP in Brazil to the national legal order, stating the order of priority according to six levels. Serving as the purpose and guiding thrust of Brazil's policy on waste, Law 12.305/10 states that the following order must be observed in the management and handling of waste:

1 non-generation;
2 reduction;
3 reuse;
4 recycling;
5 solid waste treatment;
6 environmentally-sound final disposal of refuse (article 7, section II and article 9).

With a more direct definition, the 'non-generation' of waste does not require greater interpretive clarifications. While the definitions of the other levels were provided for in article 3 of the law, the same attention was not paid to the levels of 'reduction' and 'solid waste treatment', which were not expressly defined by the Brazilian law.

Although not included as one of the levels in the order of the waste hierarchy, the NSWP highlights that technology oriented towards energy recovery from waste may be used in the management process, as long as its technical and environmental viability has been proven and that it is accompanied by a monitoring programme for the emission of toxic gases (article 9, para 1). Directly associated with the aim to follow the waste hierarchy provided by the NSWP are the objectives for reducing the volume and hazardousness of waste (section V) and the incentive to recycling industry (section VI) (Leite, 2015).

On a European level, Directive no. 75/442/EC originally stated in article 3 that the Member States should take the necessary measures to promote the prevention, recycling and transformation of waste, the obtaining of raw materials and eventual energy recovery, as well as any other method allowing for waste to be reused. They should also inform the European Commission as to any regulatory project related to the incentive to decrease, reuse, recycle or recover raw materials and energy from the waste.

It may thus be perceived that, although it did not establish a waste hierarchy to be followed by the Member States in waste management, the original text of Directive no. 75/442/EC established the Member States' duty to promote the reduction, reuse and recovery of waste materials. In 1991, the text of Directive no. 75/442/EEC was altered by Directive no. 91/156/EEC, which, while failing to establish a waste hierarchy in the terms defined here, provided further details on the Member States' duty to promote certain options in waste management. The altered text of article 3 of Directive no. 75/442/EEC stated that:

> Member States shall take appropriate measures to encourage: (a) firstly, the prevention or reduction of waste production and its harmfulness, in particular by: the development of clean technologies more sparing in their use of natural resources; the technical development and marketing of products designed so as to make no contribution or to make the smallest possible contribution, by the nature of their manufacture, use or final disposal, to increasing the amount or harmfulness of waste and pollution hazards; the development of appropriate techniques for the final disposal of dangerous substances contained in waste destined for recovery; (b) secondly: (i) the recovery of waste by means of recycling, reuse or reclamation or any other process with a view to extracting secondary raw materials, or (ii) the use of waste as a source of energy.
>
> SOURCE EU, 1991

In 1997, the European Council approved Resolution (97/C 76/01), publishing the Community Strategy for Waste Management (CSWM). Although the CSWM did not establish a waste hierarchy per se, it expressly stated that prevention in terms of waste generation and hazardous characteristics must be the top priority of any waste policy. It also invited the Member States and economic actors to pursue objectives of reducing waste quantities. Finally, it also insisted on the need to promote the exploitation of waste, particularly by means of the reuse, recycling, composting and energy recovery of waste in order to reduce disposal to a minimum (paragraphs 16, 18, 21 and 30).

Directive no. 2006/12/EC of 2006 repeated the amended text of Directive no. 75/442/EC with minor modifications which did little to alter the essence of the provisions. In its turn, affirming the need to develop a legal framework governing waste that included the entire life cycle of products, and not just waste already generated, Directive no. 2008/98/EC established in its preamble that any policy on waste must always aim to reduce the use of resources, promoting the practical application of the waste hierarchy.

Among other references to the importance of the implementation of a waste hierarchy, in its preamble the new Framework Directive clarified that such a hierarchy intends to establish an order of priorities constituting the best global environmental option, according to the legislation and policy on waste. It does, however, emphasize that in some specific cases this hierarchy can be modified if proved that certain waste streams would have a better destination following a different order.

Article 4 of Directive no. 2008/98/EC thus establishes the waste hierarchy to be followed by all Member States in the waste management process. It also states that such a hierarchy must be applied as a general principle of legislation and as part of the policy for preventing and managing waste in the EU. The order of priority explicitly given by the Framework Directive has five levels, in the following order:

1 prevention;
2 preparing for reuse;
3 recycling;
4 other types of recovery, eg energy recovery;
5 disposal.

Apart from establishing a waste hierarchy as a general legal principle for the first time, the Directive ushered in what Aragão (2003) refers to as a 'aprioristic hierarchy'. This means that although the hierarchy established by Directive no. 2008/98/EC must be seriously followed by the Member States in waste management, it is only a starting point for analyses, which must be made according to specific cases. If, upon analysis of the life cycle of certain streams of materials it may be proven that the alteration of the hierarchy initially proposed by the Directive is more advantageous in global ecological terms for a specific case, such a modification may be made as long as it is made in a complete and transparent fashion, respecting the rules on social control and public participation from the Member States in the planning of waste management.

Finally, article 4 highlights that any alteration to the order of the waste hierarchy must respect the general prohibition of poor waste management provided for in article 13. To this end, it must be made taking into account the general principles for environmental protection, precaution and sustainability; the technical workability, economic viability and resources protection, as well as the global environmental, public health and social impacts.

Unlike Brazil's NSWP, the European Waste Framework Directive defined all of the hierarchical levels provided. Despite this, the EU's waste hierarchy has been the target of criticism both in substantive and formal terms. In terms of the latter, criticism was initially directed at the fact that the measures of reuse and recovery, although defined by the Directive, did not serve as hierarchical levels of waste. In terms of prevention, the first hierarchical level of the Directive is defined as:

> [...] measures taken before a substance, material or product has become waste, that reduce: (a) the quantity of waste, including through the reuse of products or the extension of the life span of products; (b) the adverse impacts of the generated waste on the environment and human health; or (c) the content of harmful substances in materials and products (article 3.12).
>
> SOURCE EU, 2008

In defining 'prevention' as a measure designed to reduce the quantity of waste, the Directive demonstrates that its primary focus is on reduction rather than prevention. Another criticism directly related to the hierarchy's formal terms is that the Directive includes the 'reuse' of products as a measure on the level of 'prevention' and not as an autonomous level, which would be, according to Gharfalkar *et al* (2015), logically inserted after the 'preparation for reuse' level. On the same point, it may be perceived that the Framework Directive's use of the definitions of 'reuse' and 'preparation for reuse' refer respectively to 'non-waste' and 'waste'.

Although it is indisputable that 'preparation for reuse' applies to 'waste', 'reuse' is not only applicable to 'non-waste'. According to the Directive's definition of waste, it is clear that products in perfect conditions of use may be considered waste as long as the holder plans to get rid of them, even if they may be reused. According to Gharfalkar *et al* (2015), 'reuse' should be included in the waste hierarchy as an autonomous level and would be applicable to both 'non-waste' and 'waste'. Such an inclusion would also make it necessary to reassess the very definition of waste, as well as to provide clarifications on the possibilities of reuse operations.

Finally, Gharfalkar *et al* (2015) correctly criticize the fact that the Directive, just like Brazil's NSWP, uses 'recycling' instead of 'reprocessing' in its third hierarchical level. Although recycling is a type of reprocessing, it does not encompass all the types of reprocessing currently in existence, such as upcycling and downcycling. With that said, Gharfalkar *et al* believe that the ideal expression to use would be 'reprocessing', comprising the three sub-levels: upcycling, recycling and downcycling.

> **Upcycling** is the reprocessing of waste for use in other products of greater value, use or quality. **Recycling** is the reprocessing of waste for use in the same type of product before consumption, without losing the technical characteristics. **Downcycling** refers to the reprocessing of waste for use in products of a lower value, use or quality (Gharfalkar *et al*, 2015).

Although these are formal issues related to the definition and structure of the waste hierarchy, according to Gharfalkar *et al* (2015) they cause direct material impacts, reducing the effectiveness of the hierarchy as an instrument able to help improve the efficiency of the use of natural resources. On the other hand, authors such as Ewijk *et al* (2014), for example, raise questions in substantive terms on the EU's waste hierarchy, concluding that, although it meets the aim of preventing waste being sent to landfill, it is not enough to reduce the consumption of natural resources and the environmental and social impacts generated by them.

Finally, the texts of Regulation (EC) no. 1013/2006, the Basel Convention and the OECD Decision do not mention waste hierarchies (Table 4.12). Although they employ different terminology and concepts and have several distinctions in the order and levels of prevalence of options for waste management, the hierarchies legally prescribed across different territorial levels all share the *a priori* prevalence of recovery over disposal.

As Aragão (2009) highlights, serious doubts emerge in many situations as to the true nature of certain operations, as it is difficult to assess if they involve recovery or disposal. Such doubts are both legal and technical in

TABLE 4.12 Legal hierarchy of waste management under the legislation selected

	NSWP	EU	Basel	OECD
Legal hierarchy of waste	i) non-generation; ii) reduction; iii) reuse; iv) recycling; v) solid waste treatment; and vi) environmentally-sound final disposal of refuse.	i) prevention; ii) preparation for reuse; iii) recycling; iv) other types of recovery, such as energy recovery; and v) disposal.	No mention.	No mention.

nature, generating grave practical difficulties for the authorities in the decision-making process, for monitoring and judgement. The same difficulties also affect the very managers of waste themselves, given that the legal recovery and disposal regimes are significantly different.

Aragão believes that, in distinguishing between recovery and disposal, it is possible to affirm that if the operation aims to remove a substance from the economic cycle of materials, it takes on the nature of a disposal operation. It is instead a recovery operation if it is designed to maintain the substances in the same economic cycle. On the other hand, if an alleged recovery operation causes more harm to the environment or to public health than a possible disposal operation, the true nature of the former is nothing more than disposal (Aragão, 2003).

Brazil's NSWP does not mention criteria or even lists distinguishing recovery operations from waste disposal operations. Law no. 12.305 does not even use the terms 'recovery' and 'disposal'. As has been demonstrated, the term 'environmentally-sound final destination' generally encompasses both recovery and disposal operations, from specific operational standards for preventing harm or risks to public health and to safety and to minimize adverse environmental impacts (article 3, section VII). Of all the possible recovery operations, only recycling and reuse in general terms are addressed by the Brazilian NSWP. According to the Law (article 3, section XIV), the first refers to 'the process of transforming solid waste that involves altering its physical, physio-chemical or biological properties, with the aim of transforming it into inputs or new products' (Brazil, 2010).

Reuse is defined as the 'process of exploiting solid waste that does not involve its biological, physical or physio-chemical transformation' (article 3, section XVIII) (Brazil, 2010). The term 'environmentally-sound final disposal' refers to the specific operation of 'disposing' of waste in landfill (article 3, section VIII) (Brazil, 2010), therefore excluding other disposal operations – such as incineration, which is not even mentioned in Brazil's NSWP. It may thus be perceived that, apart from not using the terms 'recovery' and 'disposal', the Brazilian law also fails to provide entirely equivalent terms, effectively excluding some specific waste management operations as a result.

In the EU, waste 'treatment' encompasses all disposal and recovery operations (article 3). Article 3.19 of the Framework Directive defines 'disposal' as '[...] any operation which is not recovery even where the operation has as a secondary consequence the reclamation of substances or energy'. 'Recovery', for its part, is defined in article 3.15 as:

> [...] any operation the principal result of which is waste serving a useful purpose by replacing other materials which would otherwise have been used to fulfil a particular function, or waste being prepared to fulfil that function, in the plant or in the wider economy.
>
> SOURCE EU, 2008

It may therefore be perceived that the Framework Directive defines recovery and disposal as opposite operations and that, together these serve as the possibilities for waste treatment. The key criterion of distinction, under the terms of the Directive, is that waste, upon transformation, has a useful end, whether substituting other materials or not. Apart from such definitions, in the attempt to clarify and better distinguish such operations, the Directive provides non-exhaustive lists of what it considers to be recovery operations (Annex II) and disposal operations (Annex I). Regulation (EC) no. 1013/2006, on the other hand, does not feature any innovative stance in distinguishing between 'disposal' and 'recovery', only referring to the definitions and lists contained in the Framework Directive.

In the international sphere, the original text of the Basel Convention provides a definition of 'disposal' as 'any operation specified in Annex IV to this Convention' (article 2.4). Annex IV, on the other hand, mentions what it considers to be two categories of 'disposal'. Section A of the Annex thus lists the operations which do not allow resource recovery, recycling, reclamation, direct reuse or alternative uses, therefore encompassing all the disposal operations which take place in practice. Section B lists the operations which may result in the recovery, recycling, reclamation, direct reuse or alternative use of resources. This section thus encompasses all the recovery operations relative to materials that are legally defined or considered as hazardous waste (Annex IV).

It can therefore be seen just how confusing and inadequate the Basel Convention is in defining 'disposal'. From the outset, the Convention does not provide a specific definition of what it considers to constitute 'disposal', merely referring to the list in Annex IV. This means that such operations may only be classified if encountered in this list. In accessing such an Annex, however, it is clear that, apart from including both disposal operations and recovery operations in the general list of 'disposal', the Convention does not encompass all of the existing disposal and recovery operations, while also failing to explain if the list is exhaustive or merely provides examples.

Apart from causing serious interpretative problems for the application of the legislation, the imprecise nature of the distinction between disposal and recovery operations under the Basel Convention generates direct impacts on the very definition of waste. As has been demonstrated, unlike the European Framework Directive, which uses the expression 'discard', under the Convention materials are considered as waste when they have been disposed of or when there is a plan or legal requirement to do so. While the EU's use of the term 'discard' leaves open the possibility for a destination to both disposal and recovery, the confusion surrounding concepts in the Convention opens up space for multiple doubts and difficulties of application (see Table 4.13).

Identifying the problem, COP13 provided definitions of waste/non-waste and hazardous/non-hazardous waste as well as clarifications to assist party countries in their interpretation of the term 'disposal' in its Glossary of terms published in the 13th encounter. According to the Glossary, Section A on the list of 'disposal' operations in Annex IV refers to the category of 'final disposal', while Section B represents the 'recovery' category (UNEP, 2016).

Unlike final disposal operations, recovery operations use resources obtained from waste with useful benefits, both reintroducing them into the production process and recovering waste. The Glossary clarifies that the Convention views recovery operations as ones whose main result is the use of waste for a useful purpose, regardless of whether or not other materials are substituted (UNEP, 2016).

Such clarifications thus approximate the Convention's notion of 'disposal' and 'discard' of the European Framework Directive, while the notion of 'final disposal' is equivalent to the EU's notion of 'disposal'. The definition of 'recovery' adopted by the Glossary is also exactly the same as that given in the European Waste Framework Directive. The Glossary also highlights that the 'disposal' process may be carried out at more than one moment, including interim operations. The notions of recycling, repair, refurbishment, reuse and direct reuse are defined in the Glossary, which clarifies that, although the title of Section B of Annex IV refers to the 'operations which can result in direct reuse', the term 'recovery' does not include the notions of 'reuse' and 'direct reuse' (UNEP, 2016).

'Direct reuse' is that which occurs without any kind of repair or refurbishment. Despite the existence of recovery operations that may result in the 'reuse' of a waste material, once the waste has been recovered, it will no longer be considered as waste, with its reuse therefore no longer considered to constitute a recovery operation, as this only applies to waste. In other words, the Glossary considers operations carried out to prepare waste for reuse as 'recovery' (such as repair) but not reuse itself (UNEP, 2016).

TABLE 4.13 Conceptual differences between recovery and disposal under the legislation selected

	NSWP	EU	Basel	OECD
Recovery	Does not define recovery in general terms. Final destination: i) reuse (transformation without altering the physical, physio-chemical or biological properties); and ii) recycling (transformation altering such properties).	Any operation that renders waste useful for a purpose, upon transformation. Operations listed in Annex II of the Framework Directive.	Does not define recovery in general terms. Section B of Annex IV includes a list of recovery operations, referring to them as 'disposal' operations that may result in the recovery, recycling, reclamation, direct reuse or alternative use of resources.	Does not define recovery in general terms. Operations listed in Appendix 5.B.
Disposal	Does not define disposal in general terms. Final destination: ultimate disposal in landfill.	All operations which mean that transformed waste has no useful purpose (even if recuperated in terms of substances or energy as a secondary consequence). Operations listed in Annex I of the Framework Directive.	Does not define disposal in general terms. Any operation specified in Annex IV of the Convention. Section A: operations which do not allow for recovery, recycling, reclamation, direct reuse or alternative use of resources.	Does not define disposal in general terms. Operations listed in Appendix 5.A.

Finally, the OECD Decision does not provide any innovations on the distinction between disposal and recovery operations, nor does it offer specific definitions of such operations, merely referring to the lists provided in Appendices 5.A (disposal operations) and 5.B (recovery operations). The lists of operations provided by the OECD Decision, however, are identical to those contained in Annex IV of the Basel Convention.

Appendix 5.A highlights that the list provided aims to include all the disposal operations that occur in practice, regardless of whether or not they are suitable from an environmentally-sound perspective. Just like the Basel Convention, Appendix 5.B affirms that the list given must encompass all operations with respect to materials considered or legally defined as waste and that would be destined for disposal operations in another format.

Although the clarifications made in the Glossary of terms published in COP13 assist in an interpretation of the Basel Convention's confusing text, other difficulties of applying the legislation are generated in cases of cross-border waste transfer. As demonstrated, the OECD's new Decision was developed and approved in order to adapt to and complement the Basel Convention in a harmonious fashion, preventing the duplication of procedures. Therefore, the OECD Decision is applicable only to recovery operations, since the text of the Basel Convention gives the impression that the scope of its application is restricted to 'disposal'. According to the interpretation given by the Basel Convention Glossary, cross-border transfers of hazardous waste for recovery are governed both by the Convention and the OECD Decision. Cross-border transfers of non-hazardous waste for disposal, on the other hand, are not governed by any international instrument (see Table 4.14).

Apart from selecting the best management options according to the hierarchies provided by law, identifying the best location for waste recovery or disposal is also an essential element to just and sustainable management in ecological and social terms (Aragão, 2003). Under Brazil's NSWP, the only references made on this point are in articles 17 and 19 of Law no. 12.305/10, referring to the minimum content of state and municipal plans for the integrated management of solid waste.

TABLE 4.14 International legal regulation on transfers of different types of waste in terms of hazardousness for different final destinations

Cross-border transfer	Hazardous waste	Non-hazardous waste
For disposal	Basel Convention	Without international regulation
For recovery	Basel Convention and OECD Decision	OECD Decision

Both articles affirm that state and municipal plans must identify the most favourable areas and zones for this localization, according to the other urban planning instruments in force; especially the master plan, the ecological–economic zoning and the coastal zoning. For state plans, this information should refer to both sites for recovery and disposal, while in municipal plans, only sites for disposal. Municipal plans should identify possibilities for implementing consortium solutions or those shared with other municipalities, considering the proximity to the sites established (article 19, sections II and III). Brazil's NSWP does not provide any criteria or specific limits for the selection of sites for recovery and disposal of waste in the country.

> It must be stressed that although Brazil's NSWP makes such a provision, the majority of the country's states and municipalities are yet to develop or implement their integrated solid waste management plans.

In the EU, the Waste Framework Directive not only provides information on location criteria for identifying sites of future disposal or recovery operations as part of the minimum content of waste management plans to be developed by the competent authorities in the Member States (article 28, 3, d), but also establishes important principles directly impacting on the selection of the location in the waste management process. These are the principles of self-sufficiency and proximity which, in general terms, establish that the EU and the Member States should adopt all the measures necessary to become self-sufficient in managing waste within their territories, and that the recovery of residential urban waste and the disposal of waste in general produced in a certain area is performed in the closest treatment facilities (article 16). Such principles shall be examined in more detail in Chapter 6 of this book.

Apart from the waste management options and the location of where they are to be carried out, some management operations are specifically regulated, forming part of the juridicization (or legal regulation) of waste and its management. One of these operations is waste transfer, internationally governed by the Basel Convention and the OECD Decision; on a European level, by Regulation (EC) no. 1013/2006; and in Brazil, by the National Policy on Solid Waste, along with CONAMA Regulation no. 452 and IBAMA Regulatory Instrument no. 12 of 2013.

Given that various clauses in these legislations regulating cross-border waste transfer have a direct impact on the choice of a location for recovery and disposal operations in the waste management process, the analysis on the location element will be done in more detail in Chapter 5, researching the control procedures outlined in such instruments.

Final considerations

In carrying out a general and comparative study of the five regulatory instruments selected, the present chapter sought to demonstrate the complexity and the lack of harmony in the juridicization of waste across the various levels of global regulations. The very concept of waste has been revealed to be challenging from a legal perspective, due to the difficulty to define a material as waste or as non-waste (by-product, used good or ex-waste) in light of the ambiguities and uncertainties generated by the regulations governing the subject, which use different terms and concepts. This point is critical to defining the application of the legislations regulating waste, otherwise other legal instruments without the same attention to the socio-ecological aspect will instead be applied.

Generally speaking, interpretations made of the legal instruments have tended to consider both materials destined for disposal and recovery as waste. Legislation governing the subject also considers the three different situations in a relatively homogeneous fashion:

- when a material has already been discarded:
- when there is a plan to discard a material;
- when a material must be discarded due to a legal obligation.

Furthermore, legislators, mainly from the EU, have attempted to establish criteria to harmonize the different regulations distinguishing waste from by-products, as well as to clarify when a type of waste is no longer waste due to its subjection to recovery operations (end of the waste statute).

The classification of waste in terms of hazardousness is as challenging as framing materials as waste. The complexity inherent in the technical language of the lists of materials and potentially hazardous characteristics; the variety of materials; the unpredictable nature of the results of mixtures, alloys and the combination of different substances and materials; the impossibility of accurately predicting all the possible effects and current and future interactions of all the existing substances and materials and of those frequently emerging from new technological developments; and the very impossibility of developing exhaustive and definitive lists of waste and characteristics considered to be hazardous make this a very challenging subject. While Brazil's national legislation provides only general and brief concepts for the classification of waste in terms of hazardousness, EU legislation establishes a real classification system, whose complexity, detail and technical nature have been demonstrated. Although highly dependent on technical and specialized knowledge in its application, the European model shows great potential for providing more clarity in the classification of waste in terms of hazardousness, particularly due to its anticipation of a certain degree of discretion in the analysis of a specific case, when classification is not clear at the outset.

The international regulations on the subject have been studied by means of the two international instruments selected, revealing themselves to be less sophisticated than the European classification system. The description of the distinction made by the Basel Convention and by the OECD Decision between hazardous and non-hazardous waste (as well as between waste and non-waste) reveals how the boundary between the two is much hazier and complex than may initially be deduced. The consequences of the difficulty of classifying waste in terms of hazardousness are rather obvious, as waste classified as hazardous is more restrictively regulated than that which is non-hazardous.

It is not only the lists of hazardous and non-hazardous waste provided by the Basel Convention that may be questioned, but also the list containing the characteristics that qualify waste as hazardous. To give one example, until 2021 hard solid plastic will continue to be considered non-hazardous waste and is therefore not subject to the controls outlined by the Convention. However, recent discoveries and reports published on the enormous problem of pollution caused by plastics, notably of the oceans, and the impact on animals, ecosystems and even human health demonstrates that the inclusion of some plastics under the control of the Convention has been too slow and reactive (instead of proactive and preventive).

Similarly, the analysis of the juridicization of waste management revealed a lack of harmonization. In terms of the concept of management, it was revealed that Brazil's NSWP distinguishes between the concepts of management and handling, with the concept of handling given by the law similar to the concepts of management under the European legislation and the Basel

Convention. Both the Brazilian law and the European law demonstrate the adoption of a wider perspective on waste management, although with different approaches, reflected in the types of responsibilities outlined in the instruments. Apart from the absence of a definition of management in the OECD Decision, it was revealed that none of the legislations studied referred to the 'transfer' of waste as a stage (or operation) in waste management – a glaring omission.

In terms of subjects, the differences between the instruments became clear, as did those between the responsibilities outlined. Apart from working with different types of responsibilities, the Brazilian regulation (with the shared responsibility for the life cycle of a product) and the European regulation (with the expanded producer responsibility and the responsibility for the waste management) come together on the adoption of a more holistic perspective on the subject, encompassing both prevention and the *post factum* of waste generation. Unfortunately, the provisions on the preventive aspect focus more on encouraging the producers of waste, with no real binding power. Despite this, such an approach stands out when compared with the international legislation which adopts a more reductionist perspective, oriented towards managing waste already generated.

Both negative and positive obligations are derived from the responsibility for managing solid waste, such as prohibiting poor management and the imperative of best management. In terms of prohibiting poor management, it has been demonstrated that only the EU Regulation includes a general clause establishing such a prohibition, while the other regulatory instruments merely provide specific prohibitions or provisions derived from this negative obligation. It has also been shown that poor management may be prevented by, for example, means of administrative controls and more diffuse controls involving public participation.

In all of the administrative control instruments listed in the regulations analysed, there is an absence of specific instruments acting on the preventive aspect of waste management, for example, the product life cycle assessment, reflecting a more restrictive perspective of the waste legal regulation, focused on the *post factum* (when waste has already been generated). The instruments relative to waste transfer will be analysed in Chapter 5. In terms of diffuse control, the importance of social control and public participation in the Brazilian NSWP and European regulations stand out, unlike the international legislation, which does not even feature such a provision, thwarting the internationally recognized right of public participation in waste management processes.

> Under the Brazilian regulation, the product life cycle assessment appears in the article on the NSWP's objectives, in the sense of encouraging its implementation as one of the policy's objectives and not as a binding instrument of control in itself.

In terms of the duty of best management, it is clear that only the European legislation, albeit indirectly, addresses this obligation when referring to the need to use the best available technology. The legislation mentions the BAT at different points, including as a justification to objections made to transfer waste requests. Apart from such references, it may be seen that the duty of best management may also be deduced from the establishment of a waste management hierarchy.

As has been shown, such a hierarchy came to be established by the laws regulating waste as a form of establishing clearer strategies for overcoming the waste crisis, giving priority to recovery over disposal of waste materials. However, as demonstrated, this is a point of controversy, generating various difficulties of interpretation and implementation in the juridicization of waste.

While the international regulations do not provide a hierarchy for waste management, the Brazilian and European legislations established hierarchies that can be criticized. Generally speaking, there is a need to better clarify all the hierarchical levels, with the inclusion and exclusion of certain levels, as well as to alter the orders established. Finally, there is a complete absence of provisions on waste transfer as a hierarchical level for possible management options.

Furthermore, the analysis of the waste hierarchy also revealed the absence of a more harmonized understanding and conceptualization of the general notions of recovery and disposal, mainly in the international legal sphere. Such a problem generates serious practical doubts and difficulties for the competent authorities and for the managers of waste themselves in distinguishing between recovery and disposal operations.

In the international sphere, the confusion surrounding the concepts of these two terms has exacerbated the difficulty of framing materials as waste under the Basel Convention, as well as leading to the over-regulating of the cross-border transfers of hazardous waste for recovery and an under-regulating (or even absence) of the cross-border transfers of non-hazardous waste for disposal.

Finally, apart from defining the best option for waste management according to the establishment and fulfilment of a suitable waste hierarchy in social and ecological terms, it has been demonstrated that the defining of the location of the recovery or disposal of such waste is also a fundamental element in just and ecologically sustainable management. The international instruments do not refer to this element. In this light, the EU Regulation is more advanced than the other instruments studied, providing specific references to the principles of self-sufficiency and proximity of the generating source to the selection of waste management sites.

It is therefore evident why waste – apart from being an ecological problem (Chapter 1) and a problem of justice (Chapter 2) – may also be viewed as a legal problem. The complexity of the subject, the different concepts used, the terminology options and those structuring the legislations studied demonstrate how difficult it may be to apply and implement the legal instruments bringing the waste to the legal sphere. Such complexity has resulted in practical difficulties in the development of management systems that are socially just and ecologically appropriate, mainly in light of the current trend towards the 'globalization' of this management.

It may be affirmed that, unlike the Brazilian and European legislation, which, although susceptible to criticism, has made progress on the adoption of a more systemic and holistic perspective on the subject, the international legislation studied adopts a more reductionist and fragmented approach, hindering the approximation of the Law to its ontological bases. With such an understanding in mind, Chapter 5 will investigate in depth the legal regulation of cross-border waste transfer in itself, checking if such complexity is also reflected in this specific operation. The importance of this analysis will be revealed to lie in the fact that this operation is key to characterizing global waste management.

05

Legal regulation of cross-border waste transfer

Introduction

Chapter 1 of this book demonstrated that the international waste trade among various countries has undergone a drastic increase in recent decades, with a clear trend for waste streams flowing from the North to the South. This poses an important question: do such waste transfers represent waste trade or waste management? If such cross-border waste transfers are seen as *business* operations qualifying the international waste trade (or market), the legal regulations would accordingly adopt an economic-based perspective, with some external environmental restrictions in place at most. The focus of such a legal perspective would thus be that of a free market, reflecting the current dominant economic model.

If, however, these cross-border transfers are understood as *global waste management* operations, the legal regulations would adopt a socio-ecological perspective. The focus of this perspective, therefore, would be on protecting the social and ecological systems involved. The study carried out in Chapter 4 reveals how a socio-ecological perspective has come to predominate in the legal regulation of waste in the Brazilian national context and that of the EU. In the international sphere, however, such a conclusion is not as straightforward, prompting the present investigation to place a special focus on the existing legal regulations for cross-border waste transfer in order to verify from which perspective such instruments have been devised.

Once again, the approach taken will range from the general to the more specific, meaning that, unlike in the previous chapter, the order of the investigation into the regulatory instruments selected will shift from the global to the local. The international regulatory instruments include more general rules on cross-border waste transfer, directly influencing the local instruments, which internalized many of the regulations agreed upon internationally. Such an analysis aims mainly to assess the sufficiency and suitability of these regulations from a socio-ecological point of view, deciding if they provide a legal framework for regulating global waste management or if they instead regulate international waste trade.

After providing a description and analysis of the international, EU and national regulatory instruments selected, the final considerations in this chapter will outline the main points identified in the study. This will be done in order to offer observations on possible alternatives for the sound international legal regulation of waste (to be made in the final chapter), mainly in terms of cross-border waste transfer, as an instrument for helping tackle the waste crisis.

International regulations: the Basel Convention and the OECD Decision

Basel Convention

The Basel Convention applies to the cross-border transfer of hazardous waste and 'other waste' destined for disposal or recovery among the party states. Radioactive waste is excluded from the Convention's sphere of application and is instead subject to other international control systems (articles 1 and 2). Under the Convention, 'transboundary movements' refers to movements involving hazardous waste and other waste from one area under the national jurisdiction of one state to or through an area under the national jurisdiction of another state; or to or through an area which does not fall under the national jurisdiction of any state, as long as at least two states are involved in the movement (article 2.3).

> Although the original text of the Convention only uses the expression **disposal**, as seen in Chapter 4, the expression is currently interpreted as referring to both disposal and recovery operations. This interpretation is adopted in this book.

According to its Secretariat, the Convention's regulatory framework is based on three pillars (see Figure 5.1):

i a global control system for the cross-border transfer of waste;

ii environmentally-sound waste management;

iii minimizing waste generation.

SOURCE UNEP, 2012b

FIGURE 5.1 Pillars of the Basel Convention

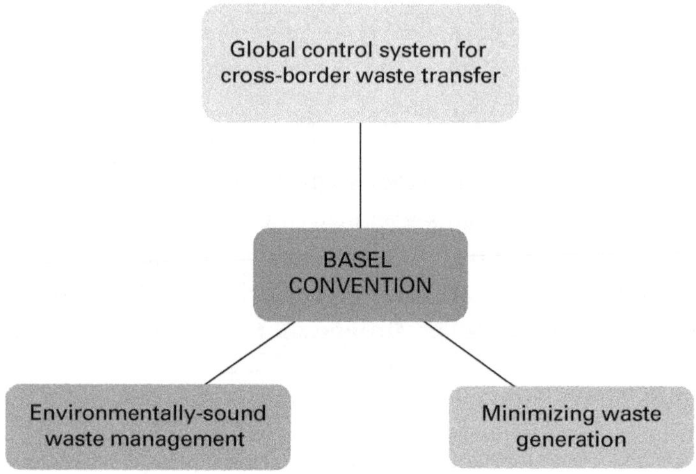

Although these pillars are highlighted in the Preamble to the Convention by means of various declarations recognizing the growing threats caused by waste and the basic need to reduce its generation, the current provisions in the instrument effectively focus on the first pillar, which is, in fact, considered by many authors to serve as more of a regulatory system for waste transfers than a control system per se.

The Convention establishes a set of general requirements imposed on the party states. Such requirements (article 4) encompass provisions related to the adoption of measures by the parties to:

- **a** reduce waste generation;
- **b** select the best location for waste treatment installations, according to the principle of self-sufficiency;
- **c** prevent the generation of pollution and environmental damage derived from waste;
- **d** reduce cross-border waste transfers;
- **e** not permit the export and prevent the import of waste that is not believed will be managed in an environmentally-sound fashion and/or whose transfer is prohibited by the states involved;
- **f** cooperate with the parties and other interested organizations.

Each party state must also (article 4):

g prohibit the export or import of waste to and from states that are not parties;

h prohibit waste export for disposal in an area south of 60° South latitude;

i guarantee waste transferred is accompanied by a movement document from the beginning to the end of the transfer;

j follow the international rules on packaging, labelling and transport for cross-border waste transfers;

k guarantee that waste transferred is managed in an environmentally-sound fashion.

Despite not featuring on the list of general requirements in article 4, the Convention also requires the party states to introduce appropriate measures to their national legislation to prevent and sanction illegal waste trade (article 9).

It is important to highlight that the Convention eases the general ban on the export and import of waste to and from states which are not parties when it lists the possibility of bilateral, multilateral and regional agreements emerging between parties and non-parties to allow for these movements. In such cases, the terms established by the agreements signed will apply, as long as they respect the general duty towards environmentally-sound waste management and are no more flexible than the Convention itself in terms of environmental protection, particularly with regard to developing countries (article 11).

The duty to transmit information stands out among the instrument's various provisions as a key element in the Convention's implementation. The parties are required to inform the other parties via the Convention Secretariat of:

i the ban on importing hazardous waste to their territory (article 4.1.a);

ii the limit or ban on exporting waste from their territory;

iii changes in the national definition of hazardous waste;

iv decisions taken not consenting, whether totally or partially, the import of waste;

v the occurrence of accidents during cross-border waste transfers that pose risks to human health and the environment (article 13).

Through the Secretariat, the party states must send the Conference of the Parties (COP) an annual report containing information which includes:

a the competent authority and focal points designated by them;

b data on the cross-border transfers of hazardous waste or other waste that they have participated in, such as, for example, the quantity of waste and other waste exported and imported; its categories, characteristics, destination, countries of transit and method of disposal as declared in the response to the notification;

c the bilateral, multilateral and regional agreements they have signed;

d the destination options for waste existing within their jurisdictional territories (article 13).

CRITERIA FOR WASTE TRANSFERS

Despite seeming somewhat out of place in the Convention text, the criteria for permitting cross-border waste transfers are also listed in the general requirements (Figure 5.2), as follows:

a that the state of export does not have the technical capacity, the necessary installations, capacity or treatment sites suitable for the disposal or recovery of the waste transferred in an environmentally-sound and efficient manner; or

b that the waste transferred is required as an input for recycling or recovery industries in the state of import; or

FIGURE 5.2 Criteria for obtaining permission for cross-border waste transfer under the Basel Convention

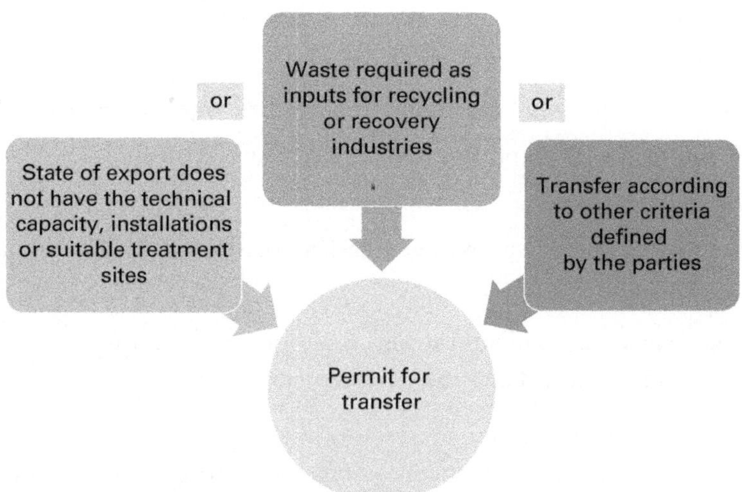

c that the cross-border waste transfer is performed in accordance with other criteria to be defined by the parties, as long as such criteria remain aligned with the Convention's objectives (article 4.9).

Finally, it must be mentioned that COP3 included a new article in the list of general requirements (article 4A), which affirms that parties which are also OECD Members must prohibit all cross-border transfers of hazardous waste destined for countries which are not OECD Members (considered as not developed countries). Such an amendment is known as the Ban Amendment, and, as shall be seen later on, has still not come into force years after its addition.

GENERAL PROCEDURE FOR WASTE TRANSFERS

Upon the fulfilment of at least one of the criteria permitting waste transfer, the Convention outlines a system with a single level of control over cross-border waste transfers regulated by this legal instrument (Figure 5.3). In general terms, the procedures to be followed under the system are founded on the instrument of prior informed consent from the recipient. The Convention (article 6) thus establishes that:

1 the state of export must provide written notification or demand this from the generator or exporter, by means of the competent authority, the competent authorities from all the states involved in the waste transfer process;

2 next, the state of import, as well as each state of transit, if any, must respond to the notifier in writing to consent to the transfer with or without conditions, to refuse it or requiring additional information on the transfer.

The information to be included in the notification of the request to export waste – to be written in a language accepted by the state of import – is listed in Annex VA to the Convention, with several examples as follows:

i reasons for the waste export;

ii designation and physical description of the waste;

iii estimated quantity in weight and volume;

iv method of disposal or recovery;

v the information transmitted to the exporter or generator by the consignee, based on which they concluded that there are no reasons to believe that the waste transferred shall not be managed in an environmentally-sound fashion according to the laws and regulations in force in the country of import.

Furthermore, the Convention affirms that the state of export may only allow waste transfers to be initiated when it has received written confirmation that:

a the notifier has received written consent from the state of import and the states of transit;

b the notifier has received confirmation from the state of import as to the existence of a contract between the exporter and the consignee, specifying the environmentally-sound management provided for the waste to be transferred (article 6).

The party states must therefore prohibit or refuse to permit the export of waste whenever there is no prior written consent from the state of import (article 4). On the other hand, the states of export may allow the waste transfer to be initiated even without prior consent in writing from the states of transit, as long as the latter do not demand prior consent for such transfers in their national legislation (article 6).

The waste movement document must be signed by all persons responsible for the transfer process, both upon receipt and delivery. Likewise, the consignee must inform the exporter and the competent authority in the state of export when they have received the waste transferred and upon finalizing the destination of the waste specified in the notification. If such information is not received by the state of export, the competent authority in the state of export or the exporter must notify the state of import. Finally, it must also be mentioned that any waste transfer made under the terms of the Convention must be covered by insurance or another type of financial guarantee (article 6).

FIGURE 5.3 General procedure for the cross-border waste transfer under the Basel Convention

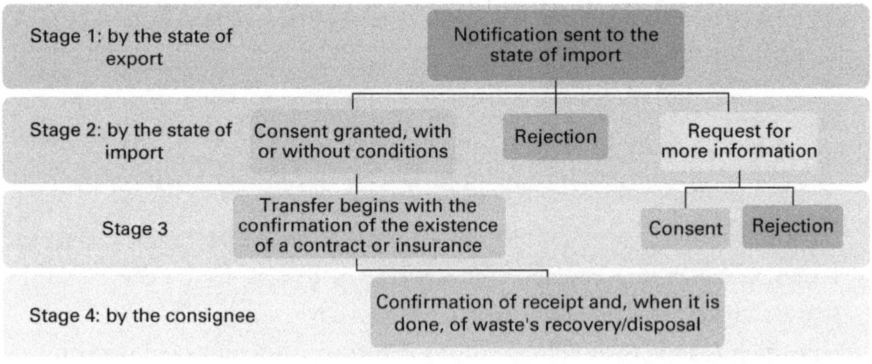

In case of disagreements over the classification of waste as hazardous or non-hazardous among the states involved in a transfer process, the demands and procedures outlined in article 6 of the Convention (provisions regulating cross-border waste transfers) shall be the responsibility, *mutatis mutandis*, of the state classifying the waste transferred as hazardous.

SOLUTIONS FOR POSSIBLE PROBLEMS AFFECTING WASTE TRANSFERS

The Convention also refers to problems which may occur in the processes of cross-border waste transfers, which may be classified in two large types of problems (Figure 5.4):

- **i** non-compliance with the terms of the waste transfer contract; and
- **ii** non-compliance with the Convention's key provisions.

The first of these refers to transfers that have obtained prior consent from the states involved, but that cannot be completed under the terms established in the contracts (article 8). For such cases, the Convention states that:

- **i** alternative agreements must be made for the environmentally-sound destination of such waste; or (if this is not possible)
- **ii** the state of export must make sure the waste is re-imported, allowing it to be returned to its territory by the exporter.

Although the Convention establishes a timeframe of 90 days for the meeting of this responsibility once the state of export has become aware of the event, such an instrument makes it possible for the states involved to agree on other timeframes. In such situations, the states of export and transit must not oppose, hinder or prevent the re-import of such waste (article 8).

The second set of problems to affect cross-border waste transfers is classified by the Convention as 'illegal traffic', specified in article 9 of the instrument. These refer to any transfer to have occurred:

- **a** without notifying all states involved; or
- **b** without prior consent from all states involved; or
- **c** with consent obtained by means of falsification, misrepresentation or fraud; or
- **d** that do not reflect the accompanying documents in material terms; or
- **e** that result in the deliberate disposal of the waste transferred, contravening the Convention and the general principles of international law.

It must be stressed that the parties consider the illegal waste traffic as a criminal practice (article 4.3).

If the illegal traffic is the result of conduct by the exporter or generator, the state of export must guarantee that the waste transferred is:

1 taken back by the exporter or generator, or, if necessary, by the state of export itself; or (if not possible)
2 redirected according to the provisions of the Convention, within 30 days of the date the state of export became aware of the occurrence or within the timeframe agreed upon between the states involved.

In such cases, the parties involved must not oppose, hinder or prevent the return of such waste to the state of export (article 9).

If, on the other hand, the illegal traffic is the result of conduct on the part of the importer or consignee, the state of import must guarantee that the importer or consignee give an environmentally-sound destination for the waste, or, if necessary, by the state of import itself within a period of 30 days or a timeframe agreed upon by the states involved as of the date the state of import became aware of the occurrence. The parties involved must cooperate, if necessary, in order to ensure an environmentally-sound destination for the waste (article 9).

FIGURE 5.4 Problems which may affect cross-border waste transfers and solutions provided under the Basel Convention

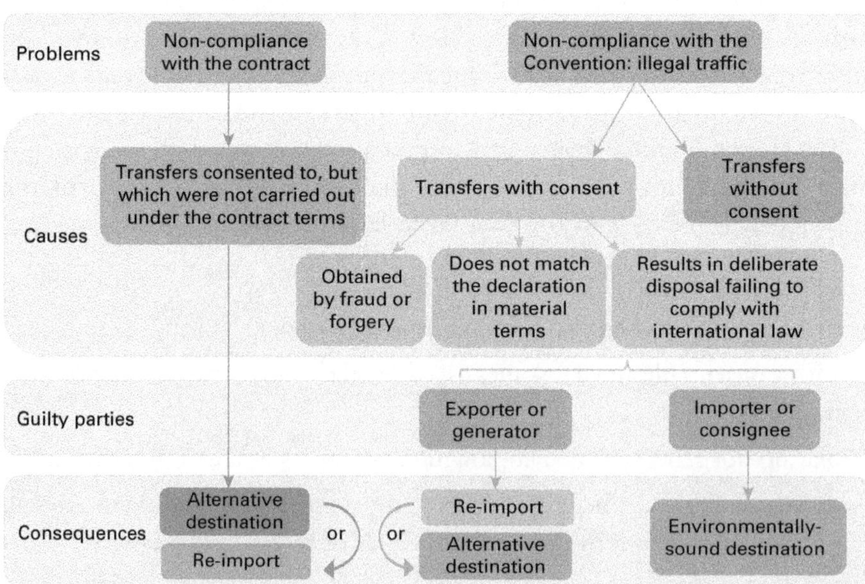

A CRITICAL ANALYSIS OF THE CONVENTION

With the provisions of the Basel Convention duly described, now is an appropriate time to reconsider the three pillars on which the Secretariat declares the regulatory framework of this international agreement is based: a global control system for the cross-border transfer of waste; environmentally-sound waste management; and minimizing waste generation.

Beginning with the third pillar of the attempt (minimize waste), although there are provisions to this end – including in the list of general requirements imposed on the party states – it can be seen that they are expressed as general objectives, without more specific, binding and procedural rules for the practical implementation of this pillar.

In this light, minimizing waste production is more a symbolic objective of what may be considered 'morally correct' in terms of waste governance, resulting in limited practical effects, if any at all. Furthermore, the very permission given for the cross-border transfer of hazardous and other waste ends up acting against reducing waste production – the result of the relative change in the exporting states' carrying capacity generated by the movement of the waste, and the transfer of the socio-ecological costs of their development processes to other territories.

As for the second pillar (environmentally-sound waste management), as may be ascertained in the description of the procedural regulation of cross-border waste transfers, the Convention lists a variety of requirements in order to guarantee that waste transferred is transported, treated and directed in an environmentally-sound fashion, marking a breakthrough in international legislation on waste transfer. Although such demands may appear adequate at first glance, they are however undermined by two key flaws which limit their application. These may seem formal flaws by nature, as they are linked to management and environmentally-sound management concepts (both outlined in Chapter 4). However, such flaws generate material consequences in the Convention's application.

The first flaw refers to the concept of waste management. It is interesting to note that the Basel Convention provides declarations in its Preamble and in article 4 (General Obligations) including cross-border waste transfer as part of the management process, affirming that each state must take the necessary measures to guarantee that such management is carried out in a consistent fashion, in order to protect human health and the environment, regardless of the location to which the waste is destined. The adoption of such a systemic perspective would be appropriate and advanced, even in comparison to the EU's current Waste Framework Directive.

However, an analysis of the Convention's text reveals that there is not a single provision or instrument in place to specifically implement this perspective. Such declarations are, in fact, isolated and are not reflected in the rest of the Convention, not even in the definitions of waste management and environmentally-sound waste management. In failing to include waste transfer as a stage in its management, the Convention effectively impedes the undertaking of a more holistic analysis in the decision-making process on the best destination for the waste in question.

All of the references made in the Convention text on reducing waste generation and transfer, and on the 'best location' and use of the 'self-sufficiency' principle for waste destination go unheeded due to the lack of inclusion of waste transfer as part of the management process. The very criteria for permitting cross-border transfers reflect this exclusion. As demonstrated, such criteria, in addition to being presented in alternatives (they do not all need to be applied together), are highly flexible and have a strong economic base. There is no doubt that the criteria for the existence of a demand for such waste as inputs for industry in the state of import is intrinsically economic, even if it does entail environmental benefits (such as a reduction in the extraction of new natural resources).

The criterion of the absence of technical capacities, installations, sites and spatial capacities to treat hazardous waste and other wastes in the state of export may initially seem to be based on environmental concerns. It does, however, still reflect an economic base, given that such absences result from a lack of investment in the field of waste management, a lack of international cooperation for technology transfer, and the development of economic activities which produce waste that cannot be managed in the same country. This means that the production of such waste is permitted in order to benefit the national economy without the infrastructure in place to deal with these negative external effects of the production process, which end up being transferred to other states.

Finally, the criterion allowing the party states to make waste transfers according to other criteria to be defined relativizes and flexibilizes the two former criteria, opening up opportunities for any and all influence of economic and political power among the states involved in such transactions.

The second intrinsic flaw limiting the application of the pillar of environmentally-sound management is derived from its very concept. As demonstrated in Chapter 4, the Convention defines environmentally-sound management as that which involves the taking of practical measures to protect human health and the environment against the adverse effects which may result from such waste. What the Convention treats as environmentally-sound management is thus any management not considered to constitute poor management, which is harmful to the environment and public health. However, the negative duties derived from the general ban on poor waste management are only one type of legal duties involved in an adequate waste management.

There are also positive duties resulting from the imperative to manage the waste in the best way possible and it is this very duty which is absent from the Convention text. Generally speaking, although the Convention attempts to prevent poor management of waste destined for transfer, it does not mention the need for the waste to be managed in the best way possible. The duty of best management should in fact serve as a criterion essential to the decision-making process on waste transfer, revealing the approach adopted by the Convention to be insufficient and economicist.

Furthermore, Anand (2004) describes the definition of environmentally-sound waste management provided by the Convention as too malleable, allowing for non-compliance with the prohibition of poor management. The Convention's affirmation that environmentally-sound management is that in which 'all practical measures are taken' seems to Anand to make it easy for those responsible for waste management to violate this requirement, resulting in poor management justified by the fact that the necessary measures have been taken by them.

Finally, the pillar of implementing a global system controlling waste transfer undoubtedly appears as the focus of the Convention's current text, which dedicates most of its provisions to addressing this subject by means of general requirements and specific procedures. Much criticism has been voiced of the Convention's handling of the subject, beginning with the definitions of 'waste' and 'hazardous waste', which, as seen in Chapter 4, have sparked confusion, doubts and discrepancies in the regulation's implementation.

Another relevant criticism made of the Basel Convention is that negotiations among the party states did not lead them to approve the ban on hazardous waste transfer, not even to developing countries, only to regulate such transactions. According to Anand (2004), developed countries considered (and still consider) waste export as a matter of free trade, arguing that such a ban would affront individual autonomy, the free market and the freedom to enter into contracts.

The costs of banning hazardous waste transfer from developed countries would also be high, given the elevated costs involved with managing and disposing of waste in these territories, added to the many legal restrictions and political pressures (as seen in Chapter 1). Such countries therefore came to adopt a double standard, given that they have high environmental quality standards, but benefit from the political and structural fragility and the low environmental standards in many of the developing countries to which they forward their waste. Developing countries, on the other hand, with an economically and politically weak position in comparison to developed countries, as well as fragile governmental and legal structures, had no alternative but to accept the regulation policy instead of an outright ban. That is to say, the reason that most developed countries did not approve the ban was mostly economic (Anand, 2004).

Although developing countries pushed from the beginning for a ban on the transfer of hazardous waste because of the lack of infrastructure and suitable resources for treating, disposing of and recovering waste within their territories, and although in theory they negotiated from a position of formal 'equality' with developed countries, their wishes were not granted and institutionalized by the Convention, which provided neither a stricter regulation, nor did it approve the ban, thus favouring the exporting countries of the North (Anand, 2004).

Despite the Basel Convention's apparent success in reaching an agreement between developed and developing countries, in actual fact it promoted the developed countries' agenda, and prevented developing countries from any significant participation in the process of outlining the regulation, thus constituting misrepresentation and socio-ecological injustice. According to Anand (2004), the Northern exporting states, at the time, took the position that they would accept nothing more than a regulation, and that if Southern countries did not agree to this, there would be no ban and no control over waste transfer at all. Anand (2004) claims that the incorporation of criteria of 'justice' in negotiation processes involving developed and developing countries, based on the criteria of responsibility, vulnerability and capacity would result in better cooperation among such countries and, therefore, a better international agreement on hazardous waste (Anand, 2004).

By the time the Convention came into force in 1992, almost 90 countries had already banned the import of hazardous waste by means of national laws and agreements. During this period, several progressive European countries such as Switzerland, Norway, Sweden and Denmark joined developing countries and several from Eastern Europe to push for a ban on the export of waste to developing countries. Along with NGOs such as Greenpeace, these countries sparked an international movement in favour of a ban, which had practical results on the subsequent Conference of the Parties (Anand, 2004).

At COP1 (1992), developing countries requested a complete ban on the transfer of hazardous waste from industrialized countries to developing countries. Two years later, COP2 (1994) approved the ban on the export of hazardous waste from OECD countries to non-OECD Members by means of Decision no. II/12, which was only adopted at COP3 (1995) as an amendment to the Convention by means of Decision no. III/1. The Ban Amendment prohibits all OECD state members, the European Community and Liechtenstein to transfer hazardous waste covered by the Convention to non-Member States of these communities (considered to be developing countries), regardless of whether the waste is destined for disposal or recovery.

Anand (2004) also states that the approval of the ban as an amendment to the Convention was received as an international benchmark in justice for socio-ecological issues. There are still many threats to the ban, however, mainly from countries wielding great political and economic power. Proof of this is the fact that many exporting countries have yet to ratify the ban to this day, such as the USA, Australia and Canada, and, even 25 years after the ban's approval, the amendment has still to enter into force due to the different interpretations of the provision on amendments of the Convention (article 17).

So far (February 2020), 98 Parties have ratified, accepted or approved this Amendment. For more information on the status of the ratifications in the Ban Amendment, see: www.basel.int/Countries/StatusofRatifications/BanAmendment

According to the Secretariat of the Basel Convention, the party states have different interpretations of article 17 of the Convention, with the majority considering it ambiguous. After many meetings held without reaching an agreement, the COP9 President invited the parties to create conditions for allowing the ban to enter into force, including initiatives led by the party countries themselves. In the wake of this declaration, Indonesia and Switzerland announced their interest in leading an initiative with this objective (mentioned in Chapter 4).

COP10 saw the adoption of Decision no. 10/3 on the initiative between these two countries to improve the Convention's effectiveness. Apart from clarifications on the terminology used by the Convention outlined in Chapter 4 (Glossary of Terms), section A of the initiative tackles the entry into force of the Ban Amendment. Although COP10 reached an agreement on article 17, such interpretation only applies for future amendment proposals, excluding the Ban Amendment. Unfortunately, therefore, such a decision merely affirmed the need for the parties and the Secretariat to continue to encourage and help other parties to ratify the amendment with no official conclusion about how many ratifications would be enough to implement the Ban Amendment (UNEP, 2011b).

With the ban not in force, the system regulating cross-border waste transfer originally approved at the Basel Convention remains intact. The description of the Convention provisions reveal that such a system is centrally founded on the instrument of prior informed consent. Such a regulation model, inspired by the principles of 'freedom of choice', 'autonomy' and 'sovereignty', remains the target of criticism. The first of such criticisms refers to the vulnerability in which developing countries may find themselves in light of the political and economic power held by developed countries. Developing countries face competitive disadvantages in the global economy and great external debts that effectively push for the acceptance of hazardous waste in their territories in exchange for immediate financial gains. This model thus presupposes a degree of relative equality among the contracting countries, which does not exist in reality (Anand, 2004).

Another criticism levelled at the regulation founded on prior informed consent is related to the presumptions of socio-ecological justice. In the description provided above, the decision-making process on cross-border waste transfer was revealed to be conducted between states, without any specific provision for public participation and information. There are also no proposals for a more systemic analysis of the impacts that such transfers generate in the ecosystems of the countries of import. This system does not take into account the fact that a state government – by means of the competent authority – accepting shipments of waste to its territory may not necessarily represent the best interests of the communities affected or ecological protection in the country. That is to say, although a system based on prior informed consent may feature positive elements for the control of cross-border waste transfers, its current format has the potential to generate socio-ecological injustices for present and future generations, both of humans and non-humans, in terms of maldistribution, misrecognition and misrepresentation.

Among other elements, criticism has also been directed at the fact that the regulation system does not consider that various developing countries are highly deficient in regulatory frameworks for environmental protection, in governmental and inspection structures, as well as lacking the scientific and technological knowledge to make well-informed decisions (Anand, 2004).

Finally, the Convention foresees two types of problems affecting cross-border waste transfers: those of contractual non-compliance and those of non-compliance with the Convention's key procedures. It is worth noting that, in terms of waste transfers which do not meet the terms of the contract, the Convention logic is to seek alternative agreements for the waste destination in the state of import, and, if not possible, to re-import such waste to the state of export. It must be noted that such a provision may mean that the best destination for waste transferred is not given, taking into account that this is not the priority set by the regulation.

Non-compliance with the Convention is characteristic of illegal traffic. In this case, when the responsibility for the non-compliance lies with the importer or the consignee the same logic for non-compliance of the contract of waste transfer must be applied. Ideally, such cases and those of contractual non-compliance would be better tackled with the use of a comparative socio-ecological analysis in order to inform the decision taken on the best destination for the waste. Furthermore, it must be highlighted that if the importer or the consignee cannot provide an environmentally-sound destination for the waste imported, such a responsibility must be assumed by the state of import, with the onus therefore falling on society as a whole – a typical case of the socialization of negative environmental externalities. If the blame lies with the exporter or the generator, the inverse logic is applied, with the state of export required to re-import the waste, or, if not possible, sign alternative agreements for the destination of such waste.

Even in including these provisions, the original text of the Basel Convention does not mention any responsibility assumed to assist in the implementation and compliance with its regulations. Article 12 mentions that the parties must cooperate in order to adopt, if possible, a protocol establishing the appropriate rules and procedures in the field of responsibility and compensation for damage. This prompts a further criticism, given that the absence of a regime of responsibility in force makes non-compliance with the Convention advantageous for waste exporters, making the Convention's implementation more difficult. In light of the problem, in 1993 a process of debate on the development of a Basel Protocol on Liability and Compensation was initiated. In 1999, the Protocol was finally adopted by COP5.

The objective of the Protocol is to establish a regime of responsibility, as well as suitable and immediate compensations for damage resulting from cross-border transfers of hazardous waste and other wastes, including incidents to have occurred in cases of illegal traffic. The document specifies who should be held financially responsible in incidents throughout the transfer process, from the dispatch of the waste to its final disposal (UNEP, 1999). Some 20 years after its adoption, however, the Protocol on Liability and Compensation – a document fundamental to better implementation of the Convention – has yet to come into force, due to the low number of ratifications. To enter into force, the document requires at least 20 ratifications, acceptances, confirmations or formal approvals. By February 2020 only 12 parties had agreed with these, Brazil, the EU and the USA not among them.

> For more information on the status of the ratifications of the Protocol on Liability and Compensation, see: www.basel.int/Countries/StatusofRatifications/TheProtocol

Despite prohibiting the transboundary movement of waste to non-party countries, the Convention created an exception in order to allow for cases featuring bilateral, multilateral or regional agreements between the countries. Criticism of this exception has been levelled at the potential abuse of such agreements, jeopardizing the very existence of the regulation outlined in the Convention (Anand, 2004).

Having described the general provisions of the Basel Convention and its main limitations and weaknesses from a socio-ecological point of view, the following section will provide an analysis of the OECD Decision, which is also an international instrument, although applicable only to transfers among developed countries. Although having a more limited scope of application, the regulation made by the OECD Decision, just like that of the Basel Convention, inspired the European and Brazilian regulations on the subject, and is therefore a relevant object of study.

OECD Decision

Given the controversies over cross-border waste transfers, the OECD adopted decisions with the aim of guiding Member States in their regulations proposing a control system for such transfers. With the majority of the OECD Member States adopting and signing the Basel Convention, including the European Union, the OECD resolved to maintain the Decision on the subject in force (Decision C(92)39/FINAL), believing itself to be providing a highly valuable framework for Member States for the control of cross-border waste transfers destined for recovery operations in a fashion that was environmentally sound and economically efficient (Preamble).

The OECD, however, revised the Decision in order to improve and harmonize its requirements and procedures with those of the Basel Convention. The OECD Decision was thus maintained as a multilateral agreement among the Member States that, according to article 11 of the Basel Convention, renders the Convention provisions non-applicable to cross-border waste transfers taking place within the agreement's scope of application (OECD Decision), as long as they are compatible with the environmentally-sound management of waste as defined in the Basel Convention (reflecting the ban on poor management) (Preamble).

The OECD Decision restricts its sphere of application to the cross-border transfers of waste destined for recovery operations within the OECD area (see Chapter 1). 'Transboundary movement' (here named cross-border transfer) is defined as 'any movement of wastes from an area under the national jurisdiction of a Member country to an area under the national jurisdiction of another Member country' (Chapter 11.A). The OECD Decision therefore only applies to waste transfers carried out between Member States and for the purpose of recovery operations.

CRITERIA FOR WASTE TRANSFERS

As criteria (or conditions) for permitting cross-border waste transfers regulated by this agreement, the Decision (Chapter II.B.1) states that:

1. the waste must be destined for recovery operations in installations which shall recover the waste in a way which is defined as environmentally sound according to the national laws, regulations and practices to which the installations are subject;
2. all persons involved in any contract or agreement for the cross-border waste transfer destined for recovery operations must possess the appropriate legal status to do so, according to the national legislation and regulations;
3. the cross-border waste transfers must take place under the terms of the applicable international agreements on transport; and
4. any transit of waste transferred performed through a country which is not an OECD Member must be subjected to the international legislation and all applicable national laws and regulations.

With such criteria fulfilled (see Figure 5.5), the OECD Decision outlines a system featuring two levels of control for these transfers (Figure 5.6):

1. the green control procedure; and
2. the amber control procedure.

FIGURE 5.5 General criteria for granting permission for cross-border waste transfers under the OECD Decision

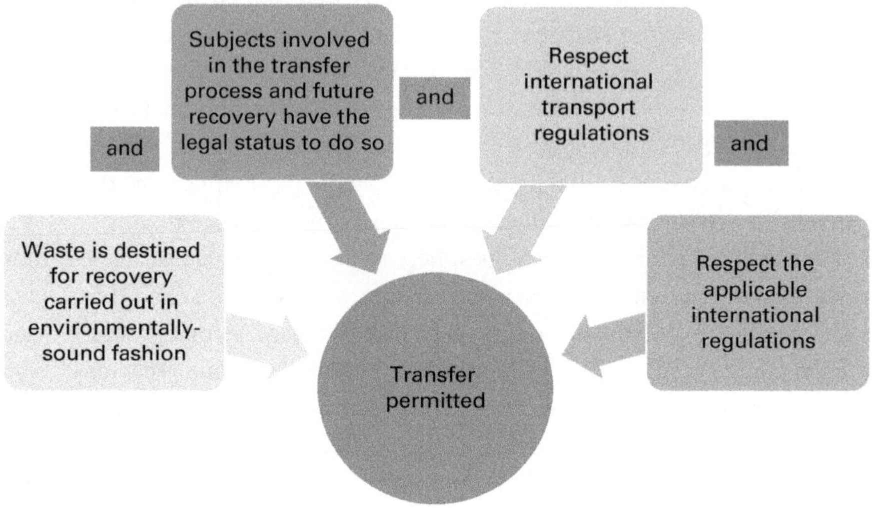

The waste to be transferred under the green control procedure is listed in Appendix 3 to the Decision, repeating the list of waste considered to be non-hazardous under Annex IX to the Basel Convention (Part 1 of Appendix 3), as well as other waste that the OECD Member States agreed upon in adding to this list, according to the criteria referred to in Appendix 6 of the Decision (criteria for the OECD risk-based approach).

Waste considered to be non-hazardous under the OECD Decision is thus subject to the green control procedure when transferred among Members for recovery purposes. This means that such transfers must adhere to the regulations normally applied to business transactions. The Decision does not therefore mention any form of specific socio-ecological control for this type of waste, treating its cross-border transfer as a simple trade operation (Chapter II.C).

This rule shall not apply to waste which, despite being listed as non-hazardous under the Decision, is contaminated by other materials to an extent that:

a increases the risks associated with this waste to a sufficient amount that it must be subjected to the amber control procedures (once again using the criteria in Appendix 6); or

b prevents such waste from being recovered in an environmentally-sound fashion (Chapter II.C).

The waste to be transferred under the amber control procedure is listed in Appendix 4 to the Decision, which replicates the waste categories requiring special consideration ('other waste') and the list of waste considered to be hazardous by the Basel Convention in its Annexes II and VIII (Part 1 of the Appendix 4), as well as additional waste that the OECD Members have also agreed upon to be subject to this control procedure, also in compliance with the criteria referred to in Appendix 6 to the Decision.

There are three conditions mentioned by the OECD Decision for the carrying out of cross-border waste transfers subject to amber control procedures (Figure 5.7). The first of these is the existence of a valid written contract, a chain of contracts or equivalent agreements between installations controlled by the same legal entity, beginning with the exporter and ending with the recovery installation. All of the persons involved in these contracts or agreements must have the appropriate legal status and, although they may be considered confidential under the terms of national legislation, the exporter must – when required by national law – provide copies of such a contract to the competent authorities both in the country of export and the country of import (Chapter II.D.1).

FIGURE 5.6 Control levels for cross-border waste transfers under the OECD Decision

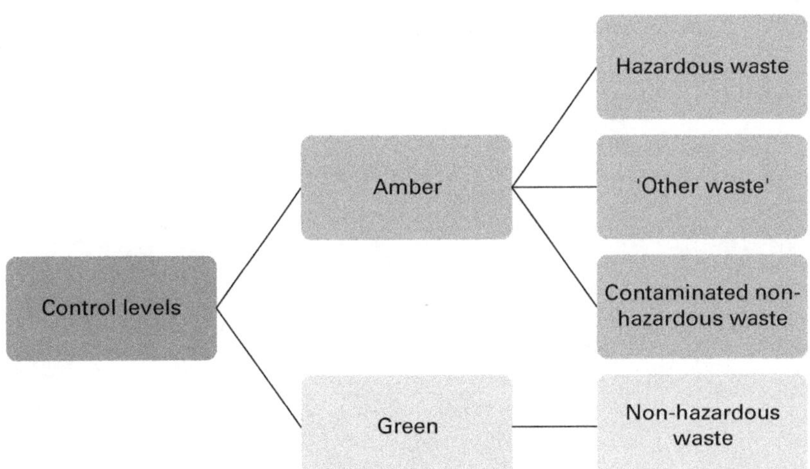

Furthermore, the contract must (Chapter II.D.1):

- **a** clearly identify the generator of each type of waste, each person that holds legal control over it and the recovery installation;
- **b** ensure that the relevant requirements in the OECD Decision are taken into consideration and are binding over all parties to the contract;
- **c** specify which parties to the contract must assume the responsibility for alternative waste management in compliance with the laws and regulations, including, if necessary, the return of the waste, as well as which parties must, when necessary, notify as to the re-export under the terms of the Decision.

The second condition for the carrying out of waste transfers under the amber control procedure is the existence of financial guarantees. According to the Decision, when applicable, the exporter or the importer must provide guarantees according to the requirements of national or international laws for alternative recovery operations to those originally foreseen for disposal or other environmentally-sound management options for the waste transferred in cases in which the agreements made for the transfer and recovery operations cannot be fulfilled under the terms originally laid out (Chapter II.D.1).

The third and final condition mentioned by the OECD Decision is, in actual fact, a provision of exception for the application of the amber control

FIGURE 5.7 Specific criteria for permitting cross-border waste transfers under the amber control procedure of the OECD Decision

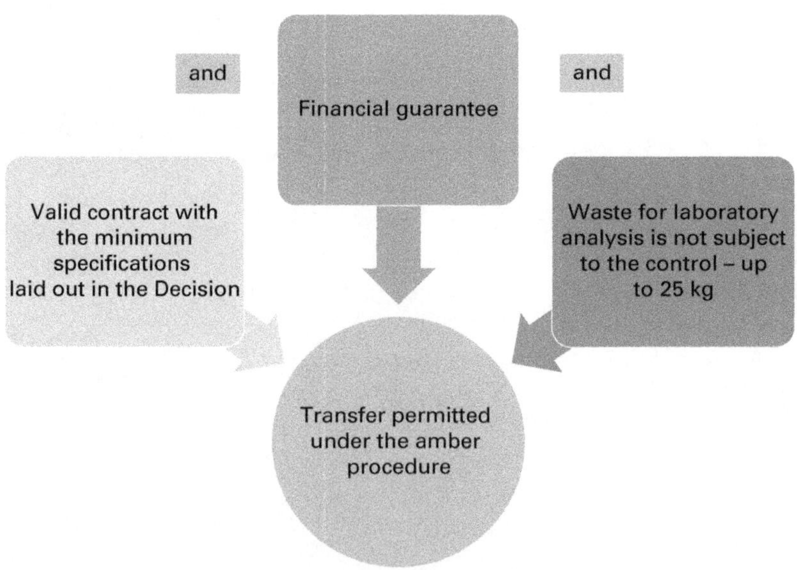

procedure to waste considered to be hazardous under the Decision. According to this provision, OECD Members are exempt from using the control if the transfer of the waste considered as hazardous is explicitly destined for laboratory analysis with the aim of assessing its physical or chemical characteristics or to determine if it is suitable for recovery operations (Chapter II.D.1).

The quantity of waste transferred with this purpose must be the minimum amount possible that is suitable for laboratorial analysis, never exceeding 25 kg. Such waste must still be packaged and labelled according to the applicable international laws. Finally, the exporter must inform the competent authorities in the countries of import and export when required by the national legislation on such transfers (Chapter II.D.1).

GENERAL PROCEDURE FOR WASTE TRANSFERS

The Decision mentions two types of procedures within the amber control procedure concerning:

- i individual or multiple cross-border transfers to recovery facilities in general; and
- ii cross-border transfers to pre-consented recovery facilities.

In terms of the former, the Decision states that before initiating each cross-border transfer, the exporter must notify the competent authorities in all states involved in writing ('single notification') (Chapter II, D.2).

The notification document must include all information listed in Appendix 8.A of the Decision (eg information on the exporter and importer, means of transport and types of waste transferred) and must be accompanied, when the national law of any of the states involved requires it, by a transfer contract for the waste to be reviewed by the competent authority. Such authorities may also request additional information if the notification is not complete (Chapter II, D.2).

Upon receipt of the complete notification document, the competent authority in the country of import (and, when applicable, the country of export), must confirm receipt within three days, with a further 30 days from the date of confirmation to raise any objections to the requested waste transfer. All objections raised by any of the states involved must be made in writing to the exporter and all of the competent authorities in the other states. If no objections are raised within the period mentioned, the Decision will consider this to serve as tacit consent, thus allowing the waste transfer to begin. Tacit consent shall be valid for one year (Chapter II, D.2).

In cases in which the competent authorities decide to consent to waste transfer in writing, such consent must be provided within 30 days, remaining valid for up to a year from the date of issue. The transfer may only be initiated upon receipt of all consent (whether tacit or written), also within the validity periods, and must always be accompanied by a movement document containing the information listed in Appendix 8B of the Decision (all information contained in the notification document, as well as extra information, such as the transporter's data and the packaging to be used for such waste) (Chapter II, D.2).

Within three days of receipt of the waste transferred, the recovery installation must send a signed copy of the movement document to the exporter and the competent authorities in the countries of export, transit and import. It should also retain the original movement document for a period of three years. Finally, the recovery installation must send a recovery certificate to the exporter and the competent authorities in the countries of import and export, as soon as possible (no later than 30 days upon finalizing the recovery of the waste transferred, and never later than a year after the receipt of such waste) (Chapter II, D.2).

The Decision considers waste which is essentially similar (eg with similar physical and chemical characteristics) and periodically transferred to the same recovery installation by the same exporter as 'multiple shipments'. In such cases, the Decision allows the competent authorities in the countries involved in the transfers to choose to accept a 'general notification' for such transfers during a period of up to one year. Although all transfers may use the same notification, each of them must be accompanied by their own movement document, under the same terms as for individual transfers (Chapter II, D.2).

Finally, the 'general notification' for multiple transfer can be revoked as long as it is made by means of an official warning to the exporter by any of the competent authorities in the countries involved. Such a warning must be delivered to all the competent authorities in all the countries involved by means of the competent authority in the country to have revoked the acceptance of the general notification for multiple waste transfers (Chapter II, D.2).

The second type of procedures outlined under the amber control procedure refers to cross-border waste transfers to pre-consented recovery facilities. This means that the competent authorities with jurisdiction over the specific recovery installations in the country of import may decide not to raise objections to cross-border transfers of certain types of waste to such (pre-consented) recovery facilities. Such decisions may, however, be limited to a specific period of time and may be revoked at any moment (Chapter II, D.2).

If opting for this type of procedure, the competent authority must always inform the OECD Secretariat upon granting and revoking a pre-consent.

In this case, it must provide details of the pre-consented recovery facility, such as the name, address, technologies used, types of waste covered by the pre-consent and their period of validity. All the rules applicable to single and multiple transfers carried out at installations without the pre-consent described above shall be applied *mutatis mutandis* to transfers carried out according to this type of procedure, with two exceptions (Chapter II, D.2). The first concerns the period granted to the countries of export and transit for raising objections, which is, as a general rule, seven days, or, exceptionally, up to 30 days, when additional information is required by the objecting countries. The second difference is found in the validity period for 'general notifications' for multiple transfers, which, instead of a year, is valid for three years for pre-consented facilities (Chapter II, D.2).

Figure 5.8 provides a summary of the procedures under the OECD Decision.

SOLUTIONS FOR POSSIBLE PROBLEMS AFFECTING WASTE TRANSFERS

Just like the Basel Convention, the OECD Decision mentions two types of problems which may occur during cross-border waste transfers in the instrument's sphere of application, as follows (Chapter II, D):

- i contractual non-compliance for any reason, including due to illegal waste traffic identified by the country of import; and
- ii non-compliance with the key provisions of the Decision on notification and movement documents or when such a transfer is considered to be illegal, identified as such by any of the countries of transit.

In the first case, the country of import must immediately inform the country of export. If alternative agreements cannot be made for the waste transferred

FIGURE 5.8 General procedure for cross-border waste transfers under the OECD Decision

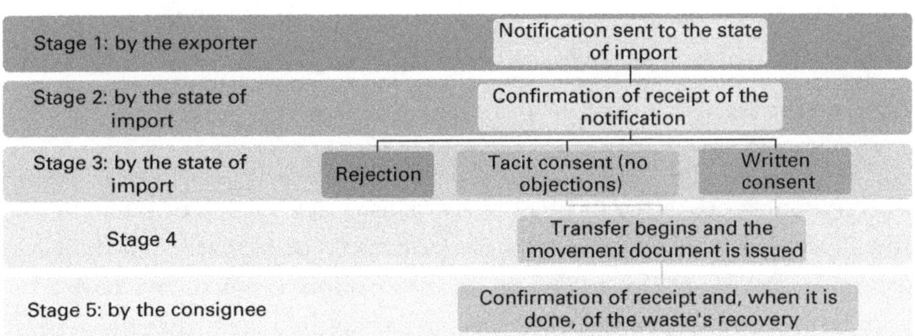

to be recovered in an environmentally-sound fashion in the country of import, the following provisions (Chapter II, D.3) must be applied:

a return of the waste by the country of import to the country of export, informing the country of export and countries of transit of the specific grounds for the return of the waste, with the countries unable to oppose or impede such a return; and

b re-export by the country of import to another country other than the one to have originally exported the waste.

The re-export may only be carried out by means of a notification sent by the exporter in the country of import to all the countries involved, as well as to the country originally exporting the waste. In such cases, all of the procedures outlined for cross-border waste transfers to recovery facilities without pre-consent (listed above) shall be applied (Chapter II, D.3).

In terms of the problem of non-compliance with the provisions on the notification and movement documents or in cases of illegal traffic identified by a country of transit, the country must immediately inform the countries of export and import, as well as the other countries of transit (see Figure 5.9). If alternative agreements cannot be made for the waste to be recovered in an environmentally-sound fashion, the country of export must accept the return of the waste to its territory, without opposing or impeding the return (Chapter II, D.4).

EXCEPTIONS TO GENERAL PROCEDURE

Although it provides the general system with two control levels (green and amber), the OECD Decision mentions three specific cases which do not theoretically follow the general rules, therefore requiring specific provisions guiding cross-border waste transfers to be carried out under the Decision's sphere of application (Figure 5.10). These are as follows:

i when specific national controls exist;

ii when waste to be transferred for recovery is not listed in the Appendixes to the Decision as hazardous or non-hazardous; and

iii when the waste is mixed.

It must be highlighted that, for the Decision, 'mixture of wastes' occurs when two or more different types of waste are intentionally or unintentionally mixed. If a transfer includes more than one type of waste, but these are separated, this is not considered to constitute a mixture (Chapter II, A).

In terms of the first case, the Decision clarifies in its text that the general control system outlined does not entirely affect the autonomy of the Member

FIGURE 5.9 Problems which may affect cross-border waste transfers and solutions provided under the OECD Decision

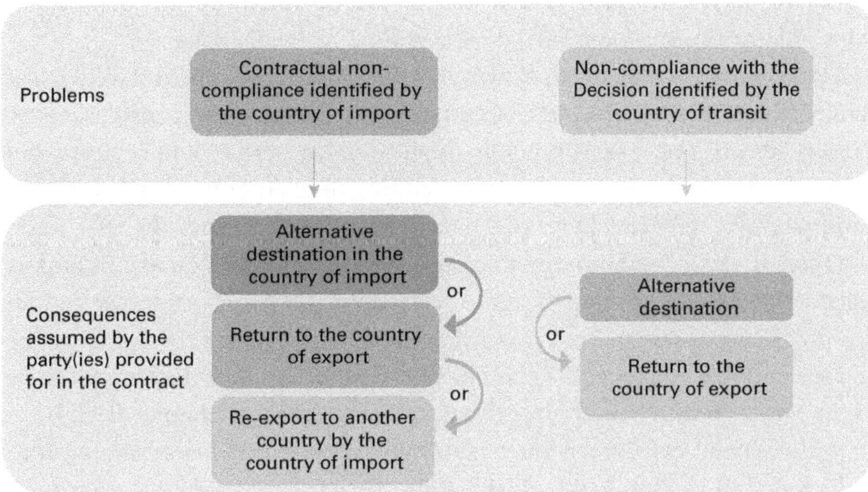

FIGURE 5.10 Exceptions to the OECD Decision general procedure

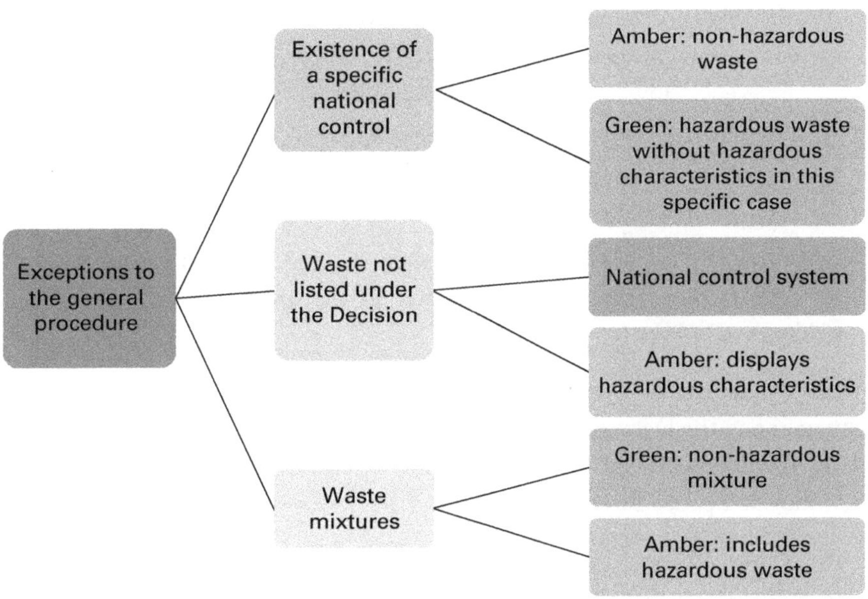

countries, which maintain the right to control, in exceptional circumstances, certain waste differently to that laid out in the multilateral agreement, according to national legislation and rules in international laws for the protection of human health and the environment (Chapter II, B.4). The Member States may thus apply the amber control procedure to waste considered to be non-hazardous under the Decision and the green control procedure to waste considered to be hazardous, but which does not feature any of the characteristics of hazardousness outlined in Appendix 2 to the Decision.

The Member States must, however, inform the OECD Secretariat when using different control levels, specifying the waste they consider to be hazardous and non-hazardous, the applicable legislative requirements and the types of analytical tests and procedures used in compliance with the national laws for checking for hazardous characteristics in the waste (Chapter II, B.5). Furthermore, in cases in which only the country of import considers waste to be transferred as hazardous, the Decision states that all the provisions in the amber control procedure applicable to the exporter and the country of export listed in the text shall be applied *mutatis mutandis* to the importer and the country of import, respectively (Chapter II, B.4).

In the second case, when waste destined for recovery operations is yet to be included in lists of waste considered to be hazardous or non-hazardous, the Decision states that it may still qualify for cross-border transfers, as long as it is subjected to the following conditions:

- **a** that the Member States identify such waste, and if appropriate, request that the Basel Convention's Technical Working Group adds it to its list of waste;
- **b** pending the inclusion of the waste in any of the lists, the control procedure must be followed for cross-border waste transfers as set out in the national legislation of the countries involved, with no country required to follow the laws of other countries; and
- **c** if the waste displays any of the characteristics of hazardousness given in the Decision, it must be subjected to the amber control procedure (Chapter II, B.6).

Finally, in the case of mixtures of wastes, in order to prevent poor waste management, the Decision states that if it is not specifically listed as a mixture under the Decision Appendixes (whether as hazardous or non-hazardous) the following procedures apply (Chapter II, B.8):

- **i** if the mixture is of two or more non-hazardous types of waste, it must be subjected to the green control procedure, as long as the composition of the mixture does not impede its environmentally-sound recovery; and
- **ii** if the mixture contains any type of waste which may be classified as hazardous, it must be subjected to the amber control procedure, as long as the composition of the mixture does not impede its environmentally-sound recovery.

A CRITICAL ANALYSIS OF THE DECISION

Compared to the Basel Convention, the OECD Decision serves as a multilateral agreement providing more explicit and simplified control mechanisms for cross-border waste transfers, facilitating their undertaking when destined for recovery among OECD Members. The regulations provided by the OECD Decision were revised in order to be aligned with the provisions of the Basel Convention, mainly in terms of the classification of waste as hazardous or not, although several procedural elements of the Decision differ from the Basel Convention, among which are the possibilities for tacit consent and the issue of notifications for pre-consented recovery facilities.

Much of the criticism of the Basel Convention therefore applies to the OECD Decision, although some specific observations must be made on the latter. The first of these refers to criteria (or conditions) provided by the Decision for permitting the carrying out of cross-border waste transfers under its sphere of application. The description of such criteria reveals all of them to be rather formal in nature (compliance with national and international regulations), referring as a whole to standards and regulations established nationally by the member countries.

Apart from the fact that the text refers at various points to the need for waste recovery (or disposal) operations to be carried out in an environmentally-sound fashion, the OECD Decision establishes such a determination as the first of the criteria for cross-border waste transfers in compliance with this agreement. At no point, however, does the Decision define what it regards as constituting environmentally-sound recovery (or management in a general sense), delegating such a task to the national law of the member countries.

Although this agreement is only applied to transfers between member countries (therefore, countries considered to be developed, theoretically excluding the 'vulnerability' factor), the absence of a clearer definition of what constitutes environmentally-sound waste management significantly weakens the control said to be exerted by the Decision, given that some countries may establish environmental standards lower than those established by the general duty towards best management.

As seen, the two-tiered control system governs cross-border transfers of waste classified as non-hazardous by the Decision by means of the green control procedure. However, this tier does not provide any kind of socio-ecological control, merely treating such transfers as business transactions. There is thus no limit on the quantity of waste transferred, the efficiency of the recovery process, the percentage of the material not recovered (and therefore disposed of), among others. The amber control procedure, on the other hand, uses the same regulatory basis as the Basel Convention, with the same pre-consent instrument in place for controlling transfers. Along with the criticisms already made of the Basel Convention must be added the questionable provision on the possibility of 'tacit consent'. This effectively means the instrument is founded on the absence of a response on the part of the state of import, and, perhaps even more seriously, the communities in such a state, who will be directly affected by such transfers.

The reference to general notifications for multiple waste shipments may also be questioned, given that it does not provide a clear and explicit definition of what wastes are essentially similar. The vague nature of this expression may prompt abuse of the use of general notifications, waiving the control of such waste transfers.

It is interesting to note that the solutions provided by the Decision for problems involving contractual non-compliance and non-fulfilment of the Decision's key provisions (on notification and movement documents), both including the illegal waste traffic, always prioritize the carrying out of alternative agreements between the countries involved for the destination of the waste transferred outside the territory of the country of export. The return of the waste to its original source appears in all cases as a second option for such problems. Such a positioning does not suggest any concern or priority for the environment, only benefits for the country of export.

The Decision provides three possible scenarios in which its control system can be relaxed. Firstly, when specific national controls are in place and if only the countries of import consider the waste to be transferred as hazardous, these countries must assume all procedural responsibilities originally belonging to the country of export. This places unreasonable demands on the country of import, which, having classified the waste as hazardous, normally has sufficient reason for doing so.

Secondly, waste which is not listed by the OECD Decision (as hazardous or non-hazardous) may still undergo cross-border transfers, and if it does not display any of the hazardous characteristics listed by the Decision, will not be subject to any international control system.

Finally, mixtures of two or more types of waste listed as non-hazardous will be subject to the green control procedure, meaning there will be no socio-ecological control for such transfers. In such cases, the Decision ignores the possibility that wastes considered to be non-hazardous when isolated may behave in an unexpected fashion when mixed, displaying recognized or unrecognized hazardous characteristics, and there is no provision in the Decision for any analysis to be done to identify the existence of these characteristics. This bends the regulations and jeopardizes environmental and human health in these types of transfers.

With the most important points of the two main international legal instruments regulating the cross-border transfer of waste now analysed, the next section will study the internalization of these rules and the regulations by the European Union and by Brazil, both parties to the Basel Convention. Although neither are OECD Members, the organization's Decision on waste transfers generated a real impact on the regulations devised for such transactions both by the EU and by Brazil. One of the possible explanations is the proximity of both to the OECD, with the EU a participant in the work carried out by this organization and Brazil a key partner.

> For more information on OECD Members and partners, see: www.oecd.org/about/membersandpartners/

Regional and national regulations: the EU and Brazil

European Union

After seeing the general outline of the legal regulation of waste in the EU under Framework Directive no. 2008/98/EC, this section now goes on to describe and analyse Regulation (EC) no. 1013/2006, which establishes procedures and control regimes for waste transfers involving EU countries.

The rules established by the EU Regulation (Article 1) feature three reference elements (see Figure 5.11 on page 295):

a spatial (transfer origin, destination and itinerary);

b type of waste transferred; and

c the type of treatment to be applied to the waste at its destination.

The Regulation works with these three elements in an interconnected fashion in order to delimit and define the prohibitions and control procedures for cross-border waste transfers established by the EU.

For the Regulation (article 2), the term 'transfer' refers to the transport, be it road, rail, air, maritime or river, of waste for recovery or disposal that takes place or is planned to take place:

- **a** between two countries;
- **b** between one country and other countries or overseas territories or areas under the protection of the former;
- **c** between one country and any area which is not part of any country under international law;
- **d** between one country and Antarctica;
- **e** from a country transited by any of the aforementioned areas;
- **f** in the interior of a country, transited by any of the aforementioned areas and that originates and concludes in the same country; and
- **g** in a geographical area not subject to the jurisdiction of any country, destined to one country.

For the spatial element of reference (transfer origin, destination and itinerary), the Regulation outlines four different types of waste transfers, providing general procedures for these transfers, with particular clauses for each case:

- **i** between EU Member States (with or without transit through third countries);
- **ii** imported from non-EU countries to the EU;
- **iii** exported from the EU to non-EU countries; and
- **iv** in transit through the EU, between third countries (article 1).

Concerning the second element of reference, types of waste, when these do not fall under any of the provisions for prohibited transfers, the Regulation applies a two-tiered (green and amber) control system following the same line as the OECD Decision. As a general rule, the green control procedure applies to waste classified as non-hazardous, which is normally subject to general requests for information and the amber control to waste considered to be hazardous, for which prior written notifications and consents must be used.

It must be highlighted that some types of waste are excluded from the Regulation's sphere of application, such as (article 1):

- the offloading to shore of waste generated by the normal operation of ships and offshore platforms;
- waste generated on vehicles, trains, planes and ships;
- radioactive waste;
- waste resulting from mining and quarrying activities;
- waste waters, among others.

Finally, the Regulation considers the type of treatment to be applied in the destination of waste transferred as a third element of reference in defining the procedure to be adopted. In such cases, the rules applied to waste destined for disposal are different to those for recovery, with the first group based on the Basel Convention and the second on the OECD Decision.

It is interesting to note that, unlike the international instruments analysed here, the EU Regulation provides specific provisions for interim recovery operations, outlining them as operations involving:

- **a** exchange of waste for subjection to final recovery operations, perhaps including preliminary operations such as dismantling, sorting, crushing, drying and repackaging; and
- **b** non-temporary waste storage for later recovery (Annex II of Directive no. 2008/98/EC).

Just like the OECD Decision, the EU Regulation outlines a two-tiered green and amber control system for cross-border waste transfers (Figure 5.12). While the first is based on the request for general information, the second is

FIGURE 5.11 Elements of reference for the delimiting and defining of prohibitions and control procedures for cross-border waste transfers under EU legislation

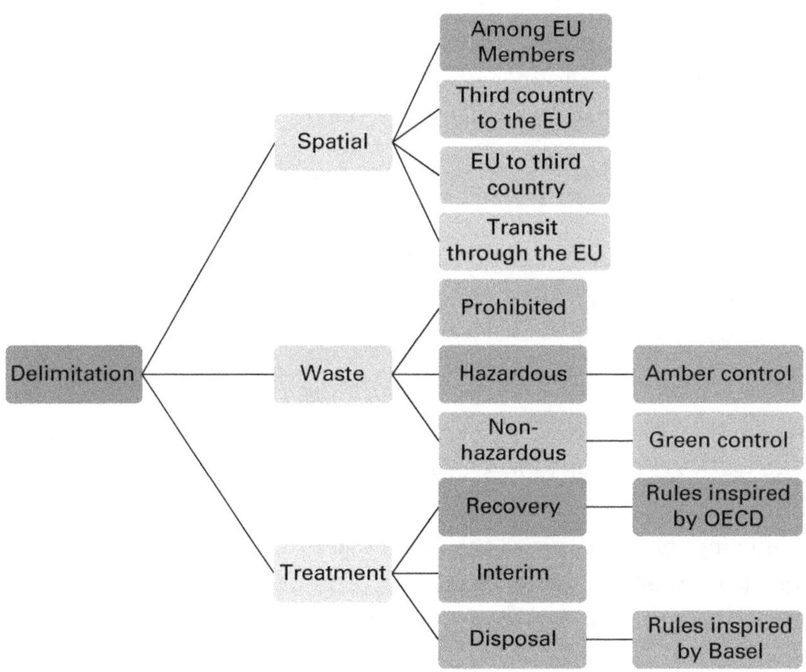

founded on the instrument of prior written notification and consent, providing much more detailed procedures. Initially, the Regulation provides a general procedural framework applied to all transfers, followed by specific procedures and exceptions for each type of transfer.

The Regulation states that wastes of the following types are subject to the prior written notification and consent procedure (amber tier):

a destined for disposal;

b destined for recovery, if:

  i it constitutes hazardous and 'other waste';

  ii it is listed in Annex IV-A to the Regulation (such waste, although classified as non-hazardous, must still be subjected to this procedure);

  iii it constitutes individual waste or mixtures that are not in any of the Regulation's lists (hazardous or non-hazardous); and if

  iv it is municipal or the mixed equivalent collected from personal dwellings (Article 3).

The Regulation highlights that, if waste classified as non-hazardous displays any of the hazardous characteristics listed under the European regulation, it will also be subject to the prior written notification and consent procedure or its transfer will be prohibited, according to the specifics of the case. Furthermore, such a situation will spark a process reviewing the lists of hazardous and non-hazardous waste under the terms of article 58 (articles 3 and 58).

Exempt from the prior written notification and consent procedure (following the green level procedure) are the transfers of wastes:

a classified as non-hazardous under the Regulation (except that listed in Annex IV-A); or

b non-hazardous waste mixtures (as long as the composition does not affect the carrying out of environmentally-sound recovery operations) listed in Annex III-A to the Regulation; and

c destined for recovery.

However, if the quantity of waste to be transferred exceeds 20 kg, the transfer will be subject to the requests for general information outlined in article 18 of the Regulation (article 3).

Following the same line as the OECD Decision, the requests for general information in article 18 for waste transfers are also applicable to waste explicitly destined for laboratory analysis for assessment of its physical or chemical characteristics or to determine its suitability for recovery or disposal. In such cases, the instruments of prior written notification and consent are not applicable as long as the maximum amount to be transferred does not exceed 25 kg (article 3).

Article 18 states that such transfers must be accompanied by the information document listed in Annex VII (article 18), which must be completed with the data of the person responsible for the transfer, the transporter, the waste generator, the importer/consignee and the states involved, as well as the type of waste, its classification, quantity, recovery or analysis operations, among others (Annex VII). It must be stressed that, just as in cases of transfers subjected to prior written notification and consent procedures (to be detailed as follows), transfers only subject to requests for general information must also be carried out by means of contracts containing provisions on responsibility in cases of non-compliance with the contract or the illegal waste traffic – illegal shipment (article 18).

CRITERIA FOR WASTE TRANSFERS

The amber control level follows the prior written notification and consent procedure. In such cases, the Regulation states that the producer of the waste to be transferred must send written notification and movement documents to the competent authority in the state of export, which will forward the notification to the states of import and transit (article 7). If the notifier is not

FIGURE 5.12 Control levels for cross-border waste transfers under the EU legislation

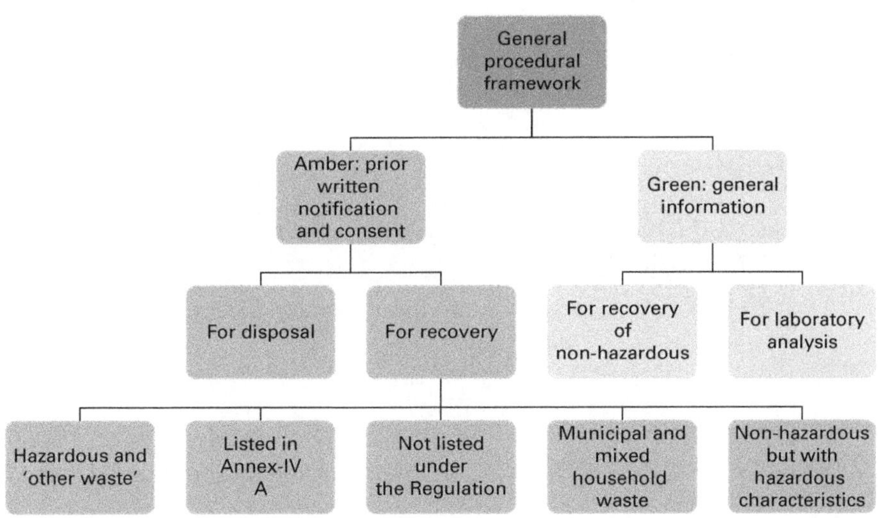

the producer of the waste, they must ensure that the producer or other persons defined as notifiers by the Regulation (such as the collector and the registered dealer, for example) also sign the notification document (article 4).

In general terms, a notification should be issued for each type of waste transferred, covering the transfer from the place of origin, including interim and non-interim recovery or disposal operations, up to the final destination. All information to accompany the notification is duly listed in Annex II to the Regulation, and, in case the competent authorities involved require further information, the notification shall only be considered complete and valid when all compulsory and additional information has been duly supplied (article 4).

Apart from the notification and movement documents, the notification must also include proof of the existence of a contract between the notifier and the consignee for recovery or disposal of the waste to be transferred, as well as proof of the existence of a financial guarantee or equivalent insurance to cover the transfer operation and the environmentally-sound management of the waste (article 4), with these the minimum conditions for carrying out a transfer under the amber control level (see Figure 5.13).

GENERAL PROCEDURE FOR WASTE TRANSFERS

The Regulation requires the signing of a contract for all waste transfers subject to the prior written notification and consent procedures. Such a contract will

FIGURE 5.13 Specific criteria permitting cross-border waste transfers under the amber control according to the EU legislation

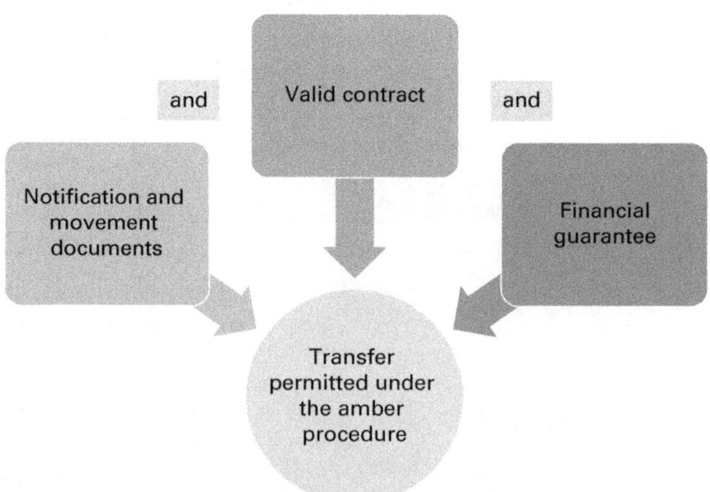

come into effect from the moment the notification is done, throughout the transfer, until a certificate is issued proving the due compliance of the final recovery or disposal operation. The contract should also provide provisions of responsibility under the terms of the Regulation (article 5).

If the cross-border waste transfer is destined for interim recovery or disposal operations, the contract signed between the notifier and the consignee should also describe additional requirements for (article 5):

a the destination installation, which should provide certificates of final recovery or disposal when complete under the terms of the contract and the Regulation; and

b the consignee, which must, upon forwarding the waste transferred for final recovery or disposal to a third country outside the EU, issue a new notification to all the states involved, including the original state of export.

The obligation to sign a contract in order to transfer waste for recovery or disposal operations subject to the prior written notification and consent procedure is only waived in cases in which the transfer is performed between two establishments under the control of the same legal entity. In this case, it may be substituted by a declaration of commitment issued by this entity to undertake the recovery or disposal of the waste transferred (article 5).

As for the financial guarantee or equivalent insurance which must accompany the notification, apart from the requirement that it is established by the notifier or another person in their name to cover cases of contractual non-compliance or illegal traffic, the Regulation (article 6) states that it must meet the costs of:

a transport;

b recovery or disposal, including any necessary interim operations; and

c storage for a period of 90 days.

It is important to highlight that the financial guarantee or equivalent insurance approved by the state of export (or importer, in cases involving imports to the EU) must be valid and cover both the transfer itself as well as the conclusion of the recovery or final disposal of the waste transferred. Furthermore, it will be released, in general terms, upon receipt by the competent authority of the certificate declaring the conclusion of the management of the waste transferred (whether involving recovery or disposal) (article 6).

Upon confirmation of the receipt of the notification by the state of import, the states of import, export and transit will have a timeframe of 30 days to take one of the following decisions:

i authorization without conditions;

ii authorization with conditions (for the notifier and for the installation of destination); or

iii objection.

Any decision made by the states involved shall be justified (ie express their reasons) and in writing. Just as with the OECD Decision, the EU Regulation mentions the possibility of tacit consent if the states involved do not raise any objections in the aforementioned timeframe and, as with the written consent, the validity of this tacit consent shall be one year (article 9).

Once the transfer has been conceded and carried out, the recovery or disposal of the waste must be concluded within a maximum period of one year as of the date of receipt at the installation. The competent authorities involved must also revoke their authorization any time that (article 9):

a the composition of the waste does not match that stated in the notification;

b the conditions established for the transfer were not respected;

c the waste was not recovered or disposed of according to the licence held by the installation to have performed the said operation; or

d the waste transfer, recovery or disposal has been carried out in a different fashion to that listed in the notification or movement documents.

The objections that may be made by the states involved in a waste transfer must be duly founded and will depend on the destination of the waste (whether disposal or recovery). The bases for objection are also divided into:

i those related to the transfer process and future waste destination; and

ii those related to the subjects involved in the transfer and future destination.

For the former group, several of the bases listed by the Regulation are applicable to both waste destined for disposal and waste destined for recovery.

The Regulation (articles 11 and 12) establishes the possibility that the states involved express objections in cases in which the waste transfer or future destination:

a is not consistent with its waste management plan;

b does not comply with the national legislation protecting the environment and human health or order and public safety; and

c is incompatible with the obligations of international conventions.

In terms of the second group of objections, concerning the subjects involved, still as common bases for both operations, objections may be raised (articles 11 and 12) if the states involved know that the notifier or consignee of the waste to be transferred:

d has previously been convicted of illegal transfers or any other illegal act related to environmental protection;

e failed to comply with the provisions of the Regulation in previous transfers; or

f does not use the best available technology in compliance with the operational permit of the facility.

The Regulation also mentions, among the first group of objections, specific grounds for objections to recovery operations and others to disposal, related to the transfer process and future destination. In terms of waste destined for disposal, the Regulation (article 11) sets out that the states involved may object to a transfer request if:

i it disrespects the principles of proximity, priority for recovery and self-sufficiency on EU and national levels;

ii its national legislation prohibits the import of hazardous waste and 'other waste';

iii the waste is mixed municipal waste collected from private households; or

iv the waste in question is subject to specific environmental regulations.

The objections based on the disrespect of the principles of proximity, priority for recovery and self-sufficiency do not apply to hazardous waste to be transferred in annual global quantities so small that the construction of new specialized disposal installations in the state of export would be economically inviable. Furthermore, if the state of import uses such grounds to oppose a transfer, it must cooperate with the state of export in order to reach a bilateral resolution on the matter, or, as a last resort, submit the matter to the European Commission (article 11).

In terms of waste destined for recovery, the Regulation (article 12) mentions that the states involved may raise objections if they consider the transfer or recovery planned:

a to be inconsistent with the waste hierarchy, the ban on poor management, the permit rules, the prohibitions and other provisions, all set out in the European Waste Framework Directive;

b to be inefficient from an economic and/or environmental perspective due to the estimated value of the waste material to be subjected to recovery or the costs of recovery and disposal of the non-recoverable fraction of the waste;

c not to handle the waste transferred according to the legally-binding environmental protection standards for recovery operations.

Finally, the Regulation mentions that the state of export may oppose a transfer request if it does not comply with its national legislation on recovery (especially if the regulations in the state of import are less exacting). These grounds for objection shall have no effect, however, if (article 12):

a EU legislation is in place on the subject and the legislation in the state of import is as exacting as the EU legislation;

b the recovery operation in the state of import is carried out in conditions that are roughly equivalent to those established under the national law of the state of export; and

c the state of import has not been notified of the technical standards and regulations adopted by the state of export.

A summary of the grounds for objections to waste transfer under EU legislation is provided in Table 5.1.

TABLE 5.1 Grounds for objections to waste transfers and their future destination according to EU legislation

Objections to waste transfers and their future destination	Process	Subjects (notifier or consignee)
For both operations	i) Does not comply with the national waste plan; ii) Is not consistent with legislation protecting the environment, public health, safety and order; iii) Is incompatible with the international convention it is party to.	i) Previous conviction for illegal traffic or illegal environmental acts; ii) Did not comply with the Regulation in previous transfers; iii) Does not use BAT.
Disposal	i) It does not respect the principles of proximity, priority for recovery and self-sufficiency; ii) Mixture of municipal household waste; iii) Waste with specific regulations.	–

(continued)

TABLE 5.1 (Continued)

Objections to waste transfers and their future destination	Process	Subjects (notifier or consignee)
Recovery	i) Is not consistent with the waste hierarchy, the ban on poor management, permit rules or other rules in the EU Waste Framework Directive; ii) Is environmentally and economically inefficient; iii) Treatment does not comply with environmental regulations; iv) Does not comply with national legislation.	–

Following the terms of the OECD Decision, the European Regulation allows for the possibility of 'general notifications' and 'pre-consented recovery facilities', repeating the requirements and timeframes outlined by the Decision, except for the demand for the use of the same itinerary for all transfers carried out under a single general notification (articles 13 and 14).

On the transfer of waste destined for interim recovery and disposal operations, the Regulation mentions several provisions additional to those previously described (Figure 5.14). The first of these is the need to include information in the notification document on all the installations in which operations subsequent to the interim operations will be carried out, which will also be taken into consideration in the analysis leading to the issue of consent or objection from the states involved (article 15).

The installation undertaking the interim operation must issue a certificate marking the conclusion of the operation under its responsibility, as well as obtain a certificate of final conclusion from the installation(s) that will carry out the subsequent recovery or disposal operations (article 15). This same final conclusion certificate for the recovery or disposal operation must be presented for transfers to (final) non-interim operations (article 16).

As general requirements to this procedural framework, the EU Regulation lists:

a the prohibition of the mixture of waste during transfer (article 19);

b the preservation of documents and information on transfers by all subjects involved for at least three years (article 20); and

FIGURE 5.14 General procedure for cross-border waste transfers under the EU legislation

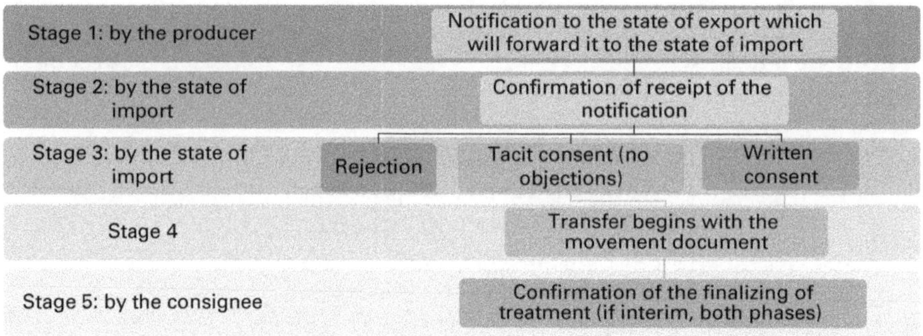

c the possible publicizing by the states involved of information on the notification for transfers authorized (article 21).

SOLUTIONS FOR POSSIBLE PROBLEMS AFFECTING WASTE TRANSFERS

Just like the international instruments analysed here, the European Regulation foresees two types of problems which may occur in the waste transfer processes (Figure 5.15, page 307), derived from:

a non-compliance with the notification and movement documents or the contract for the waste transfer and destination; and

b illegal waste transfers (illegal shipments).

In both cases, the Regulation refers to the duty to re-collect the waste, with specific provisions for each case.

In the first case, the Regulation sets out that, as a general rule, the state of export must guarantee that the waste is taken back by the notifier, or, if this is not a viable possibility, by the state itself or any other person acting on its behalf. There are, however, cases in which such a duty does not apply, as follows (article 22):

a when the states involved agree on alternative means of recovery or destination for the waste transferred; or

b when the waste transferred has been irreversibly mixed with other waste during the recovery or disposal operation at the installation.

It is important to note that if the alternative solution agreed upon for the two exceptions above is for the destination of the waste transferred outside the original state of import, a new notification must be issued by the original notifier if appropriate, or by the other persons responsible, or, if not possible, by the original state of export (article 22). All of the costs involved for the take-back or the alternative recovery or disposal operations, including the transport and storage costs, must also be charged (article 23) in the following order of priority:

1 to the notifier;

2 to other persons on their behalf;

3 to the state of export; or

4 as agreed upon between the parties involved.

The second type of problem that may affect cross-border waste transfers – cases of illegal transfer (or illegal trade) – is defined by the European Regulation (article 2) as any waste transfer carried out:

a without notification issued to all states involved;

b without consent issued from all the states involved;

c by means of falsification, misrepresentation or fraud;

d so as to materially contradict the information contained in the notification and movement documents;

e in violation of EU and international rules on waste recovery or disposal;

f in contradiction with the bans on transfers outlined in the Regulation;

g misrepresented under the exemptions to the prior written notification and consent instruments listed beforehand or in non-compliance with the requirements for such exemptions.

For such cases, as with the Basel Convention, the Regulation outlines different consequences depending on the subject responsible. If the responsibility lies with the notifier, the waste transferred must be, in the following order (article 24):

i taken back by the notifier *de facto*;

ii taken back by the notifier *de jure* (if no notification was issued);

iii taken back by the state of export;

iv disposed of or recovered in an alternative fashion in the state of import or export; or

v disposed of or recovered in an alternative fashion in another state by the state of export, if all of the states involved agree to do so.

On the other hand, if the responsibility for the illegal waste traffic lies with the consignee, such waste must be recovered or disposed of by the consignee, or, if not a viable possibility, by the state of import, in an environmentally-sound fashion. If the illegal transfer cannot be attributed to the notifier or the consignee, the states involved must cooperate to ensure such waste is destined in an environmentally-sound fashion (articles 24 and 25).

Finally, in cases involving a transfer exempt from the use of the prior written notification and consent instrument and non-compliance with the Regulation provisions, the subject to carry out the transfer will assume the responsibilities attributed to the notifier. The costs for take-back of the waste or for the alternative recovery or disposal options will be charged in the same order as the responsibilities described above (articles 24 and 25).

For both types of problems affecting cross-border waste transfers, the Regulation (articles 22 and 24) establishes that:

i the state of import must store the waste until alternative re-collection, recovery or disposal may be arranged;

ii the requirement for the notifier (or, alternatively, the state of export) to re-collect the waste or provide alternative recovery or disposal operations will terminate when the installation issues a certificate confirming the conclusion of the non-interim disposal or recovery;

iii neither of the states may oppose or impede the return of the waste when applicable; and

iv the assigning of the costs and responsibilities under the Regulation does not exclude the application of the EU or national provisions on responsibility.

In finalizing the general procedural framework for cross-border waste transfers, the Regulation outlines provisions for:

i disagreements on classification issues among the states involved;

ii general administrative costs; and

iii transboundary agreements on specific transfers.

In terms of the first point, it has been established (article 28) that, when disagreements are on:

a the classification of certain materials as waste or non-waste, these must be classified as waste;

b the classification of waste as hazardous or non-hazardous, this must be classified as hazardous;

c the classification of the treatment operation to be performed on waste to be transferred, this will be considered to constitute disposal.

FIGURE 5.15 Problems which may affect cross-border waste transfers and solutions under the EU legislation

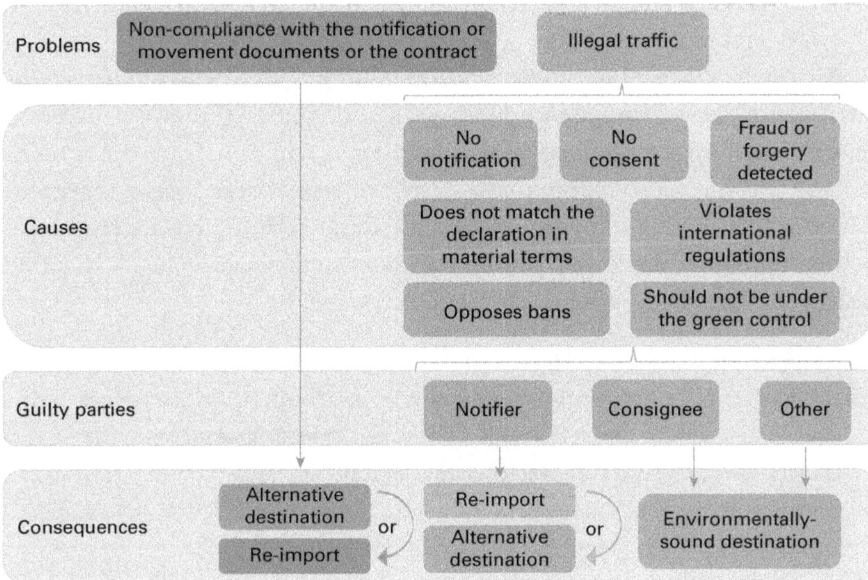

On the general administrative costs, the Regulation states that these are attributed to the notifier as a general rule, encompassing the notification, monitoring, analyses and inspection procedures, if necessary (article 29). Finally, the Regulation mentions the possibility that the Member States, in exceptional cases, may conduct among themselves and other party states of the Agreement on the European Economic Area (EEA), transboundary agreements in order to simplify the notification procedure for installations suitable for certain special waste streams in places closer to border areas (article 30).

> The **European Economic Area** (EEA) is a geographical area created by European institutions to allow for the free circulation of goods, services, persons and capital within the internal market of the European Union.

After providing a general procedural framework for the undertaking of cross-border waste transfers involving countries in the European Union, the Regulation outlines specific details on transfers involving third states, ie those that are not members of the European Union. The first of these refers to transfers performed between EU states transiting through third states. In such cases, transfers for waste destined for disposal will require written consent from the third state (article 31). For waste destined for recovery, prior written consent will be required only if the state of transit is not an OECD Member (article 32).

Waste exports from the European Community to third states and overseas territories and countries are prohibited when destined for disposal, except if undertaken between Member States of the European Free Trade Association (EFTA) which are also parties to the Basel Convention. Such exports are also banned if the EFTA country nationally vetoes them or if there are reasons to believe that such waste will not be managed in an environmentally-sound fashion in the state of destination (articles 34 and 40).

> Overseas countries and territories are countries and territories which are not independent and are linked to EU Member States. The **European Free Trade Association** (EFTA) is a European economic bloc of free trade, currently consisting of only four countries: Switzerland, Liechtenstein, Norway and Iceland.

Two specific elements of the procedure for this case are worth highlighting: a transfer may only be initiated when the waste management is guaranteed to be carried out in an environmentally-sound fashion, and the waste must be sent to installations in working order that are authorized under the terms of the national legislation applicable in the country of import (article 35).

If such exports are destined for waste recovery, on the other hand, following the provisions of the OECD Decision, the Regulation provides distinct rules for OECD Member States and non-Member States. If the state of import is not an OECD Member or the country or territory is overseas, exports are banned if (Figure 5.16, page 310):

- **a** the waste is classified as hazardous or 'other waste' under the terms of the Regulation;
- **b** the waste is a mixture of hazardous waste;
- **c** the state of import classifies the waste as hazardous;
- **d** the state of import has prohibited the import of such waste; or
- **e** the state of export has reason to believe that the waste will not be managed in an environmentally-sound fashion (articles 36 and 40).

The Regulation provides for the possibility that the Member States, in exceptional cases, may adopt different provisions to those set out in the instrument, if they are able to prove that the waste in question is hazardous or not, based on the characteristics of hazardousness described in the Regulation and in EU legislation. In such cases, the Member State of export must inform the state of import before making a final decision on the matter (article 36).

If the waste to be transferred for recovery in states which are not OECD Members and in overseas countries and territories is classed as non-hazardous or mixtures of non-hazardous waste and does not fall under the possible prohibitions in article 36 (described above), the EU Commission must contact such states requesting written confirmation that the waste may be exported from the EU for recovery in their territory, apart from providing information on the eventual control procedure (articles 37 and 40).

If the non-OECD Member State does not issue a response or has not been contacted, the prior written notification and consent procedure shall be applied. If, on the other hand, information is sent that it does not have a control system for this type of transfer, the request for general information will be applied (according to the general procedural framework described above). Just as with waste exported for disposal, waste transferred for recovery must be destined for installations in working order authorized by the state of import under the terms of national legislation (article 37).

The Regulation allows for waste exports destined for recovery in OECD Member States and applies the general procedural framework to these transfers, regardless of whether the waste is classified as hazardous, non-hazardous or not included in any of the classification lists in this regulatory instrument. Some specific details are added, however (article 38). The first of these is the need to use the prior written notification and consent system, if:

- **a** non-hazardous waste mixtures are destined for interim operations, with subsequent operations (whether interim or not) to be carried out in non-OECD Member States; and
- **b** waste considered to be non-hazardous awaits inclusion on the Basel Convention's list of non-hazardous waste.

In cases of waste classified as hazardous, the Regulation sets out the need to guarantee that it will be managed in an environmentally-sound fashion, with the destination installations authorized in accordance with the national regulations in the state in which they are located (article 38). Finally, the Regulation expressly prohibits waste exports for both disposal and recovery from the EU to Antarctica (article 39).

The Regulation (article 41) bans all imports from countries outside the European Community when the waste is destined for disposal, except when:

a originating in states which are parties to the Basel Convention;

b governed by other exceptional agreements in force (when the waste cannot be managed in the state to have produced it in an environmentally-sound fashion and for exclusive disposal in the Member State which is part of the agreement) or;

c from areas in situations of crisis or war.

The import of waste to the EU destined for recovery is also prohibited. The same exceptions related to waste destined for disposal are applied to this case, adding the transfers coming from OECD Member States (article 43).

Finally, the Regulation authorizes the import of waste from overseas countries and territories regardless of the destination (article 46). Imports for both disposal and recovery are, in all three cases, subject to the rules of the general procedural framework, with some specific details; among these, the guarantee that the transfer will only take place if it may be proved that the waste will be managed in an environmentally-sound fashion (articles 42, 44 and 45).

Transfers originating in and sent to countries outside the EU to transit through the EU follow the same rules in the general procedural framework, with the same specific details on waste imports to the EU (articles 47 and 48). Finally, in terms of transfers exclusively taking place within EU Member States, the Regulation establishes that such states must create a monitoring and control system in compliance with the rules laid out by the Regulation (article 33).

FIGURE 5.16 Cross-border waste transfers prohibited under the EU legislation

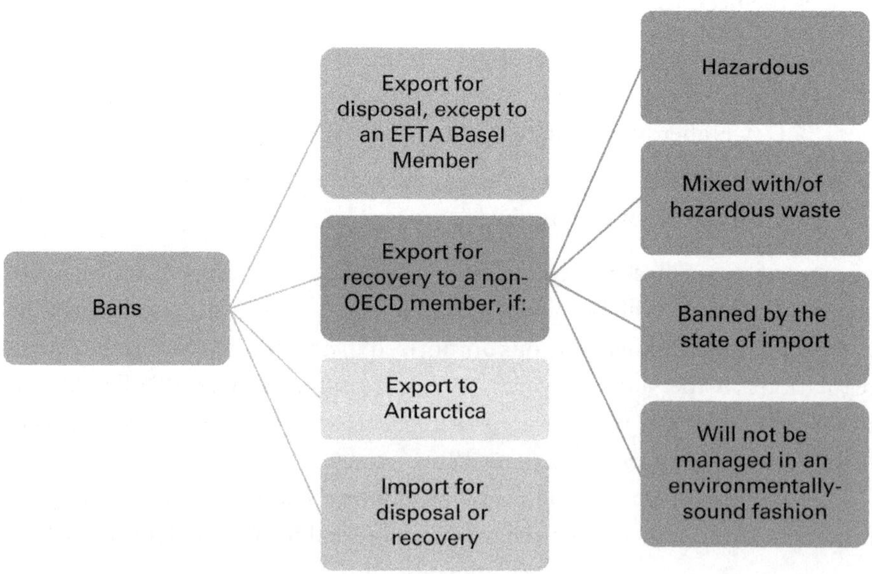

Within the context of the general procedural framework and the specific details applicable to the different cases of waste transfers involving EU Member States, the Regulation mentions several further obligations worth highlighting. The first of these concerns environmental protection, establishing that all those involved in a cross-border waste transfer and its subsequent recovery or disposal must guarantee that all the measures necessary are taken to ensure that the management does not jeopardize human health and is carried out in an environmentally-sound fashion (article 49).

For cases both of waste import and export within the EU, the competent authorities in EU countries must always demand and guarantee that the waste transferred is managed in an environmentally-sound fashion during the entirety of the transfer, including the recovery or disposal operations, prohibiting transfers that are not believed to comply with this general requirement (article 49).

For Member States, this means following the provisions of the Regulation, as well as those in the EU Waste Framework Directive. Environmentally-sound management will be assumed in non-EU Member States of import if the destination installation operates under the national regulations protecting human and environmental health equivalent to those in EU legislation. Such a presumption may however be reversed if a global assessment is carried out on the management of the waste transferred (article 49).

Another general obligation established by the Regulation is the need to carry out controls on compliance with the rules on European waste transfers by the Member States. In this respect, the Regulation sets out that the Member States must establish rules on the sanctions applied for infractions of the provisions described above, which must be effective, proportionate and dissuasive, taking the measures necessary to guarantee their application (article 50).

The Member States must also take the measures necessary to conduct inspections in establishments and companies performing waste treatment, as well as in waste transfer sites or those offering recovery or disposal. The transfer controls may be carried out in the place of origin, at the destination, at EU borders and/or during the transfer within the EU, including document inspection, confirmation of the identities of those involved and, if necessary, physical checks on the waste (article 50).

A CRITICAL ANALYSIS OF THE EUROPEAN REGULATION

Given the main strands of the EU Regulation on cross-border waste transfer, it can be seen that a more socio-ecological and less economicist approach is being taken to the subject. The list of grounds for objection to waste transfer requests provides sound preventive instruments in socio-ecological terms for states which are in fact politically committed to ecological integrity and social well-being. The respect for the principles of proximity and self-sufficiency, the coherence with management plans and the hierarchy of waste management, and the insufficiency of environmental and economic gains upon analysis of the values and costs of transferring, recovering and disposing non-recoverable waste reveal the holistic treatment and the ecological bias with which the European legislation handles the subject.

It is interesting to note that the Regulation text reaffirms the need to guarantee that waste will always be managed in an environmentally-sound fashion, whether it is transferred or not. Special attention is given to this point in the provision imposing the duty of environmental protection. This European regulation defines 'environmentally-sound management' in a very similar fashion to the Basel Convention, stating that it constitutes in 'taking all practicable steps to ensure that waste is managed in a manner that will protect human health and the environment against adverse effects which may result from such waste' (article 2) (EU, 2006b).

Furthermore, in Annex VIII, the Regulation links several documents establishing directives on the environmentally-sound management of certain waste (such as biomedical waste) and the provision of healthcare under the Basel Convention and the OECD's references to used, obsolete personal computers. As demonstrated, this definition has been the subject of criticism, effectively restricting itself to the banning of poor waste management. Annex VIII does not add much in this respect, listing directives effectively non-binding.

However, the text directives exceed the definition of sound management, revealing themselves to be more holistic through the inclusion of the waste transfer process as part of the analysis of management and proposing initiatives to guarantee best management by means of a global assessment of the management and the use of the best available technology, not only to prevent environmental damage and risks to human health, but also to provide a high level of protection to both. It would be better if the Regulation established such instruments as indispensable to the decision-making processes on waste transfer. Despite this failure, it does offer the possibility for the Member States to do so by means of the internalization of the regulation at national level.

A more preventive approach may also be identified at various points, such as in the provision on the possibility of disagreements over waste classification, which may occur within the states involved in transfers. The precautionary principle and *in dubio pro natura* ('when in doubt, favour nature') reign, always prioritizing the use of more rigid controls when there are doubts over the classification of a material as waste, its classification as hazardous or non-hazardous and the operation it is to be subjected to (recovery or disposal).

Of course there are elements of the Regulation susceptible to similar criticism as that made about the international instruments analysed here, such as the control system founded on the prior written notification and consent instrument, which, although showing great potential, requires modifications able to guarantee that the most vulnerable states and their populations are in fact in positions of equality in political and informational terms in the decision-making processes on waste transfers. The fact that alternative agreements are prioritized over the duty of re-collection when problems arise over contractual non-compliance or the terms of the Regulation also reveal it to have repeated errors of the international legislation.

In light of the general terms of the Regulation on cross-border waste transfers within the EU, the next and final section of the present chapter will provide an analysis of the legal regulation on this operation in the Brazilian national sphere. The elements of greatest interest to the various analyses made here will be highlighted at the end of this chapter.

Brazil

Upon the approval and signing of the Basel Convention by various countries in 1989, in 1992 Brazil signed the instrument, which came into force in the country in the same year. As a subsequent act, Brazil promulgated the Convention internally in 1993 by means of Decree no. 875, reaffirming its position as party to this international multilateral agreement, declaring that the Convention must be complied with in its entirety within Brazilian territory, except for the reservations expressed upon deposit of the instrument of ratification (article 1).

Although the Brazilian government stated that it believed the Convention text to be positive, the declaration of its reservations expressed concern over certain points of the instrument it referred to as 'deficiencies'. This declaration highlighted the absence of solutions for a concrete reduction in the generation of hazardous waste and more rigorous controls for the movements of such waste, considering the provisions of the Convention, notably article 4, paragraph 8, and article 11, too malleable (Brazil, 1993).

The Brazilian government thus affirmed its signing of the Basel Convention as the first step of a negotiation process for:

a reducing cross-border waste transfers to the minimum levels possible for efficient and sustainable management;

b minimizing the quantity and toxic content of hazardous waste;

c ensuring the environmentally-sound disposal of hazardous waste as close as possible to the location in which it was produced; and

d assisting developing countries in managing the hazardous waste they produce in an environmentally-sound fashion.

SOURCE Brazil, 1993

Almost 20 years after the internalization of the Basel Convention to the national legal order, Brazil finally promulgated its first National Policy on Solid Waste, expressly prohibiting the import of hazardous waste and refuse, as well as any waste whose characteristics may harm the environment, public, animal and plant health, even if for treatment, refurbishment, reuse, recycling or recovery (article 49).

The NSWP thus banned the import of (Figure 5.17):

i hazardous waste destined for both disposal and recovery;

ii waste destined for disposal (refuse); and

iii other waste that, although not classified as hazardous or as refuse, has characteristics which may harm the environment as a whole or specifically human, animal or plant health, regardless of whether it is destined for disposal or recovery.

Although the NSWP established such prohibitions, it did not provide further details on the specifics of the control of cross-border waste transfers, nor did it define 'hazardous waste' and 'non-hazardous waste' (as seen in Chapter 4). It was in this context, therefore, that CONAMA devised and published Resolution no. 452 in 2012, regulating the control procedures for waste imports based on each country's sovereign right to regulate the entry and exit of foreign waste in its territory recognized by the Basel Convention (article 1).

Apart from the definitions of hazardous and non-hazardous waste, CONAMA Resolution no. 452 defined controlled waste as non-hazardous waste, whether inert or non-inert, subject to control by IBAMA, which may restrict its import (article 2, section VI). In compliance with the Brazilian NSWP, the Resolution affirmed the ban on imports of waste classified as hazardous (class I); and waste destined for disposal (refuse) (article 3).

It should be noted that waste listed in Annex II of the Basel Convention (waste collected from households and the refuse resulting from its incineration),

FIGURE 5.17 Levels of control for cross-border waste transfers under Brazilian legislation

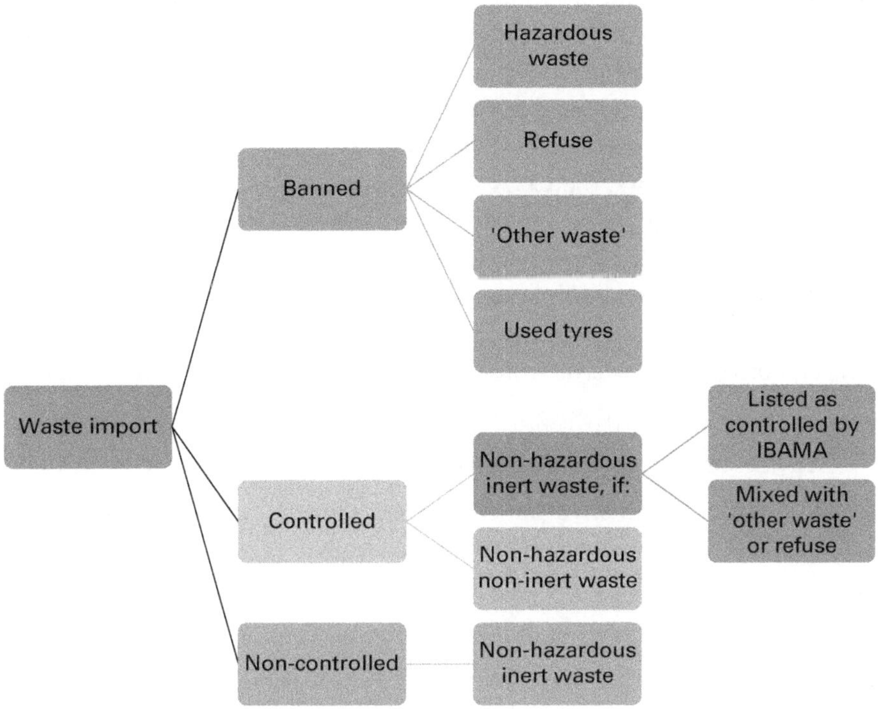

following the Convention's line of reasoning, is defined as 'other waste' by Resolution no. 452. As mentioned, such waste is encompassed by the definition of 'hazardous waste' in the Resolution, with the general ban on imports thus applicable. The single paragraph of article 4, however, states that the ban may be lifted if bilateral agreements signed by Brazil are in place.

The import of waste destined for disposal (refuse), hazardous waste, 'other waste' and waste from used tyres is prohibited across Brazilian national territory. The import of non-hazardous waste is permitted and controlled by IBAMA, whether inert or non-inert. As a general rule, waste classed as non-inert (class IIA) is always controlled, while waste classed as inert (class IIB) is not subject to import restrictions, although with some exceptions. This means that, even if classed as inert non-hazardous waste, import is controlled if: i) it is on IBAMA's 'controlled waste' list; or ii) if it is mixed with 'other waste' or refuse (article 6).

The list of inert non-hazardous waste controlled by IBAMA may be modified upon the identification of potential harmful effects, whether to the environment or human, animal or plant health, by means of a reasoned, sole decision (article 6, paragraph 1). Such waste will be governed by the general procedures applicable to 'controlled waste' (article 6, paragraph 2). Examples of waste considered to be non-hazardous and subject to import control by IBAMA include: battery residue and refuse; unserviceables; electric parts of machines and devices (Annex VI, IBAMA, 2013).

On such procedures, the CONAMA Resolution establishes that only the consignees may carry out controlled waste imports, whether directly or by means of third parties contracted, as long as they have duly licensed installations for such purposes, with prior authorization and approval granted from IBAMA (article 7), and the waste is destined for recycling. IBAMA, for its part, only authorizes imports if the consignee meets the following requirements:

I good standing on the Federal Technical Register of Potentially-Polluting Activities or Users of Environmental Resources (CTF), managed by IBAMA;

II presentation of an environmental licence on the part of the consignee, issued by the competent environmental body;

III technical report attesting to the classification of the load of waste imported, except in cases with grounds for exemption according to IBAMA;

IV meeting of national and international regulations on packaging and transport, as well as special handling during transit, including in internal operations, apart from the provision of emergency procedures for each type of waste;

V compliance with the conditions established by federal, state and municipal legislation on environmental control for the storage, handling, use and reprocessing of the imported waste, as well as any waste generated in this operation, including in terms of final disposal.

SOURCE CONAMA, 2012

It is worth emphasizing that, apart from these demands, imports carried out by third parties under contract will require the presentation of a copy of the contract signed with the consignee (article 7, paragraph 4). Furthermore, IBAMA may, if necessary, request other documents and information from importers (article 7, paragraph 5), as well as establish complementary control and consignment regulations aligned with national laws and the Basel Convention (article 10). Finally, the technical report mentioned in section III of article 7 must be issued by laboratories accredited by INMETRO or by foreign laboratories accredited by accreditation bodies, to have signed a mutual recognition agreement to which INMETRO is a party (article 7, paragraph 6).

Apart from the requirements mentioned above, importers of non-inert non-hazardous waste must also meet the prior notification procedures set out in the Basel Convention (article 8). The Resolution also specifies that non-hazardous waste from countries which are not parties to the Basel Convention, regardless of whether it is inert or non-inert, may only be imported by Brazil if bilateral, multilateral or regional agreements or arrangements are in place with such countries (article 9).

In terms of the duty to monitor compliance with the environmental regulations and those on waste in Brazil, it is the responsibility of the environmental bodies across all federal levels to inform IBAMA of any breach of the regulations at any stage of the management of imported waste. Apart from the duty to monitor, IBAMA shall take the necessary measures laid out under the Basel Convention whenever it detects or is informed of non-compliance (article 11). Apart from the measures outlined under the Basel Convention, offenders will also be subject to sanctions set out in the Brazilian Environmental Crimes Act and its regulations (article 13).

Complementary to Decree no. 875/93, the NSWP and CONAMA Resolution 452, is IBAMA's further publication of Normative Instruction (NI) no. 12 of 16/07/2013 for regulating control procedures for waste imports. Under the same terms as the CONAMA Resolution, IBAMA's NI adopted the methodology for classifying waste in terms of hazardousness developed by standard ABNT NBR 10.004/2004, reproducing practically the same definitions and rules as the CONAMA Resolution. IBAMA's NI, however, features a list of waste whose import is prohibited or controlled according to the MERCOSUR Common Nomenclature (MCN), based on the Harmonized Commodity Description and Coding System (HS).

Furthermore, IBAMA's NI no. 12 also specifies that authorization for imports of 'controlled waste' must be requested from this body by the waste importer by means of the presentation of a form suitable for the type of import to be carried out (article 5). Before requesting this authorization, the importer must ensure that the notification procedure for the waste export as defined under the Basel Convention has been initiated by the exporting company (article 5, paragraph 1). If the importer proves that the product to be imported – although listed as 'controlled waste' by the NI – is not waste, there shall be no need for authorization as laid out in the NI.

With the request received, it will be IBAMA's responsibility to issue authorization for the import within 15 working days (article 6). It must be highlighted that such authorization will be subject to revision whenever modifications or updates are made to the information or documents presented by the importer, which must be immediately communicated to IBAMA under penalty of cancellation or suspension of the authorization in cases of omission (article 6, paragraph 2).

Also constituting causes for cancellation or suspension are the following:

- **I** non-compliance with the regulations governing waste import;
- **II** the omission or forgery of information subsidizing the authorization act;
- **III** cases of serious environmental harm or risks to public health;
- **IV** non-compliance with specific conditions, which, if existent, shall be established by IBAMA under the authorization act; and
- **V** non-compliance with the general conditions mentioned in the NI (article 7), as follows:
  - **i** maintenance, during the period authorized for the import, of the validity of the environmental licence for the operations conducted by the consignee, issued by the competent environmental body;
  - **ii** compliance with the conditions established by federal, state, district and municipal legislation on environmental control for storage, handling, use and recycling of the waste imported, as well as any further waste generated during this operation, including its final disposal;
  - **iii** compliance with the national and international regulations on packaging and transport, and with the special care to be taken handling the waste during transport, including internally, as well as the provision of emergency procedures for each type of waste;
  - **iv** maintenance, during the period authorized for the import, of the good standing of natural and legal persons involved in the import process on the Federal Technical Register;
  - **v** undertaking of the recycling of the waste imported only in the industrial installations listed on the authorization document;
  - **vi** maintenance of the authorization, and notification and cross-border waste movement documents of Basel Convention along with the cargo imported during the waste management operations;
  - **vii** sending to IBAMA, within 10 days upon receipt of the waste cargo at the recycling installations of the Transboundary Waste Movement document that accompanied the cargo, with lines 17 (Transfer received by the importer) and 18 (Transfer received at the recycling installation) duly completed (article 7, IBAMA, 2013).

Apart from the cancellation or suspension of the authorization of the waste import, non-compliance with the regulations governing waste imports will also constitute illegal traffic under the terms of the Basel Convention (article 9). In such cases, those responsible for the import will be subject to the sanctions listed in the Brazilian Environmental Crimes Act, notably in articles 56 and 70 (article 12).

Finally, IBAMA NI no. 12 outlines that it may carry out inspections and monitoring of waste cargos imported, at any time (article 10), as may it permit cargo transfer by means of the shipping authorization, making the definitive approval of the import licence subject to the terms of its prior physical inspection of the goods (article 9).

A CRITICAL ANALYSIS OF THE BRAZILIAN REGULATION

An analysis of the Brazilian regulation on the cross-border transfer of waste reveals that the national legislation focuses on waste imports, with no specific regulations on exports. This is no more than a legal reflection of the geopolitical reality which sees waste streams mainly flowing from North to South (as seen in Chapter 1 of this book). Despite adhering to the Basel Convention, it is also due to this global logic that Brazil, in light of the shortcomings identified since the Convention's formulation and signing, positioned its own signing of this international multilateral agreement not as an end, but rather as the beginning of a necessary process of negotiation between the various parties in order, if not to interrupt waste transfers, then at least to minimize them and the waste generation.

With the reservations incorporated into its instrument of participation to the Basel Convention, Brazil, just like other Southern countries which are consignees of waste, declared its unsatisfaction with the Convention's malleability and insufficiency at dealing with the problem of cross-border waste transfers. With the promulgation of the NSWP, Brazil took a step beyond the Basel Convention, banning the import of hazardous waste, other waste and waste destined for disposal (refuse). Apart from the positive elements and although the import of non-hazardous waste may be controlled, gaps and insufficiencies can be identified in the national regulation.

Positive elements of the Brazilian regulation include the fact that only persons whose activities are duly licensed to carry out recycling are able – whether directly or by means of third parties – to carry out the import of non-hazardous waste to Brazil. Apart from this, in order to receive import authorization from IBAMA, such persons must comply with all the environmental and information regulations, guaranteeing that the type of waste imported, the packaging, the transport, the handling during transit and the recycling process are carried out according to national and international regulations. Finally, authorization once given is not definitive in nature; its continuation depends on compliance with specific and general constraints, as well as with national and international standards applicable to the import process and recycling, with suspension and cancellation applied in cases of infraction.

However, the requirements for granting and maintaining authorization for waste imports do not guarantee that cross-border waste transfers to Brazil are the best option for managing such waste in ecological and social terms. There is no provision for the need to undertake a comparative analysis on the life cycle of these materials or a systemic analysis of the environmental impacts to better define the best option for their destination; nor is the possibility of public participation in the decision-making process for waste transfer in Brazil contemplated.

The national legislation specifically regulates waste imports for recycling, excluding all other possibilities for waste recovery. Such inadequacy may generate confusion and an inappropriate application of the regulation. There is also no reference to the limits of quantities of waste to be imported, much less on the material efficiency of the recycling process to be carried out on such waste. In other words, a certain quantity of waste imported might not cause significant impacts on the local environment in which it is to be reprocessed, but greater quantities may have the potential to generate serious environmental damage and to public health (ecological rucksack).

Apart from this, although the regulation mentions the possibility of refuse generation after the process of recycling the waste imported, it does not impose any limits on this generation. Even though it bans refuse imports, the law permits part of the waste imported to be destined for final disposal. If the recycling process is not efficient, such a provision of the Brazilian regulation, combined with the absence of a limit or percentage on refuse production, may allow large quantities of waste imported to be effectively destined for disposal (ecological rucksack).

The specific criteria used by IBAMA in forming its list of controlled non-hazardous waste are not clear, nor is the process for monitoring the loads of the waste imported. The regulation merely mentions that IBAMA may conduct inspections and monitoring of the loads of waste imported. It is, in fact, known that the monitoring of loads imported to Brazil is carried out on a sampling basis, meaning only a small percentage of the loads are actually inspected by the competent national authorities.

Despite the fact that all of the loads imported to Brazil are submitted to customs (request for import clearances), the Brazilian regulation mentions that once a risk analysis and the cross-referencing of information between federal databases has been carried out, three levels of monitoring are conducted:

i green, when a documentary inspection and the checking of the goods are not required;
ii amber, in which only the load documents are examined; and
iii red, in which both the load documents and the goods are inspected.

For more details, see Complementary Law 87/1996; Decree no. 6.759/2009; Ministry of Finance Ordinance no. 389/1976; Federal Revenue Office NI no. 680/2006; and Cosit Normative Act no. 7, of 22 August 2014.

Therefore, only a small percentage of the loads imported to Brazil are actually inspected, with the selection made mainly on the basis of declarations provided by importers and exporters.

To this must be added other problems which lead to poor control in Brazil, such as the insufficient number of professionals trained for such operations, the immense size of the national territory and issues linked to corruption. All of these factors end up weakening and sometimes hindering the implementation of the Brazilian regulation and a more efficient control of cross-border waste transfers.

Final considerations

Notwithstanding the fact that many other elements of the regulatory instruments studied here could be questioned, it must be considered that the emphases made in this chapter have the greatest relevance to the object of analysis in this book. The significance of the Basel Convention in beginning to regulate what was previously only left to the market's 'invisible hands' cannot be denied. However, the points raised reveal the Convention's shortcomings in attempting to control cross-border hazardous waste transfers and, more profoundly still, in contributing to a real reduction in the generation and transfer of such waste, and to its environmentally-sound management – the three pillars on which the Convention Secretariat states the instrument rests.

The entry into force of the Ban Amendment, along with the Protocol on Liability and Compensation, will no doubt make a decisive contribution to addressing the problem of cross-border waste transfers, although many other issues must also be faced, with a more in-depth reformulation of the current international regulation seeming inevitable.

The lack of more specific rules which are binding and procedural for reducing waste generation and its transfer; the lack of the inclusion of 'transfer' as a stage in waste management; the lack of provisions grounded on the duty of best management; the excessive malleability and deep economic foundations of the criteria permitting transfers; the prioritizing of an alternative destination instead of return to the country of export in cases of contractual non-compliance or illegal traffic due to the fault of the importer or consignee without a more systemic analysis of the socio-ecological impact in order to identify the best destination for the waste; and the absence of provisions in instruments for public participation and for the carrying out of more systemic analyses on the social and ecological impacts in the decision-making processes on the cross-border waste transfers regulated by this legislation – all of these factors demonstrate the insufficiency of this approach from a socio-ecological point of view, revealing the economistic perspective adopted to regulate the subject, even if its preliminary and general provisions affirm the contrary.

Furthermore, the original text of the Basel Convention violated socio-ecological justice criteria by misrepresentation of the Southern countries, due to its failure to incorporate their demands and needs in the actual development of the text. Although they participated in the negotiations, the positions of economic and political inequality between developed and developing countries remained clear in the original text approved by the Convention. Such a violation is perpetuated and exacerbated over time due to the fact that the Ban Amendment and the Protocol on Liability and Compensation have not entered into force, even after several years have passed. Additionally, organized civil society and non-governmental organizations, which could have helped in ensuring wider representation for current and future generations of humans and non-humans, were excluded from the process of negotiating and approving the regulation.

It is also clear that the OECD Decision adopted an economistic approach, with some limitations of an environmental nature. In its very Preamble, the Decision reveals this bias in affirming that the exploitation of materials and energy from waste is seen as an integral part of the international economic system and that well-established international markets already exist to collect and process these materials. The Decision also affirms that environmentally- and economically-sound waste recovery justify the practice of cross-border transfer of waste between Member countries. However, apart from failing to define what environmentally-sound waste management entails, the Decision does not mention the fact that the duty towards best management and social issues must form part of the decision-making process on waste transfer, going beyond the economic element and the prevention of poor management.

Apart from reproducing many of the limitations of the Basel Convention, from a socio-ecological point of view it may be affirmed that the OECD Decision makes the economic bias even clearer throughout the course of its text. Issues such as the complete absence of control for transfers of waste considered to be non-hazardous; the possibility of tacit consent; the malleability of the regulation for multiple shipments; the absence of international control for non-listed waste and the negligence on possible negative ecological interactions between wastes considered to be non-hazardous demonstrate that this regulatory instrument also falls short from a socio-ecological point of view.

Unlike the international documents studied here regulating cross-border waste transfers, the European Regulation clarifies its emphasis on environmental protection, to the detriment of economics. This emphasis is highlighted in the Preface to the Regulation, which categorically affirms that the principal and predominant objective and strand of the EU Regulation is to protect the environment, with its effects on international trade merely secondary. It declares that, if such an objective cannot be sufficiently met by the Member States, the EU may take measures to ensure it is fulfilled on a European level (Preface).

The provisions of the EU Regulation reflect such an objective very well, going beyond mere declarations of intent. While the international instruments seem to provide empty declarations on the need to reduce waste production and the transfer of such waste, the European Regulation and the European Waste Framework Directive establish clearer paths to do so, notably on grounds for raising objections to requests for cross-border waste transfers.

It has been demonstrated that the legal regulating of the subject in the Brazilian national sphere also incorporates many elements of the international regulatory structure, despite adopting a socio-ecological rather than economicist perspective in legal terms. In this context, it must be highlighted that just like the legislation, Brazilian jurisprudence also clearly adopts a socio-ecological perspective on the subject. Iconic precedents in this respect are the rulings on the import of used tyres, with the ban on such cross-border waste transfers upheld in the country based on the constitutional principles of protecting public health and an ecologically balanced environment, such as in the ruling made in Special Appeal no. 1.129.785-PR (2009/0144125-4) by the Brazilian Superior Court of Justice.

Despite the positive elements of the national legislation, such as the firm position taken in banning waste imported for disposal and on hazardous waste, there are regulatory limitations which weaken the control system for cross-border transfers to the country, opening up loopholes for illegal traffic and unjust management that is ecologically unsustainable for waste classified as 'non-hazardous'.

Although not exempt from criticism, generally and in comparison to the international and Brazilian regulations on the subject highlighted here, the EU Regulation appears much more coherent, detailed and effective at governing cross-border waste transfers, and may be used as a basis for the creation of a new model of governing waste transfers in an international context, in order for the subject to be treated as global waste management rather than international waste trade.

This Regulation also expresses the European Union's responsibility and commitment to global socio-ecological justice, revealed in the many bans on waste exports to outside the Community, for example, which may, and should, be reproduced in the international sphere. Based on such conclusions, and on the theoretical developments carried out in the first part of this book, the next and final chapter will outline several strategies for legally governing global waste management as a political–legal instrument which may help overcome the waste crisis.

06

Global governance and the management of waste

Introduction

Throughout its course, this book has sought to demonstrate how waste has come to serve as a socio-ecological problem in recent decades, rendering it an important target for the environmental movement and also a problem of justice. As a result, a process involving the juridicization (legal regulation) of waste has been sparked in order to attempt to solve such problems. As the legal regulation of the issue has so far fallen short in dealing with the crisis, waste has also become a legal problem due to the limits of the legislation on the subject (studied in Chapters 4 and 5 of this book).

These limits reflect problems in the juridicization of waste by the regulatory instruments assessed, directly impacting on the regulating of its international management. Although it is clear that the way in which the law regulates waste is key to overcoming the waste crisis, the form such regulation should take remains the subject of debate. As demonstrated in Chapter 3, the process by which Nature is regulated in law should occur alongside the process by which the Law is ecologized, in order for the legal regulation of the subject to supplant the mechanistic paradigm. A more suitable means of regulating waste must therefore be sought, emerging as a path towards the adoption of a new legal paradigm that is more reflective of the complex reality.

To contribute to such a construction, the present chapter will work with the notion of 'material flows' from the revolutionary thesis of Material Flows Law by author Alexandra Aragão. Her proposal incorporates theoretical frameworks from other areas of knowledge, such as the social metabolism and thermodynamics, in order to create a new legal–ecological paradigm that seeks to overcome the mechanistic model. In this sense, the central focus of this chapter will be the development of strategies for regulating global waste management in the most effective way possible according to a socio-ecological standpoint.

The rationale will narrow from presenting wider strategies to those which are more specific. It will begin with the notion of 'governance of material flows' to then concentrate on the notion of 'global waste management', with strategies for negotiation, development, general content and specific content of the international regulatory instrument devised here. The strategies outlined will not be presented as singular, complete and definitive, but as possibilities which are open to dialogue and engagement in the aim of establishing a form of global waste management that is compatible with the assumptions of socio-ecological justice.

Governance of material flows

Social metabolism and law

An analysis of the legal concepts of waste in the legislation studied in Chapter 4 reveals that all of them adopt a static perspective, according to which waste is considered as tangible, appropriable objects, discarded due to the lack of interest they represent to their holders. This treats waste as physical objects, whether in a solid, liquid or gaseous state, as long as they are appropriable (thus excluding emissions, whether thermal, visual, acoustic, light or radioactive, for example) (Aragão, 2006).

This static approach distinguishes waste from other things in a legal sense not by physical characteristics, but by the holder's attitude towards it. The material towards which the holder demonstrates a lack of interest or rejects is considered waste. In referring to waste as 'things thrown away', Aragão (2006) clarifies that she chose this expression as it is rather suggestive within the static vision of waste, reflecting as it does an attitude displayed by someone with special responsibilities towards the object rejected. This individual is seen by the author as the 'natural parent' of the object rejected, given that they produced or acquired it, or that they were in some way responsible for its emergence as waste.

According to the static perspective, the natural destination of waste, or the thing thrown away, is to be discarded. The owner of the thing thrown away does not therefore have the intention to pass it on to someone else, regardless of the purpose. And it is for this very reason that Aragão prefers the expression 'throw away' to 'abandon', given that it is closer to waste's destination, which is to be discarded. According to the author, while 'abandon' is more linked to the voluntary renunciation of the possession of interesting things, 'throw away' better reflects the discarding of uninteresting things, with the owner not caring how the objects are thrown away or where to. To summarize, from a static perspective, waste is appropriable things, which, due to the lack of interest they represent to their owner, are thrown away or have already been thrown away (Aragão, 2006).

In the legal system, the static view of waste generates an approach singly focused on the controlled destination of these materials in the environment. Known as the 'classic waste law', this identifies the (local) public authorities as the main subject of waste management in the legal regulations, making them responsible for the collection, processing and destination of these materials by means of the adoption of techniques considered to be suitable in terms of the environment and public health. In this sense, such an approach grounded in the static perspective is restricted to a 'waste disposal law', as it is only concerned with the controlled destination of waste (Cipriano, 2015).

As Aragão (2006) highlights, the static perspective fundamentally orients the law towards finding suitable destinations for waste, thus exerting a limited influence over any possible shifts in social behaviour, given that it ignores its origin and only considers it once it has been revealed to be a risk, whether potential or real. It may thus be perceived that the static perspective is founded on the logic of Mechanistic Law, treating waste as isolated objects, without a context and disconnected from ecological principles.

According to the logic of metabolism (outlined in Chapter 1 of this book) and the complex models (developed in Chapter 3), it is essential to understand that, apart from being a rejected static object, waste must be seen as part of the material flows permeating ecologic, social and economic systems. Known as dynamic, this new perspective serves as another approach to the subject of waste, viewing it as an output of the social metabolism. According to Aragão (2006), using the language of physics makes it possible to affirm that waste – just like light – behaves both like particles and waves.

The use of biological theories to consider the social process of production, consumption and waste elimination, summarized in the idea of social metabolism, demonstrates that the social and economic systems receive matter and energy from the ecosphere (or ecosystem), which are then processed and returned to the ecosphere in the form of waste and pollution, forming a stream of matter and energy. From the dynamic perspective, therefore, waste is found in the final stage of a material flow when it is transferred from the economic and social systems back into the ecosphere.

As they are open, the social and economic systems have entries and exits by which they interact with the ecosystem. The environment influences the social and economic systems by means of the inputs, with such systems influencing the environment by means of the outputs. As demonstrated by the laws of thermodynamics, the time element is also a variable and must be taken into consideration in an analysis of the systems before and after the exchanges of matter and energy; before and after the processing and transformation of these elements during the flow, as the process of entropy is inevitable and irreversible in this metabolism.

Guided by this dynamic perspective of waste, Aragão divides the social metabolism into anabolism and catabolism, identifying two great moments of legal intervention in material flows, which she refers to as Anabolic Law and Catabolic Law. Based on a time analysis of material flows, anabolism refers to the period prior to the production of waste, when materials leave the sphere of the natural and enter the spheres of the social and the economic. The anabolic period of the social metabolism ranges from the extraction of the raw materials in the ecosystem to the transformation of these materials into products, and finally, the use or consumption of these products within the social and economic systems. In this context, the legal regulation of access to ecological goods, the preservation of Nature, industry, commerce, services and consumption are all the object of Anabolic Law, which seeks to suppress and reduce material flows by means of precaution and prevention (Aragão, 2006).

Catabolism, on the other hand, refers to the period in which materials leave the social and economic systems (technosphere) and re-enter the ecosphere, ie the moment after waste is produced, upon its management for disposal. It is therefore the object of Catabolic Law (or the waste law in a strict sense), which seeks to reduce and redirect catabolic material flows selecting final destinations that are more suitable for waste that cannot be avoided, promoting sound catabolic management (Aragão, 2006). According to Aragão (2006), while the static perspective focuses on catabolism in the process of material flows of the social metabolism, the dynamic perspective spans both moments, placing equal or greater emphasis on the first stage, the anabolic one. In this sense, such a perspective does not ignore the origin of waste, considering the material flows among the different systems before they are transformed into a harmful anti-ecological phenomenon.

Unlike the law of waste management, the law of material flows would therefore primarily seek to prevent waste from occurring in the first place, before then locating suitable destinations for waste which is inevitable. Furthermore, the more efficient Anabolic Law is, the less the need for Catabolic Law, with this ideally disappearing. The law of material flows therefore entails two ecological problems recognized to be growing in seriousness and complexity, and whose message for all of us is clear: resource preservation and waste management' (Aragão, 2006).

In this light, Gharfalkar *et al* (2015) highlight that the number of researchers to have come to describe waste as 'resources' has grown significantly, following the trend by which the concept of 'end-of-pipe waste management' has migrated to a more holistic approach of 'resource management', such as the standout 'zero waste' management model. The importance of focusing on waste as a resource has been widely promoted by models such as that of the Circular Economy, which reinforces the paradigm based on resources instead of the paradigm based on waste.

According to the law of material flows, the best waste is therefore that which does not exist. Prevention thus assumes a key role in this perspective. The notion of prevention founded on a metabolic approach is complex and extends beyond the mere proposal to reuse and recycle waste: according to the dynamic perspective, preventing waste encompasses (first) preventing and (then) reducing material flows. Avoiding waste means consuming less, and, in order to consume less, we must produce less, which in turn means reducing the use of raw materials from Nature (Aragão, 2006).

Among other measures, controlling production conditions and the conditions of goods consumption, and even prior to this, the conditions in which raw materials are extracted, is part of preventive action. Within this expanded vision of the legal regulation of waste, the law of waste flows and the unification of a set of subjects – which were traditionally approached by different disciplines in a compartmentalized fashion – reveals the most logical path (Aragão, 2006). Aragão's thesis therefore connects with the proposal of ecologizing Law to an overcoming of the reductionism, isolation and fragmentation of the mechanistic paradigm.

The notion of governance

Based on the dynamic perspective outlined by Aragão, an analogy may be made by strategically working with the notions of governance and management. In business terms, the notion of governance refers to the form in which an organization is managed at the highest level and the systems used to do so. Governance offers a wider view of the organization, decisions on strategic management and the development of policies and regulations defining the purpose, values and the structure of a certain organization, offering a guide for its management. Management refers to the activity of organizing and implementing definitions made on higher levels, that is, to routine decisions and administrative work on the organization and implementation of goals and values defined on the level of governance (Cambridge, 2011).

The notion of 'governance' adopted here emphasizes that the subject of waste must be treated in terms of material flows, and, following Bosselmann (2016), not only by the nation states, internally and internationally, but also by other actors that participate in global environmental governance – such as international organizations, non-governmental organizations and organized civil society itself – engaged in socio-ecological protection. Furthermore, the selection of this term aims to highlight the potential for multi-level action (across various levels) of environmental governance (global, regional, transnational, state and local).

Finally, the choice of the notion of governance is justified by its underlining of an action that moves beyond the legal regulation of the subject, also being expressed in other forms, such as by the power of geopolitical, economic and cultural relations, among others. In this light, governance is revealed to be a complex process, in which different systems interact and are mutually modified in circles of retroactivity and recursion, with Law only one of them. Of course, it is also true that the governance of material flows will be influenced and reaffirmed by the domestic and international legal regulation of the subject, at the same time as it influences the juridicization of waste by means of the theoretical approach to material flows.

Sass and Melo (2014) stress that the notion of governance has been strengthened in light of evidence as to the shortcomings of the (Westphalian) nation-state model in confronting issues of a global impact. For the authors, global governance refers to the set of instruments (both formal and informal) used in decision-making processes on subjects of an interest that transcends national borders, without the establishment of a single system or a central government. In short, the main characteristics of this notion are the existence of multiple levels, pluralism, geometric variability and a complex structure.

Alier (2003) claims that the majority of the world's current environmental problems are complex and feature many contradictory scientific, economic and social facets. The nation states have thus not revealed themselves to be capable of adopting procedures to integrate the different points of view – tripping up on environmental agendas which they, in actual fact, have not mastered and have not fully assumed. With that said, nowadays the adoption of a strategy to migrate to a form of governance defined as the capacity to bring together a wide range of opinions, both from specialists and from the sectors affected, may lead to a decision-making model in which different valuation discourses are adopted, and such decisions, even if not optimum, will be legitimate for having been agreed upon.

There is no doubt that waste has come to represent a problem with a global impact and that the international structure focused solely on the nation states has been ineffective in overcoming the waste crisis. Such a fact has been proven by the results of the Basel Convention itself, from its negotiation to its current incomplete implementation. In fact, the international structure focused merely on the actions of the nation states has solely acted on the treatment of cross-border waste transfers more as a business than a socio-ecological issue. These are some of the reasons why the notion of governance seems to offer an alternative model in the search for better ways to govern the subject.

The priority objectives of the governance of material flows

The proposal here is for governance not only of waste, but of material flows. Although the framing of materials as waste or non-waste has remained nebulous under legal regulations, it is possible to affirm, in a simplistic fashion, that materials requiring recovery for their reuse or that have been irreversibly degraded and must be disposed of may be considered as waste. All other materials, are, by contrast, non-waste.

> This general definition of **waste** – 'materials requiring recovery for their reuse or that have been irreversibly degraded and must be disposed of' – is used here merely with the aim of the theoretical development proposed. In practice, the distinction between waste and non-waste is much more complex, as outlined in Chapter 4, requiring more detail and clearer criteria in the legislation regulating the subject.

According to the dynamic perspective on material flows, governance is required both over waste and non-waste. In other words, governance of material flows requires actions and regulations throughout the flows' entire metabolic cycle, working in a coherent, systematic and holistic fashion in order to reduce flows of matter and energy passing through social and economic systems as much as possible. This would be the first priority goal for the governance of material flows, and the strategies for achieving such a target may be summarized by the ideas of 'de-production' and 'de-consumption' presented by Aragão (2006).

Aragão (2006) meticulously develops each of these strategies and how the Law can internalize them. Although it is of fundamental importance to the subject of waste, the details of these theoretical developments will not be analysed here. Based on brief points made on the Circular Economy (in Chapter 3), the following strategies to reduce flows of matter and energy can be highlighted: virtualizing (or dematerializing), optimizing systems in order to reduce matter and energy (or eco-efficiency), non-consumption, consumption reduction, sharing, prolonged use and direct reuse (without the need to recover in any way) of products.

The second priority goal of the governance of material flows would be to unite all of the measures, strategies and efforts in order to get as close as possible to closed cycle of these flows. Taking into consideration the entropy law, as seen in Chapter 3, such cycles may never be completely closed, as there will always be some loss of matter and energy in the socio-economic transformation processes. Such an aim will be sought for flows that cannot be avoided by the actions previously mentioned, with Aragão's terms once again relevant to summarize the strategies for this goal: 'lasting production' and 'lasting consumption'.

Again, this book will not delve into the meticulous details explored by Aragão. However, an example of a strategy that reflects this second aim is the use of a cascade logic (cascading): the return of the material to the production cycle at the end of each of its uses, in order to allow for its reuse with the least interference possible from recovery processes, thus maintaining its highest degree of usefulness and, as a consequence, the lowest degree of entropic loss.

Generally speaking, these two priority objectives may be said to seek to prevent the extraction of new natural resources, from one end of the social metabolism, as well as the disposal of new waste into the ecosphere, from the other end. There is also a third objective that may be listed as a priority for the governance of material flows, and this is to spark the development of actions, measures and strategies to reduce as much as possible the negative impacts generated in the social and ecological systems by flows of matter and energy that cannot be avoided.

This is also a preventive objective; however, the prevention it seeks to engender is of the negative impacts of flows that cannot be avoided or closed within the socio-economic system. Examples of measures for the implementation of this objective include a maximum reduction in technical cycles with the gradual migration of the production system to a biological cycle (using materials based on plants instead of plastic, for example), the use of renewable instead of non-renewable sources, the reintroduction of materials recovered in the ecosphere and the non-production and use of toxic materials, among others.

Other elements of the governance of material flows

In terms of actors, Sass and Melo (2014) highlight the asymmetry of power specifically occurring in the current experiences of global governance due to economic and political powers (typical coloniality of power), which means that in practice only some actors develop the terms of governance – the 'governance makers' – while others are simply positioned as 'governance takers'. Such an asymmetry of power was in fact revealed in the results of the negotiations at the Basel Convention, as well as in its implementation to date.

In this sense, it may be understood that, in order to express democratic values and those of socio-ecological justice, the governance of material flows must recognize this reality and seek strategies to meet the demands made by highly diverse actors who, apart from having their voices heard, must have their demands considered and included. Such strategies, linked to the 'how' element of justice in Fraser's theory mentioned in Chapter 2, will be developed in more detail in the following section.

The empowerment of global organized civil society, in the sense defended by Bosselmann (2016), has great potential for a form of governance oriented towards the targets of socio-ecological justice, given that state actors do not always represent the real interests of their representatives. This does not of course refer to the exclusion of these traditional actors, but instead, an amplifying of possibilities for democratic dialogue in the making of decisions essential to the future of humanity; more specifically, to the very continuity of the human race on this planet.

Furthermore, added to these strategies for expanding democracy in terms of participation in the governance process are strategies for expanding democracy in epistemological terms that should also be developed, for example, by means of the ecology of knowledges proposed by Boaventura de Sousa Santos. This means that the governance of material flows must seek to transcend actions founded purely on technical–scientific knowledge, opening up to the dialogue between different types of knowledge, represented by different actors. Such a form of governance must therefore prompt a re-appropriation of Nature by individuals and communities that, by means of and in dialogue with various actors, shall seek the best solutions, measures, actions and strategies in terms that are not just ecological, but also social, in an open and democratic fashion.

The governance of material flows has the potential to transform the preventive aspect related to waste. Instead of treating the subject in terms of 'encouragement', an effective and binding model may be consolidated by means of a more systemic and holistic approach. In general terms, this governance should seek to balance the entropic and negentropic processes of the socio-economic and ecological systems, with an increase in the effectiveness of the governance acting on the anabolic phase of the social metabolism, gradually reducing the need for governance to act on the catabolic stage as a result.

Empowering communities linked in a network in a transnational fashion – organized civil society – seems to increase the chances of migrating from a dichotomized approach between the market and state sovereignty to an approach founded on protecting the commons, in which other discourses are legitimized in the processes of pondering different interests (such as the discourse of socio-ecological justice). Furthermore, such empowerment may help make part of the humanity and Nature marginalized under the mechanistic abyssal logic more visible, prompting a greater socialization of power. In this sense, it may be believed, for example, that if other actors had had access to the negotiation process of the Basel Convention (apart from the nation states), the results achieved would perhaps have been quite different.

Waste management within the governance of material flows

After briefly sketching the general outlines of the governance of material flows, the next step is to consider the second notion of fundamental importance to this new proposal for international waste regulation: management.

As demonstrated, management is contained in governance. The governance of material flows encompasses a series of different types of management, all interlinked and oriented towards the practical implementation of the general aims of governance, such as the management of natural resources, the management of production processes and waste management (see Figure 6.1). While governance would function in a holistic fashion in planning how to deal with material flows (from non-waste to waste) and guaranteeing that this planning will be implemented, each type of management would be dedicated to the organization and implementation of the plans in specific subjects.

According to this logic, waste management would tackle the organization and implementation of planning carried out across the different levels and by the different actors in the governance of material flows. In general terms, waste management must therefore push for a more direct implementation of the final two priority objectives of the governance of material flows (the closure of the flows and the reduction of the socio-ecological impacts) for all materials destined for recovery (and, thus, remaining within the socio-economic system – in the anabolic stage), as well as those destined for disposal (and thus returned to the ecosphere – the catabolic stage).

FIGURE 6.1 Governance of material flows

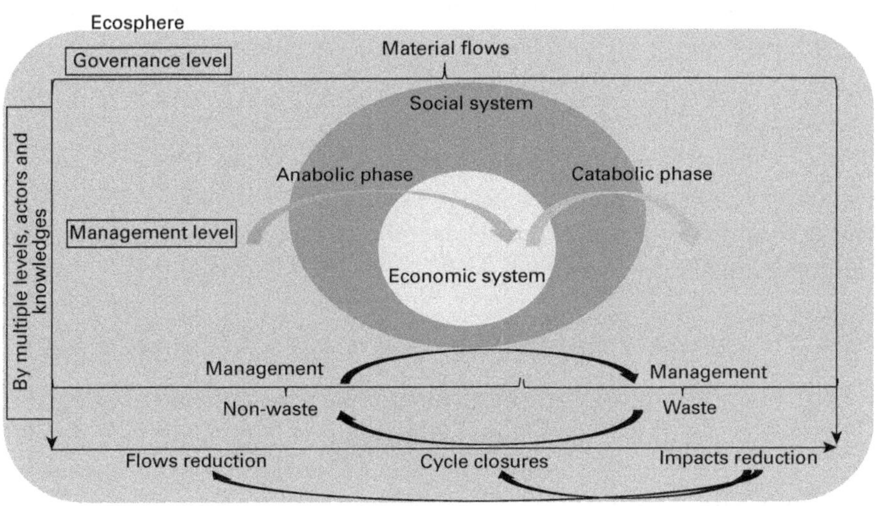

For the organization and implementation of plans made as part of the governance of material flows, waste management must consider the following:

- **a** the use of the cascade logic (cascading), with the defining of a waste management hierarchy that is socially and ecologically sound;
- **b** the techniques used in waste treatment;
- **c** the existence of an ecological rucksack in waste destined for recovery;
- **d** the location of waste destination facilities;
- **e** the involvement of the interested parties (stakeholders);
- **f** the correct analysis and decision between the possible destinations for waste identified by means of instruments such as life cycle assessment and material flow analysis, among other matters.

Due to the recurrent practice of its cross-border transfer, waste management should not only be limited to state borders. Such a practice alone establishes the need not only for the global governance of material flows (also determined by the practice of the international trade of natural resources and goods), but, also, global waste management. However, as proven in Chapter 5, the current international legal regulation of the subject does not provide tools for the establishment of true waste management, but only for a waste trade, with few, insufficient environmental limitations.

For this reason, and given the general outlines of a model for the global governance of material flows, the following section will be dedicated to formulating a proposal for the legal regulation of global waste management in the most suitable fashion from a socio-ecological perspective, which will be informed by the general priority objectives of the governance of material flows. To do so, the notion of socio-ecological justice developed in Chapter 2 will be returned to, both for the process of shaping/designing the regulation and in terms of its content. Finally, methods and criteria in harmony with this logic will also be sought out.

Global waste management: strategies to overcome the global waste crisis

The complexity of action

In turning its attention to the current waste crisis, this book has sought to investigate its deepest underlying causes by interconnecting the points of contact between different areas of knowledge, as well as with the epistemological grounds for modern Western thought (here referred to as mechanistic). Chapter 3 highlighted several epistemological, economic and legal models to have been developed in the attempt to overcome this paradigm. Such models have sought – some more explicitly than others – to reflect on the complexity of life in their theoretical frameworks to somehow locate stronger responses to the strong questions of today.

The great challenge is to materialize such models in light of the fact that the field of action remains dominated by the mechanistic paradigm. This means that action is mostly dictated by reductionist, fragmentary, objectivist and determinist practices that seek to guarantee social stability. An example of this is Law, which, in its actual implementation, always seeks solid bases in order to offer stability. As highlighted by Morin (2008), however, action is the material and often vital sphere of complexity. That is, thought and reflection are intrinsically linked to action, within the domain of complexity.

Although life in society demands that its social members act as trivial machines, in moments of crisis, when the making of important decisions becomes a matter of emergency, the machine becomes non-trivial, meaning its actions cannot be predicted. The emergence of the new is never trivial and can thus never be a predicted fact (Morin, 2008). The crisis represents a perturbation (whether internal or external) that causes disorder in social organization. If the social system has adequate instruments, it will have the conditions to spark the emergence of a new organization from the interaction between order and disorder, becoming more complex and thus evolving and adapting to the new reality.

If the social system does not have adequate instruments to deal with the crisis, on the other hand, it will march towards its own collapse. For Morin (2008), the use of a simplified and linear vision may work very well and prove useful and necessary in times of stability, where order predominates, and organization is not perturbed. In times of crisis, however, such a vision may become completely mutilated. Complex thought appears as an instrument for dealing with the crisis – not for predicting the unexpected, but for making everyone more prudent and attentive.

Complex thought facilitates an understanding that no form of organization lasts forever (and history has infinite examples of this), just as the most important events on this planet have occurred unexpectedly. This thought does not reject clarity, order or determinism, but merely understands them to be insufficient, unable to programme discoveries, knowledge or actions. While the mechanistic paradigm directs action by means of programmes, the paradigm of complexity requires strategies (Morin, 2008).

A programme establishes a series of actions that were previously decided upon and that work in an invariable sequence. Programmes work well when the environmental conditions remain constant and in the absence of perturbations. Strategy, in turn, is a scenario of action that may be modified whenever new information or changes in events occur. In this light, strategy is placed as the art of working with uncertainty: in terms of thought, it means the art of thinking with and in light of uncertainty, and, in terms of action, it translates into the art of acting in uncertainty (Morin, 2008).

Complexity needs strategies, as these can deal with the unexpected or the uncertain. Complex thought does not solve problems emerging in times of crisis on its own, but is rather a form of assistance, a key instrument to a strategy that can help solve them. Complexity is thus a starting point for a richer and less mutilating form of action, requiring the abandonment of programmes and the creation of new strategies for overcoming crises, letting go of solutions that have worked in past crises in order to prompt the development of new solutions (Morin, 2008).

Complex thought reveals that all action represents a decision, a choice and also a bet, given that, regardless of what the action is, as soon as an individual acts, that action will not necessarily result in what they originally intended. In entering the universe of interactions, action is taken over by the environment so that, ultimately, its consequences may in the end represent the opposite of the initial intention of the individual to have acted. Unlike the simplistic reductionism of mechanistic paradigm, complexity recognizes the 'bet' character of action, that is to say, it is conscious of risks and uncertainties. This is why, under complex thought, action entails complexity, and, thus, strategies for dealing with it (Morin, 2008).

In this sense, strategies assume the risk, danger, initiative, decision and the awareness of derailments and transformations. They require vigilance and innovation and recognize the bet character of action, in order to allow for the prediction of a certain number of scenarios of action based on an initial decision that may be modified according to the emergence of new information and shifts that perturb the initial intention. Strategies battle against chance, seek information and use chance to their favour. The domain of action is, thus, risky and uncertain (Morin, 2008).

The waste crisis, part of the ecological crises, as a perturbation of the order and the organization of the social system, demands an overturning of the mechanistic paradigm and the adoption of the paradigm of complexity, both in the domain of thought and in that of action. Strategies must be developed in the sphere of governance in order to overcome the waste crisis, generating a new form of social organization. The proposal of the governance of material flows seems like a suitable approach to tackling this challenge and the strategies developed on this level may be consolidated in the form of action on a management level.

Recognizing the bet character of action, the strategies outlined here – just like the models developed in Chapter 3 – should not be interpreted as singular, complete and definitive proposals, but rather as a starting point for the interaction between order and disorder for the possible defeat of the waste crisis by the emergence of a new form of organization. Plurality and dialogue must once again be favoured in this process. Based on these premises, this and the following sections will present several possible strategies for better regulating global waste management as an instrument for overcoming the waste crisis, approaching both the collective process designing the regulation and its content.

Structure for drafting and negotiations of the international legislation

The process of developing the regulation should be carried out in the sphere of the governance of material flows. Apart from meeting the priority goals of closing the materials cycles and reducing the socio-ecological impacts generated by flows that cannot be avoided (regulation content), the process must develop and utilize strategies ensuring that the different demands (the 'what' of justice) of all those subjected to the governance (the 'who' of justice) are heard and considered in the formulating of the regulation of global waste management (the 'how' of justice).

At this point, the third element of Fraser's theory of 'abnormal justice' comes into play: the 'how' of justice, described in Chapter 2 of this book as a proposal of justice for abnormal times, ie times of crisis, in which mechanistic simplistic discourses are not able to give strong responses to the strong questions perturbing the organization of systems. In light of the general growth of democratic expectations, the debate on the 'how' of justice thus represents the contestation of the institutions and the procedural frameworks granting the prerogatives of states and elites in determining the grammar of justice (Fraser, 2008).

In the case of global waste management, the contestation of the 'how' of justice represents the challenge to the process of negotiation and design of the international legislation regulating the subject, as the means by which they are brought about and the legitimized actors to actively participate in it (and not only as listeners, as was the case of the NGO Greenpeace in the negotiations over the Basel Convention). Thus, the development of strategies for the process to shape and create a regulation of global waste management that meets the assumptions of the notion of socio-ecological justice represents the search to overcome the current hegemonic premises of the 'how' of justice, requiring the invention of new global democratic institutions.

Initially, therefore, the 'how' of justice must deal with issues of the 'who' of justice – who are the subjects legitimized to have their demands heard and considered in the decision-making processes at the level of governance. The normative principle favoured here has been outlined by Fraser (2005; 2008), and that is the principle of the 'all-subjected' which, in the case of socio-ecological justice, may include present and future generations of humans and non-humans. The Westphalian perspective, by which only the nation states have the legitimacy to participate in these decision-making processes, reveals itself to be increasingly insufficient in times of globalization, as it does not always represent the real demands of all those who are subjected to this governance.

These are disputes at the meta-political level of framing the justice in abnormal times, requiring the creation of new, counter-hegemonic procedures. Apart from seeking new procedures moving away from the hegemonic model, Fraser (2008) claims that the model she refers to as 'scientistic' should also be avoided. That is to say, the 'how' of justice must seek alternatives that, apart from being counter-hegemonic, do not commit the error of leading to the definition of the frame of justice using the normal (mechanistic) social science, which believes itself to be in possession of incontrovertible facts on who is affected by injustices and why.

According to Fraser, science tends to reflect the perspectives of the privileged, omitting the blind points of their theories. Therefore, the use of mechanistic science, as well as the Westphalian model, as a strategy for defining the frame for who is legitimized to participate in the decision-making processes carries the risk that the grievances of the disadvantaged will be rejected (Fraser, 2008), as did in fact occur in the Basel Convention process. For Fraser (2008), both the hegemonic and scientific models propose to solve framing issue (the 'who' of justice') in a monological way, appealing to an authority (one power, the other science).

Rejecting the monological approach characterizing these models, Fraser (2008) suggests treating framing issues in a dialogical way, as meta-political conflicts whose solution requires inclusive public debates, without barriers and that reject appeals to the authorities. Dialogism is thus the first characteristic of the 'how' of justice. The author stresses that dialogue is not a solution in itself – apart from promoting dialogue this process should establish the means for creating executable solutions, forms of linking contestations and the legitimacy of decisions for their implementation.

The search for executable solutions led Fraser (2008) to propose a second track in the dialogical process: a formal institutional track, which must be positioned in a dynamic and interactive way in relation to the first track – that of organized civil society. By means of a bidirectional communicative process, the formal institutional track must dialogue with the track of organized civil society. The structure must contemplate two aspects: on the one hand, just procedures and a representative structure to guarantee the democratic legitimacy of its deliberations; and on the other hand, the representatives must be able to guarantee their decisions will become binding requirements in terms of the 'who' of justice, realized in a dialogical form.

Furthermore, Fraser (2008) highlights that the 'how' of justice must also be equipped to deal with uncertainty and indefiniteness, as disputes over justice may not be susceptible to definitive and final solutions. The author suggests that this new model of governance, structured on the two tracks in dialogue (organized civil society and formal institutional), must handle its disputes as perennial issues of political life in a world in a process of globalization. One possibility, for Fraser, would be the creation of a democratic forum for provisory solutions for disputes in permanent dialogue with transnational civil society.

The 'how' of justice combines dialogical and institutional requirements. If, on the one hand, dialogue can provide validation of disputes not previously considered, the existence of a formal institution outside civil society guarantees the implementation of decisions made in a democratic and dialogical fashion. In this light, this approach has the potential to provisionally resolve framing conflicts (on the 'who' of justice) in abnormal conditions of justice, and, by unravelling such meta-political problems, this proposal will open the way to prompt solutions to the first-order disputes about justice, such as those of maldistribution, misrecognition and misrepresentation.

This therefore delimits the third element of Fraser's theory for a notion of justice suited to abnormal times, here adopted for the construction of the notion of socio-ecological justice. The two initial elements – the 'what' and the 'who' – were developed in Chapter 2 in order to overcome the hegemonic, now insufficient, Westphalian–distributive model. The 'what' was developed with a multi-dimensional social ontology (distribution, recognition, representation and others), with normative monism (development of the basic capabilities for full functioning). The 'who' was constructed in a reflexive and determinative way with the use of the principle of 'all-subjected'. Now, the element 'how' is presented as both dialogical and institutional, proposing the creation of new global representative institutions in which deliberative and democratic procedures are used in the decision-making processes (Fraser, 2008).

Using Fraser's lessons, the final element in the theoretical framework of the notion of socio-ecological justice proposed here – the 'how' of justice – is delineated so as to potentially assist in the shaping of strategies in order for the governance of material flows to develop a suitable legal framework from a socio-ecological point of view on global waste management. By means of this framework, governance must happen supported by the tracks of a formal international institution and that of civil society, both in constant dialogue. The track of civil society may be formed by organized civil society in transnational networks, as well as by international and non-governmental organizations that may represent the interests of Nature and of future generations. In turn, the formal institutional track will guide the decision-making processes with representatives from all the nation states and in constant dialogue with the track of civil society, in order that their decisions may be supported, even if provisionally, by the other track.

Fraser's theory thus outlines sound strategies for moments of crisis, refuting the reliance on past solutions that do not serve current times (distributive–Westphalian model of justice), at the same time as they avoid the celebration of abnormality in itself, without a specific proposal to overcome it. Fraser (2008) highlights that her proposal is not the renormalization of justice, which would risk prematurely closing new paths for contestation, creating refuges for new forms of exclusion and establishing a fixed set of presumptions of a specific historical period. For the author, the circumstances of justice are in a fluid state and require flexibility, combining elements of normal and abnormal discourse oriented towards a conclusion, necessary to political arguments, but treating each conclusion as provisory and subject to questioning.

Fraser proposes a form of 'reflexive justice' in which the three elements of justice (what, who and how) are viewed as perennial elements, without treating them as a licence for paralysis in terms of the remediation of injustice (Fraser, 2008). Such a proposal is suitable both in moments of stability and order as it is in moments of disturbance and disorder, providing strategies in order that, in moments of crisis, the social system will find new forms of organization (that are always temporary) in constant processes of complexification and evolution. Issues of – at the very least – distribution, recognition and representation must guide the dialogue, always with the ultimate aim of guaranteeing the development of basic capabilities for the full functioning of individuals and communities to promote the dignity of life and the ecological integrity of all those subjected to decisions made by the governance of material flows (see Figure 6.2).

In short, the process for developing the regulation on the global waste management must include not only the nation states but also civil society, international organizations and non-governmental organizations, guaranteeing that all those subjected are fully recognized, may be duly represented, and are thus guaranteed fair distribution in terms of the waste problem. These serve as possible strategies for tackling the disturbances generated in the economic, social and ecological systems by the waste crisis.

Apart from the strategies on the process for negotiating and developing the legal regulation, reflections must also be made on the content of the regulation, to be tackled in the next section.

General content of the legislation

LEGAL CONCEPTS AND CLASSIFICATION OF WASTE AND ITS MANAGEMENT

Chapters 4 and 5 of this book demonstrated the limitations and inconsistencies of the current international regulations on cross-border waste transfers. The first, and perhaps most elementary, refers to the legal concept of waste. In terms of governance, the distinction between waste and non-waste is not as relevant due to the approach oriented towards material flows; on a management level, on the other hand, such a differentiation is fundamental to defining the sphere of management the materials will be subject to, even if all types of management are interconnected. It is important to develop clear criteria for differentiating waste, by-products and used goods, as well as waste which is no longer waste. Of all legislation studied, the EU's Waste Framework Directive was the most effective at distinguishing between such materials, a distinction that could be replicated in a new legislative framework, with the addition of a better definition of used goods.

Apart from clearer criteria for framing a material as waste or non-waste, it is important that the regulation allows for questioning and debate on materials subject to doubt, and this may be performed at a governance level. The legal concept of waste may be developed based on the fulfilment of criteria established by the regulation. Its importance is highly relevant to the

FIGURE 6.2 The ‘how’ of socio-ecological justice

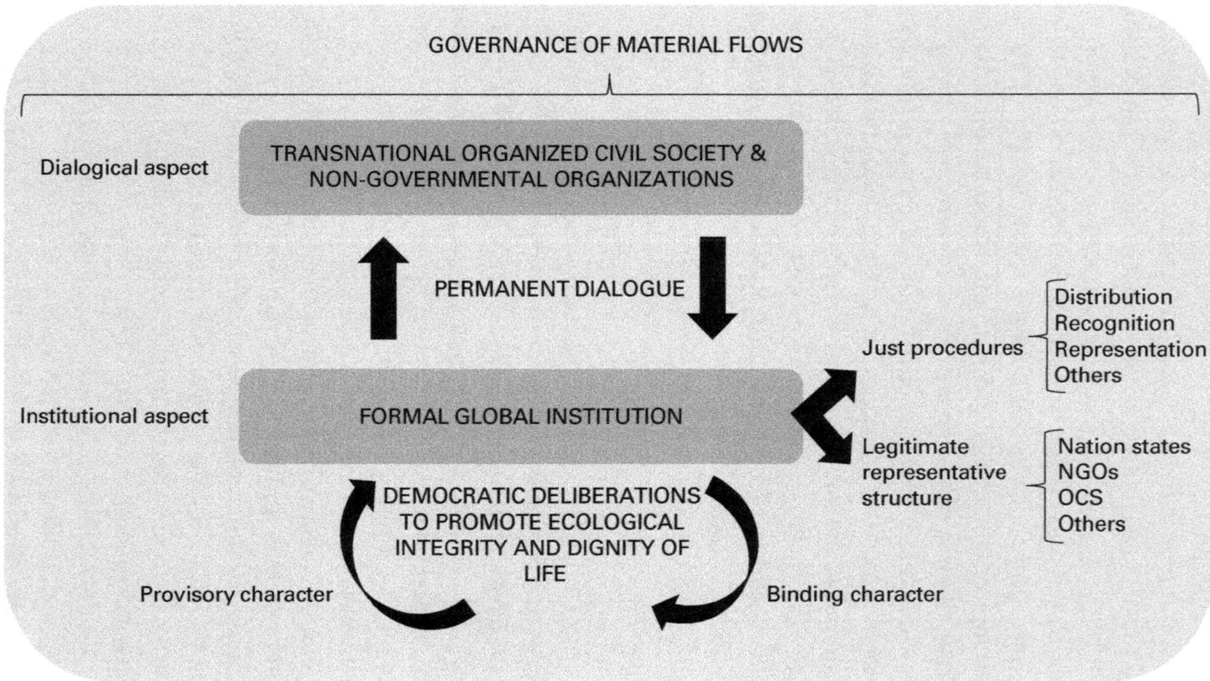

very definition to which management (and thus legal regulations) the material will be subjected. Generally speaking, this framing process must follow the precautionary principle, with any doubt leading to the classification of the material assessed as waste and subject to the most rigid environmental control measures.

In terms of the legal classification of waste according to its hazardousness, the European regulation stood out once again as the model to be copied, despite its complexity. As demonstrated in Chapter 4, in order to define waste as hazardous or non-hazardous in the EU, a five-step route must be followed:

1. classification of the material as waste;
2. selection of the appropriate entry on the EWC;
3. analysis of the waste's composition;
4. checking for hazardous qualities;
5. checking the waste for POPs over the limit values.

All of these steps are guided by the precautionary principle, determining that in case of doubt over how to classify a material under assessment, the material must be classified as hazardous and subjected to the most rigid legal environmental controls. Similarly, doubts over the classification of waste and non-waste mean the material must be classed as waste. The application of *in dubio pro natura* to the classification process emphasizes the adoption of an ecological perspective instead of an economicist approach, conceding a certain discretionary power to the public authorities who, in light of uncertainty, can always classify materials as waste in the first case and as hazardous waste in the second case.

In terms of the concept of waste management, waste transfer does not figure as a management stage in any of the regulations studied, constituting an important omission. Cross-border waste transfer is a key stage in qualifying waste management as global – and it is from this that waste is managed by two or more countries, transcending state borders. Although all the regulations include 'transport' as a stage in waste management, this should not be confused with 'transfer', given that, without treating waste at its source, transport will always be required, regardless of whether a transfer is performed or not.

The inclusion of 'transfer' as a possible management stage has, firstly, a symbolic value: such transfers instead of waste trade will become part of global waste management. That inclusion also has a practical value, given that laws with a socio-ecological approach may be developed to regulate global waste management, including in order to formulate a more suitable hierarchy of waste management, as will be demonstrated later on.

RESPONSIBILITY SYSTEM AND CONTROL INSTRUMENTS

In terms of responsibility, the European and Brazilian regulations stand out for their use of a more holistic approach and for touching on themes of prevention and post-waste generation. Unfortunately, it must also be noted that the preventive aspect of such regulations is located on the sense of 'encouragement', without any real binding power.

On the other hand, the international regulations demonstrate the adoption of an eminently *post-factum* approach, failing to dedicate due attention to the aspect of prevention. The absence of binding regulations on prevention is also true of the international instruments, notably the Basel Convention. Here the absence of binding power even affects the *post-factum* responsibilities, due to the long wait for the Ban Amendment to enter into force, as well as the Protocol on Liability and Compensation.

Adopting the approach of material flows will help create a legal structure that is more systemic and holistic with more binding power. The proposal of the governance of material flows has as priority objectives the reduction of material flows, the closure of the cycles of these flows, and the prevention of the socio-ecological impacts from flows which cannot be avoided. While governance encompasses these three objectives, waste management focuses on the final two. In this sense, and in making a connection with the developments of Chapter 4, the ideal scenario to completely close the material cycles within the socio-economic system would represent compliance with the duty of best management (Figure 6.3). On the other end, the prevention of socio-ecological impacts would represent compliance with the ban on poor management.

The responsibilities for material flows – whether the expanded responsibility of the producer added to the responsibilities for management or the shared responsibility for the life cycle of products – must be outlined in the legal regulations on all managements encompassed by the governance of material flows in order to act in a binding fashion at every stage of such flows, from extraction to the return to Nature. Once more, the importance of dialogue and interconnection between all types of management of the material flows cannot be stressed enough, as well as clear and binding provisions on liability and compensation.

FIGURE 6.3 Metabolic analysis for the completion of the concepts of best, sound and poor management

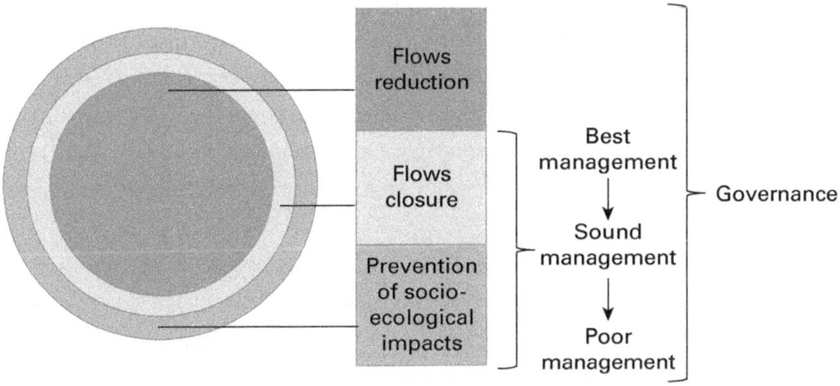

A direct consequence of the approach oriented towards the *post facto* in the regulations analysed in Chapters 4 and 5 is the absence of the provision of control instruments specifically for preventing the forwarding of waste into the ecosphere. There is a remarkable absence of instruments under international regulations for carrying out systemic analyses in order to prevent socio-ecological impacts in the decision-making processes on cross-border waste transfers. It is extremely important to provide instruments that directly assist the consolidation of aims to reduce and close material flow cycles, as well as in the prevention of socio-ecological impacts in the legal regulations on all managements of these flows, including waste management.

Life cycle assessment (LCA) and material flow analysis (MFA) are examples of instruments that should be used in the legislation governing all management of material flows, including that of global waste management, in order to effectively prioritize prevention. In the case of waste management, this refers to the priority for recovery made by the closing of cycles, as well as to the prevention of socio-ecological damage by materials that cannot be reintroduced into the production cycles.

Just like other terms in the environmental sphere, 'life cycle' and 'life cycle assessment' have been used in a variety of ways. The definition most commonly used by researchers and professionals from different fields is the official ISO one, defining LCA as '[...] the compilation and evaluation of the inputs, outputs and the potential environmental impacts of a product system throughout its life cycle' (Heijungs *et al*, 2010:424). The life cycle, on the other hand, is defined as the 'consecutive and interlinked stages of a product system, from raw material acquisition or generation of natural resources to the final disposal' (Heijungs *et al*, 2010:424).

This definition works with the notion of products as systems – from the extraction of the natural resources via the processing and use of these materials to their final destination. Although it may be the target of criticism, this definition has helped establish vocabularies and with the delimitation of a central structure, providing a context for the development of more elaborate guides in order to increase productive dematerialization and eco-efficiency (Heijungs *et al*, 2010).

According to Aragão (2003), LCA reveals that waste is generated from the process of extracting materials from the ecosphere, to the transport, processing and consumption, which demonstrates the relationship of direct proportionality between resource extraction and waste generation. In this light, LCA is a sound instrument for basing the decision-making process on the most suitable options for waste destination from an ecological point of view, directing it towards the best management and location alternative for each specific case.

Material flow analysis, on the other hand, is a descriptive and systematic approach taken to assess the metabolism of a defined system and is based on the principle of conservation of mass. Using this instrument, Turner *et al* (2016) presented an analysis of the waste management system being operated in Cardiff, the capital of Wales, demonstrating that recycling targets set by local public policies could not be achieved unless there was a reduction in waste generation. This revealed the importance of MFA to the adoption of a more preventive perspective on waste management. The authors also stress the need to use MFA along with LCA in order for the environmental impacts of the system described by MFA to also be assessed.

According to Turner *et al*, if used in a combined fashion, these two instruments may provide valuable information on the performance of the waste management carried out. Such an approach may thus be used to identify the most effective strategies from an ecological point of view (Turner *et al*, 2016). Such instruments, however, are more oriented towards the technical and biophysical aspects of the analyses carried out by experts.

In terms of diffuse control, there is clearly a complete absence of provisions on public participation in the international instruments investigated, although mention is made in the Brazilian and European legislation. Previous sections of this chapter revealed how public participation may be amply incorporated into the governance of material flows, as well as into the process of devising regulations under this governance. As a question of socio-ecological justice, it is important to guarantee due representation for all those subjected to this governance in order to promote ecological integrity and the dignity of life during all of these forms of management of material flows.

On the level of waste management, therefore, legal regulations must guarantee that the communities directly subjected to the management have the opportunity to participate in decision-making processes, including those referring to cross-border waste transfers. This may encompass participation from representatives of international organizations, non-governmental organizations, associations (such as those for waste-pickers, for example), as well as organized civil society, all representing the interests of present and future generations of humans and non-humans.

Nowadays, there are countless creative ways to allow for such participation, mainly by means of the use of technologies, the internet and social networks. An example of this is the online tools offered by the UN's Special Rapporteur on the human rights implications generated by the management and disposal of waste and hazardous substances. The platform allows for and facilitates the posting of claimed human rights violations by individuals, communities and organizations. The structure of the 'how' of justice proposed by Fraser is also suitable for guaranteeing dialogue and participation from all-subjected, along with the existence of a global formal institution for legitimizing and implementing the decisions made in a democratic fashion between these two tracks.

THE DUTY OF BEST MANAGEMENT AND WASTE MANAGEMENT HIERARCHY

Although there are no provisions expressly addressing the subject in any of the regulations highlighted, respect for the duty of best management may be deduced from the provision on the use of the best available technology in European legislation, as well as that for the establishment of waste management hierarchies, both under the European and Brazilian regulations. Even so, such hierarchies have been the target of criticism and reveal the need for better restructuring and clarifications.

Furthermore, the identification of the absence of provisions for guaranteeing the best management and for the establishment of hierarchies between the management forms coupled with the absence of a harmonized understanding of the distinctions between recovery and disposal operations in the international legislation analysed demonstrates how such limitations further entrench the difficulty of framing materials as waste or non-waste, as well as the consequent hyper-regulation of cross-border transfers of hazardous waste for recovering and the hypo-regulation of non-hazardous waste transfers for disposal.

It is therefore highly important that the regulation of global waste management provides clearer criteria for a differentiation between recovery and disposal operations. The European legislation on this subject also stands out, due to its defining of recovery and disposal as opposite operations comprising the possibilities for waste treatment. Under the terms of this legislation, the key criterion is that materials, upon their transformation, must serve a useful purpose. Although lists of operations may help in the definition, these are insufficient, with a clear definition of criteria required for purposes of differentiation. These should always be guided by the precautionary principle, meaning in case of doubt, an operation should be classed as disposal and therefore subjected to the most rigid environmental controls.

According to Aragão (2003), two premises must always be present in order to ensure the best management: the impossibility of applying single, abstract solutions to all decision-making processes in waste management; and the need for an integrated perspective. The first premise requires that an analysis is made of all possible waste management alternatives in a specific case, as a solution may be ideal for one specific case and disadvantageous in others.

In this sense, Aragão affirms that a realist vision does not fit with an abstract, inflexible hierarchy of management objectives valid for all waste in all circumstances. To the contrary, the most suitable strategy for dealing with the waste problem may only be established according to each specific case (Aragão, 2003).

For the author, the ideal legal regime for waste management establishes an *a priori* hierarchy of management options that is therefore not immutable, merely serving as a starting point for an analysis of each specific case (Aragão, 2003). In this light, after clarifying the distinction between recovery and disposal, the legal regulation of global waste management must present a clear and suitable waste hierarchy for the priority general objectives of closing the materials cycle and preventing socio-ecological impacts. Based on these premises and taking into consideration the criticism in Chapter 4, the waste management hierarchy in Figure 6.4 is proposed.

Initially, it may be observed that although waste management (in the model presented here) addresses the aims of closing the cycles and preventing socio-ecological impacts more directly, as well as carrying out waste governance (leaving non-waste for other material flow management systems), it is

FIGURE 6.4 Proposed waste management hierarchy

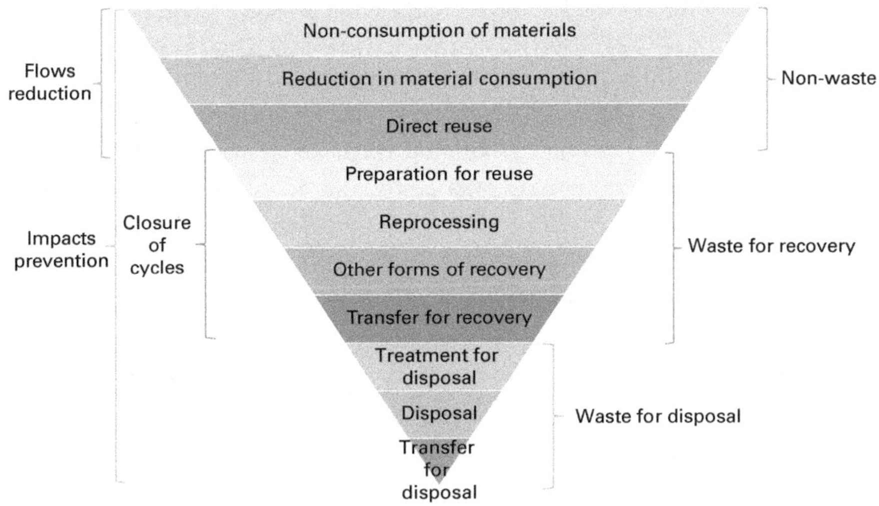

important that the aim of reducing material flows is also featured in the management hierarchy, reflecting that this is the primary main objective sought by the governance of material flows, as well as the interconnection between waste management and other forms of management, which must be in constant dialogue.

With such a clarification made on the operations suggested, the first level, at the top of the hierarchy, refers to the reduction of material flows and is composed by the operations of:

- **i** material non-consumption;
- **ii** reduction of material consumption; and
- **iii** direct reuse.

While the first represents a complete de-consumption, the second operation encompasses different actions, such as the virtualization of consumption, shared consumption and the prolonged use of products.

The third operation on the first level of the hierarchy is direct reuse, referring to used materials with perfect and full functionality and conditions or that require small interventions without introducing new material flows to the product (such as cleaning). That is why this operation is on the first level of the hierarchy, as it features the reuse of used goods. The preparation for reuse involves all of the operations required for the material to be reused with the introduction of new material flows, such as repackaging and parts exchange. For this reason, this operation has been included in the second level of the proposed hierarchy – cycle closure – as it is considered a recovery operation.

Reprocessing is also placed at the recovery level. This stage encompasses three sub-operations: upcycling (reprocessing which increases the purpose and/or value of the material); recycling (processing which maintains the same function and/or value of the material); and downcycling (reprocessing which results in the loss of purpose and/or value of the material) (Gharfalkar *et al*, 2015). The category of 'other forms of recovery' encompasses all the further existing recovery operations and opens up space for possible technologies that may be developed in the future.

The possibility of transferring materials for recovery appears as a final option within the second level of the *a priori* hierarchy given, because, even if it fulfils all the proposed criteria permitting the transfer, it must be noted that it will always add new management stages, with longer transport times and spaces, increasing the material and energy flows to the recovery process, as well as heightening the risk of producing socio-ecological damage (which must be avoided).

The third and final level of the hierarchy – whose main aim is to prevent socio-ecological impacts – encompasses three operations related to disposal, giving preference for waste to be treated before its disposal, even with energy recovery technologies. Disposal itself must always prioritize methods which cause the least socio-ecological impacts, in order for incineration, for example, to be avoided, if not banned altogether. And, finally, waste transfer for disposal only appears in the hierarchy due to the fact that some territories do not have other alternatives for managing their waste.

Generally speaking, transfers for disposal should be prohibited, mainly concerning flows in the direction North to South. On this point, Bartl (2014) highlights that targets for achieving 'zero waste' to landfill that are pursued by many developed countries may generate the redirecting of waste to less sustainable and just management forms, involving incineration and transboundary transfers. This reveals the importance of including this operation in the waste management hierarchy, in order for it to be factored into the global waste management system (Gharfalkar *et al*, 2015).

The final point with regard to the hierarchy is that the use of the best available technology for all operations, whether involving recovery or disposal, is essential if best management is to be achieved.

THE LOCATION OF WASTE TREATMENT

Finally, the defining of the location in which the waste will be treated (whether recovered or disposed of) is fundamental to just and sustainable management. Once again, the international regulatory instruments lack references to the best location, although, on the other hand, there are useful principles for guiding the decision-making process in the European legislation, such as the principles of self-sufficiency and proximity.

In general terms, the principle of self-sufficiency means that responsibility for the waste produced in a given territory cannot be transferred to subjects outside it (Aragão, 2009). According to Aragão (2006), such a principle is founded on ecological and not social issues and will gain fruitful content if interpreted as a reflection of the principle of a High Level of Environmental Protection (HLEP) in stipulating an adequate number of waste treatment facilities.

Currently, in the EU, this principle has come to refer only to general waste disposal and the recovery of municipal household waste. It also holds two meanings – that of EU self-sufficiency and the self-sufficiency of each Member State (Aragão, 2009). In this context, the principle of self-sufficiency is interpreted in the sense that a territory cannot create too many facilities, such as a 'mini landfill' in the yard of every factory, or too few facilities, such as 'mega landfills' that concentrate the majority of waste produced in the territory (Aragão, 2006).

In applying this principle in the EU, the adequate level of waste disposal will be subject to the creation of the necessary number of waste disposal facilities, where they are lacking in ecological terms. In this light, and according to Aragão (2006), the sphere of application of the principle of self-sufficiency is simultaneously global and local – ie *glocal* – establishing the exact dimensions of a territory's disposal facilities required to form an integrated network. Understanding the sense of application of such a principle in the sphere of the European Community means understanding that such a principle must be interpreted in a different fashion on an international level.

Based on the premise that international political organization is still based on the Westphalian model, ie the sovereignty of the nation states, in terms of global waste management, the principle of self-sufficiency will apply to state or community territories, the latter when this is an express provision in the treaties creating such communities, such as in the EU. In this sense, the principle of self-sufficiency would point to the nation states' responsibility or that of the state communities to guarantee that the number of waste treatment facilities existing in their territories is proportional to the number of waste producers in terms of their treatment capacity.

Although the international recovery market can generate economic and ecological benefits, it must be remembered that the carrying out of cross-border waste transfers always adds new stages to the management process, increasing the material and energy flows used, as well as the risks of socio-ecological impacts. In this sense, the use of this principle as a general rule also for cases of waste destined for recovery is coherent. It is thus evident that, just as the demand for self-sufficiency can be bent in certain situations when waste is destined for disposal, its application will also not necessarily be absolute in terms of waste destined for recovery.

Following the *a priori* hierarchy logic, territorial self-sufficiency may be provided for in the regulation of global waste management as a general rule. It shall, however, be made possible by the use of global and systemic analysis instruments to initially check ecological reasons for not following the principle. The social element may then be verified, in order to guarantee that the waste transfer, even if beneficial in ecological terms, does not generate negative social impacts in the receiving country, mainly in the communities directly affected.

In general terms, under the principle of proximity waste must be treated as close as possible to the location in which it is produced, avoiding its transfer (Aragão, 2009). According to Aragão (2006), before being a legal principle, the principle of proximity reflects common sense, ie is almost intuitive in its application. There is nothing more coherent than expecting waste produced by a certain source to be treated as close as possible to this source, for reasons of socio-ecological justice.

Apart from avoiding the addition of new material and energy flows and increasing the risks of socio-ecological impacts with waste transfer, waste treatment carried out near the source of its generation avoids the transfer of the social and ecological costs (ecological rucksack) which should be borne by the source of production. Not transferring the waste reveals the real carrying capacity of the territory in which the waste was produced, prompting measures to reduce waste generation by means of decreasing flows or closing the cycle of such flows.

The application of the principle of proximity should be called upon when choosing the site for treating waste by a particular generator centre, which must take into account the existing facilities acceptable in ecological terms due to the procedures and technologies used for the treatment. The closest facility to the site of production should always be selected. On this point,

Aragão (2006) emphasizes that proximity is not measured in a straight line, but according to the existing communication channels, and their characteristics shall be taken into consideration in the assessment of the possible risks inherent to the transport.

She uses the notion of 'gravitational field' as a means of delimiting the sphere of influence of a certain treatment facility. This space is seen as a circle not necessarily circular in shape, in the centre of which is the treatment facility, attracting all the waste generated within this space, if appropriate for treatment in such facility. Just like the principle of self-sufficiency, the principle of proximity also promotes the HLEP, to the extent that it represents the weighing up of two possible waste treatments: one which is closer and the other further away from the generation site (Aragão, 2006).

Aragão (2009), however, raises the issue of the possible clash between the principles of self-sufficiency and of proximity; that is to say, the proximity required must respect the borders of the territory in which certain waste was produced. The author claims that, as a general rule, the principle of self-sufficiency at state level must prevail over the principle of proximity, although with two exceptions. The first of these is when the distance between the site in which the waste was produced and the treatment site within the same national territory is clearly greater than the distance between the production site and a treatment site outside this national territory.

The second exception is when there is excess treatment capacity in another national territory, referring to when the waste treatment facilities in a certain national territory can handle all the waste produced in their own territory and still have extra capacity to take on materials from other territories. According to Aragão, the principle of proximity should prevail over the principle of self-sufficiency in both cases. Thus, while the principle of self-sufficiency raises borders to the circulation of waste between territories, the principle of proximity ignores state borders in order to prioritize ecological issues (Aragão, 2009).

Like the principle of self-sufficiency, the EU uses the principle of proximity for waste destined for disposal and recovery – in this last case, for municipal household waste. In terms of global waste management, this principle should also be applied, for the same reasons outlined for the principle of self-sufficiency, not in an absolute form, both for disposal and recovery operations for all waste.

Returning to the two premises that Aragão (2003) refers to as essential for the best waste management, possible strategies for management options were identified and arranged in an *a priori* hierarchy, as well as options for the location of the waste treatment in question with the application, although not absolute, of the principles of self-sufficiency and proximity. Such strategies mainly refer to the first premise, the impossibility of the use of singular, abstract solutions for all decision-making processes in management – in this case, on both the type of management and the location of the treatment.

However, Aragão's (2003) second premise –the need to use an integrated perspective – allows all the alternatives to be analysed, considering and duly pondering all of the possible ecological and social consequences. In the majority of cases, such a perspective points to the need to use not only one management option, but the simultaneous use of various options for a single type of waste.

According to the International Solid Waste Association (ISWA), integrated sustainable waste management (ISWM) is an excellent framework of analysis for waste management systems. So much so that it was one of the analytical frameworks used by the Global Waste Management Outlook (GWMO) published by the UNEP and by the ISWA as a result of debates held at Rio+20, given that environmentally-sound waste management is an internationally established sustainability target (ISWA, 2016).

In this report, UNEP (2015b) affirmed that, for a system to be sustainable in the long term, three basic elements must be taken into account:

i the infrastructure of a waste management system (physical elements);

ii the actors involved (stakeholders); and

iii all of the strategic aspects, such as political, institutional and environmental.

The proposed sustainable integrated waste management takes these three dimensions into consideration. The first, physical, element is considered by the ISWA as the hardware of the waste management system, providing the management infrastructure required. This element is formed of waste collection, treatment, recovery and disposal (UNEP, 2015b).

The majority of analytical frameworks for waste management only touch on the physical elements of management systems. A management system which is truly integrated and sustainable must, however, extend beyond this, also touching on aspects of governance, which would in turn encompass further elements related to the stakeholders and strategies. These elements thus form the 'software' of the management system and focus on including the stakeholders, on financial sustainability, on suitable institutions and on proactive public policies (UNEP, 2015b).

In this sense, it may be perceived that the proposed governance of material flows containing global waste management is aligned with the analytical framework of ISWM proposed by the UNEP and ISWA, taking a further step with the adoption of a systemic approach also in relation to the very object of governance and management (material flows).

A systemic approach, therefore, involves both the 'hardware' and the 'software' of waste management, and has the potential to facilitate the sound functioning of the system in a sustainable fashion in the long term. This approach entails an understanding that each 'hardware' organization only works with the specific 'software' instruments and vice versa (ISWA, 2016). This means that, in order to successfully modify or improve the physical infrastructure of a waste management system, the components of governance – the system software – must be reformulated and vice versa. Therein lies the importance of reformulating the process and content of the international legislation regulating global waste management in order to achieve the best management.

Criteria and procedures for waste transfer requests

CRITERIA FOR WASTE TRANSFERS

Given some of the general elements of the regulation's content in terms of cross-border waste transfer operations, it has been demonstrated how the criteria for permitting transfers, the procedures and provisions for solutions to possible problems in waste transfer operations, as provided for in the international legislation studied, reflect an overwhelmingly economicist perspective, with few environmental limitations oriented towards preventing poor management.

However, the means of regulating global waste management in order to overcome the waste crisis must seek a migration from an economicist perspective to a socio-ecological perspective. The proposals developed on procedures for negotiating and developing the regulation, as well as its content, move in this direction. Also, in alignment with such a target is the necessary adoption of strategies permeated by this perspective for the establishment at the least of criteria, procedures and instruments facilitating the specific decision-making process on cross-border waste transfers. Here, once more, the assumptions of socio-ecological justice may help guide the process by which global waste management is performed in a just and sustainable fashion for present and future generations of humans and non-humans.

Initially, in terms of criteria, it has been demonstrated that the Basel Convention has three alternative criteria (or conditions) for allowing cross-border waste transfers, which may be summarized as follows:

i that the state of export does not have the technical capacity, the adequate facilities or sites with the sufficient capacity to treat the waste; or

ii that the waste is required as a raw material for recycling or recovery in other states; or

iii other criteria to be defined between the countries involved.

The OECD Decision, for its part, establishes four general criteria applicable to all types of waste and two specific criteria for hazardous waste subject to the amber control system, to be fulfilled in tandem. These criteria may be summarized as follows:

i the waste may be recovered in an environmentally-sound fashion;

ii the subjects responsible for recovery have the legal status to do so;

iii the transfer complies with the international transport regulations;

iv other regulations applied to the subject are respected;

v there is a valid contract between the parties;

vi there is a financial guarantee.

Finally, the EU establishes three criteria for hazardous waste transfers subjected to the amber control procedure, which may be summarized as the existence of:

i notification and movement documents;

ii a valid contract;

iii a financial guarantee.

Added to these criteria under the European regulation is a list of bans which must be considered as criteria for permitting or prohibiting a transfer. In short, the following are banned:

a export for disposal;

b export for the recovery of hazardous waste, hazardous waste mixtures, waste banned by the country of import and waste not believed to be handled in an environmentally-sound fashion;

c export to Antarctica; and

d waste import to the EU.

Based on the analyses and criticisms made in Chapter 5 and adopting a socio-ecological perspective on global waste management, the following general criteria are proposed to ensure that the transfer must (Figure 6.5):

i be ecologically advantageous in a global sense;

ii not impact the ecological integrity of the place of receipt; and

iii not impact the dignity of life of individuals and communities in the place of receipt.

These proposed minimum criteria for allowing a cross-border waste transfer are based on the premise that these transfers are only justified when they offer ecological advantages on a global level while promoting the local ecological integrity and the dignity of life in the place where the waste is received, all elements of the normative core of the concept of socio-ecological justice seen in Chapter 2.

The regulation of global waste management could thus establish general rules, such as a ban on waste transfers destined for disposal or on hazardous waste for recovery, just like the EU Regulation. Such rules must, however, always meet the criteria described above, meaning a specific request may be permitted if all stakeholders agree that it is ecologically advantageous in

FIGURE 6.5 General criteria permitting cross-border waste transfers of the global waste management system proposal from the perspective of socio-ecological justice

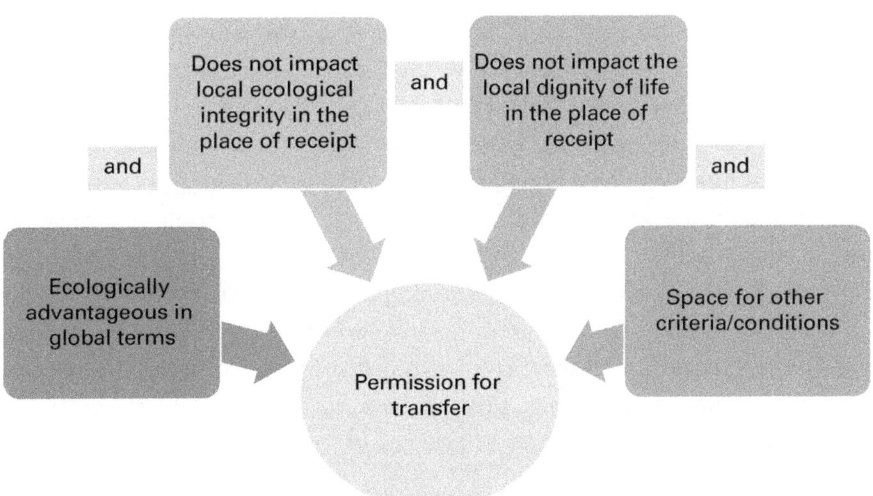

global terms when compared to the possibility of treating such waste in the territory in which it was generated, and, simultaneously, will not impact the local ecological integrity and dignity of life in the places receiving the waste.

Such an assessment is not, of course, simple, as it must take into account factors based on distant notions of values escaping the classic and reductionist economic cost–benefit analysis. This must consider issues such as:

- material and energy flows involved in the options considered;
- the ecological rucksack to be transferred with such waste;
- the ecological efficiency of the waste treatment process to be carried out, with the selection of the best available technology;
- the adherence to the waste management hierarchy, with an analysis of the best options for each specific case;
- the compliance with the principles of self-sufficiency and proximity;
- how the country of import, and mainly, the communities will be directly affected and their positioning on the matter.

GENERAL PROCEDURE FOR WASTE TRANSFERS

The procedure for permitting cross-border waste transfers must also be revisited. It has been demonstrated that all of the regulations studied in this book use the prior notification and consent instrument as the general procedure for cross-border waste transfers classified as hazardous: the Basel Convention, because it only regulates this type of waste; the OECD Decision, due to not applying any environmental control to non-hazardous waste; and the EU Regulation, as it uses the request for general information for non-hazardous waste.

Although it is understood that waste classed as hazardous requires greater ecological control, it is not plausible from a socio-ecological point of view that waste classified as non-hazardous does not have any form of control, as is the case nowadays with the international regulations. Thus, following the logic of the European regulation, there is a need for control procedures also for non-hazardous waste, mainly due to the fact that a percentage – depending on each specific case – of this waste will end up being disposed of in the country of import, with this constituting its ecological rucksack.

The prior consent instrument may (and should) be used both for hazardous and non-hazardous waste, even with different control levels (more rigid for that classified as hazardous). After the notification has been issued, all the information relevant to the specific case must be provided for the decision on the transfer request to be taken in the global formal institutional sphere (the 'how' structure of justice), with the transfer consented to or rejected. In order to promote better-informed decisions in socio-ecological terms, the multi-criteria analysis model must be used.

On the subject, Alier (2003) highlights that public decisions on the environment currently use authoritative power in the decision-making process, mostly prioritizing the focus on cost–benefit analyses, thus opting for commensurable monetary valuations established by 'scientific experts' with specialist knowledge. Sometimes, such decisions also employ cosmetic assessments on environmental impacts, with the resource of monetary compensation for negative externalities. Even so, these are marked by the use of strong comparability–commensurability, revealing themselves to be reductionist and authoritarian and excluding values which cannot be monetarily measured.

The **cosmetic environmental assessments** mentioned by Alier refer to assessments carried out only to meet formalities, adopting a reductionist bias to validate decisions eminently economicist.

Strong comparability of values – or commensurability of values – refers to the application of a single 'supervalue', whether monetary or of another type, reducing all the various values to this supervalue in order to make possible a (strong) direct comparison between them. With this reduction, the values become commensurable. The alternatives under analysis, in this case, could be easily hierarchized according to the values presented (Alier, 2003). For example, in an analysis on waste transfer using the monetary values (supervalue) needed for treatment in the territory in which the waste was produced versus in the territory to which the waste is planned to be transferred, the decision is made by prioritizing the less costly option in monetary terms (as demonstrated in Chapter 1).

If such an authoritarian and reductionist assessment is opened up to public debate, however, the supervalue chosen may be questioned by other values (such as that of material and energy flows involved or the social impact generated by each option). Other alternatives may also emerge (such as the redirecting of the waste for recovery instead of treatment or transfer for disposal), or the entire procedure undergone as part of the decision-making process may be criticized (for instance, the instruments used for assessment or the absence of direct public participation). When taken into consideration, all of these possibilities would represent a weak comparability of values approach, in which different values would be assessed for the making of decisions in a multi-criteria fashion.

In this sense, such a method admits the irreducibility between different value forms (the incommensurability of values), leading to weak comparability. The incommensurability of values – referring to the absence of a common unit of measurement in light of a plurality of values – implies a rejection not only of monetary reductionism, but also of any reductionism (Alier, 2003). The multi-criteria method, in this sense, is not so much a determinist formula for decision-making, but a strategy that can lead to unpredictable results in each specific case.

Alier (2003) thus defends the use of a non-compensatory multi-criteria focus that may accommodate a plurality of incommensurable values in the environmental decision-making processes. This is a proposal for an 'integrated assessment' considering the complex physical, biological, chemical, social and human relationships, recognizing the legitimacy of the various points of view on the same problem. The multi-criteria method helps prompt a (weak) comparison of alternative situations taking into consideration incommensurable plural values, among them the economic, social, ecological and cultural.

This method considers simultaneously monetary values and physical and social indicators of (un)sustainability. It is important to highlight that contradictory tendencies may be observed in the different parameters, sparking the need to establish priority targets for such parameters, implementing the most affordable (or cost-effective) solution for such targets to be reached. Indexes combining several indicators in a single figure, such as that of the 'ecological footprint' may also be employed (Alier, 2003). Despite the difficulties of calculating physical indicators, these are fundamental to ensuring that socio-ecological decisions meet the criteria of socio-ecological justice.

Although the use of the multi-criteria analysis method does not exclude the monetary valuation, it treats it merely as one of the values to be considered, along with other physical and social values, as well as the ecological impacts of the human economy. The opting for values which are purely monetary for analysis conceals other non-chrematistic discourses by individuals and communities directly subjected to these decisions and that make non-equivalent (or distinct) descriptions of reality (Alier, 2003).

Cost–benefit analyses, for example, must be recognized as representing one perspective of many, legitimate as a point of view and as a reflection of the existing power structures, guided by the commodification of Nature.

According to Alier (2003), there is thus no way of proposing a common (universal) unit of measurement that is totally reliable, which does not mean that alternatives cannot be compared on a rational basis by means of a multi-criteria analysis. Relegating the logic of monetary valuation to where it belongs, ie simply as another point of view, opens up political space for other discourses, quashing any attempt to simplify complexity.

As highlighted in Chapter 2, the current socio-ecological conflicts transcend economic demands for just distribution, including demands of a social and political character, such as those for full recognition and due representation. It is in this context that the multi-criteria method's potential to implement the multi-dimensional perspective of socio-ecological justice is revealed. The recognition followed by the representation of current and future generations, of humans and non-humans, for a fair distribution that allows for the development of basic capabilities for the full functioning of individuals and communities, promoting ecological integrity and the dignity of life, are possible with the use of this method.

In pondering all of the values and interests involved in a decision-making process, some should take priority over others. The ecological advantages of a waste transfer must serve as the first parameter, essential to the consideration of whether or not the transfer should be permitted. According to Aragão's understanding (2006), there is no way to justify the carrying out of a cross-border waste transfer unless it is ecologically advantageous in global terms, ie involving all the territories part in the management process.

The analysis must therefore take into account all the management stages of the waste to be transferred, including transport, treatment in the receiving country, the ecological rucksack of the waste and its final destination. As mentioned, such an analysis may be undertaken by combining the use of different instruments, such as the LCA and the MFA. This assessment must be made by comparing it with the waste management options within the territory in which the waste was generated, in order to apply the self-sufficiency principle. If it is proved the transfer – with all its stages – is ecologically advantageous when compared with the internal management, the analysis procedure may proceed.

Next, and based on the notion of socio-ecological justice, the preservation of the local ecological integrity and dignity of life of the communities and ecosystems in the place receiving the waste should be placed as indispensable elements for the transfer to be permitted. With that said, however, although ecological integrity does not completely encompass the notion of the dignity of life, as human beings, for example, developed capabilities (and needs) that surpass Nature, it is a basic precondition for promoting the dignity of life. In other words, without preserving ecological integrity, there is no dignity of life. Ecological integrity will therefore represent the major priority for the achievement of socio-ecological justice for present and future generations, both human and non-human.

Furthermore, the orchestrating of different forms of knowledge is an essential characteristic of the multi-criteria method. Here, non-specialists are included as part of the extension of assessment beyond experts, inducing participative methods for decision-making and even deliberative democracy (Alier, 2003). Morin *et al* (1999) highlight that, even if technical competency is outside the immediate reach of non-specialists, such knowledge may be communicated and discussed in a language common to all.

The authors also highlight the importance of the adoption of measures recognizing the complexity of current problems, understanding the damaging effects that decisions based on compartmentalized and linear knowledge may produce (Morin *et al*, 1999). Santos' proposal of the ecology of knowledges seems like another sound instrument or strategy for articulating knowledges derived from the participation of various actors, specialists and non-specialists. For Alier (2003), the alternatives and the criteria relevant to the application of the multi-criteria method may even be determined based on the interaction between the affected population and the external specialists.

In this sense, it is possible to consider the procedure of informed and participative prior consent, in which not only the states involved but also organized civil society and other organizations, are able to participate. The creation of a global formal institution (the second track proposed by Fraser) to make possible and mediate the dialogue between these different actors would be an option for transfer decisions to be made not only by the states involved.

Furthermore, the use of this governance and management platform may facilitate a preliminary analysis of the possible use of other options of management operations, placed at higher levels of the waste management hierarchy when compared to waste transfer, and the selection of other locations, according to the principles of self-sufficiency and proximity. The platform may also open up the necessary space for debate and the establishment of all of the criteria the actors involved believe to be relevant, apart from the minimum criteria described above for analysing waste transfer requests.

With these clarifications in mind, the next step on the notification of the transfer request is for the party interested in exporting the waste to provide information including (among others):

- the characteristics of the waste (whether hazardous or non-hazardous);
- the destination (recovery or disposal);
- the reasons for thc transfer;
- if the transfer represents a business transaction or not (with the values involved, who is paying, etc).

Based on this preliminary information, the institution of governance and management, after dialogue with the interested actors (stakeholders), will be able to assess the existence of alternative options, according to the *a priori* hierarchy, which better reflect the order of the priority objectives of the governance of material flows.

In the absence of alternatives meeting the *a priori* hierarchy, such an institution must verify the existence of alternative sites for the chosen waste treatment closer to the source of production in relation to the destination requested (principle of proximity). With such information, a comparative analysis can then be made between the national and/or closest options and the destination requested for the waste transfer, using the minimum criteria described, along with additional criteria established by the actors involved.

In global terms the option which must be chosen is the most ecologically advantageous one. However, if this option is to perform the transfer, this should only occur if it does not impact the local ecological integrity and dignity of life of the communities and ecosystems in the place receiving the waste, to be evaluated by means of physical and social indicators with the use of the multi-criteria analysis method. With such criteria met and the transfer approved, responsibilities and guarantees must be established in order for the management process to take place as authorized and planned, and monitoring indicators devised for applying the regulation in general and the decision more specifically (see Figure 6.6).

SOLUTIONS FOR POSSIBLE PROBLEMS AFFECTING WASTE TRANSFERS

If problems arise during the transfer operation, unlike those currently provided for under the legislation studied, the global waste management regulation should instead establish an *a priori* hierarchy of actions, stressing the return of the waste to the country of export as a priority, followed by the selection of an alternative destination that is both socially and ecologically suitable (Figure 6.7). Once again, this general rule is malleable. If it is proven that an alternative destination would be more ecologically advantageous than returning the waste to its country of origin, such an action should be permitted as long as it does not affect the local ecological integrity and dignity of life in the place in which the waste will be alternatively treated.

FIGURE 6.6 General procedure for cross-border waste transfers of the global waste management proposal from the perspective of socio-ecological justice

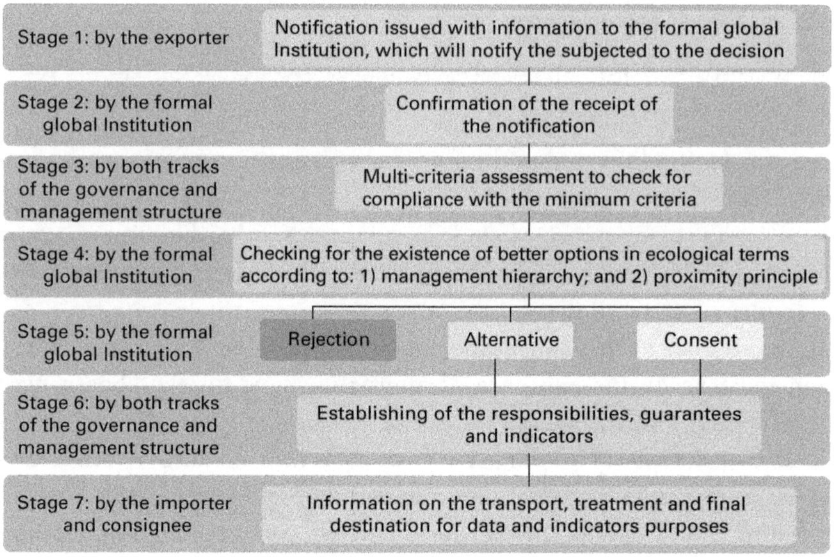

FIGURE 6.7 Problems that may occur during cross-border waste transfers and solutions in the global waste management system proposal from the perspective of socio-ecological justice

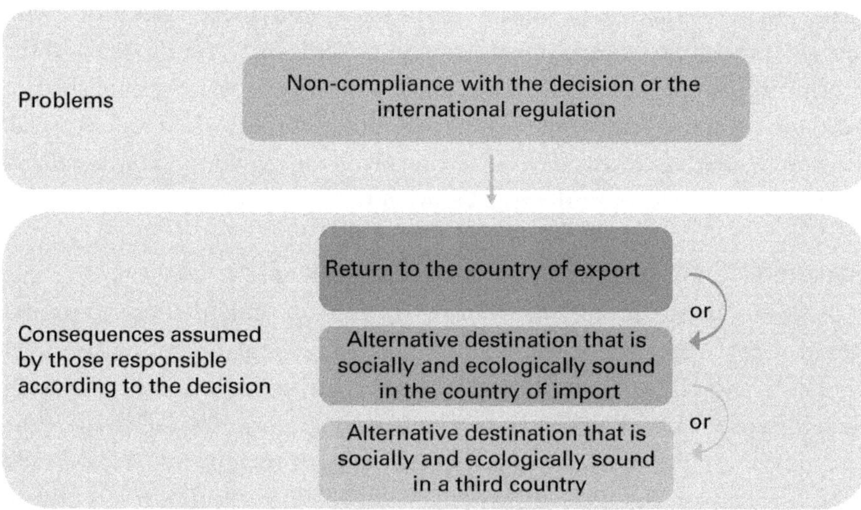

These are the elements essential for creating a regulatory framework for global waste management which has truly adopted a socio-ecological perspective with the aim of promoting socio-ecological justice. Many other elements could be addressed and debated, and it is for this very reason that this proposal is not said to be singular, complete and definitive. The governance of material flows and global waste management proposed here should be viewed as possible reflective and pragmatic strategies that, whenever challenged by events of disorder, will seek new possibilities for a new form of organization emerging from dialogue, with the aim of overcoming the current waste crisis.

Final considerations

Inspired by the theoretical developments made throughout this book, this final chapter has sought to outline some of the possible strategies for establishing global waste management as an instrument for overcoming the current waste crisis. To do so, it began by proposing the governance of material flows, encompassing both the anabolic and catabolic periods of these flows, thus addressing both waste and non-waste. This model of governance selected three priority objectives:

- **i** reducing material flows;
- **ii** closing the flows of these cycles; and
- **iii** minimizing the socio-ecological impacts generated by unavoidable flows.

The governance of material flows should also provide strategies for a democratic expansion in terms of participation and representation, as well as in epistemological terms. The first by the inclusion of all those subjected to this governance in its decision-making processes, and the second by the dialogue and engagement among these actors, permitting the use of different types of knowledge. While seeking to achieve an entropic–negentropic balance in the technological and ecological processes, this governance should also help promote the empowerment of organized civil society, by means of different representatives working together to treat Nature as a common and for the socialization of power. As a key role in developing a complex approach to the subject, waste management in particular should focus on the two latter priorities (closing cycles and reducing socio-ecological impacts), implementing general strategies defined in the sphere of governance.

In terms of the negotiation and development of the legislation regulating global waste management, the model of the 'how' of justice, founded on the two tracks of organized civil society and a global formal institution, both in constant dialogue, shows great potential for facilitating the adequate definition of the actors with legitimacy to participate in the decision-making processes (the 'who' of justice), as well as to encompass different demands (the 'what' of justice).

In terms of the general content of this regulation devised for global waste management, a focus has been placed on the elements of the concept of waste, classification in terms of hazardousness, the concept of management, the responsibilities, the control instruments for the preventive aspects, the diffuse control, the distinction between recovery and disposal operations, the duty of best management, the establishment of a hierarchy of management options and location criteria that are all more suitable for the adoption of a socio-ecological perspective on the regulation. The precautionary principle has also been highlighted in order to ensure *in dubio pro natura* (when in doubt, favour nature) is always applied when uncertainties arise in the framing, classification or destination of waste.

Finally, on the specific content of the regulation related to cross-border waste transfers, strategies were outlined for redefining the criteria, procedures and solutions to the problems to have arisen in these operations, based on the premise of socio-ecological justice. The use of the instrument of informed and participative prior consent for both hazardous and non-hazardous waste was thus proposed, with the analysis of the management options hierarchy and the principles of self-sufficiency and proximity, as well as, at the very least, the criteria for global ecological advantages, along with the preservation of the local ecological integrity and dignity of life, by means of the use of a multi-criteria method of analysis.

The solutions to the problems that have arisen in transfer operations may also be established in an *a priori* fashion, prioritizing the return of the waste to the country of origin, with other solutions accepted if they meet the minimum criteria described. This is understood to be a possible minimum structure for establishing global waste management that fulfils the premises of socio-ecological justice.

Conclusion

The existence of the current waste crisis has become increasingly clear. The lack of adequate management of the planet's growing amounts of waste of increasing hazardousness is manifested in various ways; among them, in cross-border waste transfer, an important topic of this book. The perspective of social metabolism has been harnessed in Chapter 1 to demonstrate how the practice of cross-border transfer has deepened the crisis, all the while manipulating the carrying capacity of the territories involved and displacing the resulting socio-ecological costs of the waste produced by the countries of export to the countries of import.

It has also been demonstrated how the increase in the number of transfers carried out in the North–South direction has sparked unequal ecological exchanges by means of ecological rucksacks, forcing the global South – with less political and economic power – to bear the brunt of the negative effects generated by this waste crisis disproportionately when compared with the global North. Furthermore, empirical data on the subject indicates the potential ineffectiveness of the international regulations currently governing the subject, notably the Basel Convention. In this context, the need for the use of the social metabolism approach by the social and human sciences (such as Law) has been highlighted in order to facilitate the development and implementation of more concrete measures for a just and ecologically sustainable regulation of the metabolism of the social and economic systems, and ultimately, for overcoming the many ecological crises, including the waste crisis.

Apart from assisting in an understanding of the biophysical reasons, the importance of unveiling the deepest origins of these crises became clear. In this sense, it has been demonstrated how modern Western thought has brought about an epistemological paradigm known as 'mechanistic', whose logic has extended to practically all domains of life, influencing both global geopolitical relations and the theoretical formulations and practical applications of Economics and Law. The mechanistic paradigm, therefore, by means of reductionism, isolation, dichotomization, determinism, universality and other characteristics, has distanced human beings from the complexity of life, generating inordinate appropriation and destruction of Nature in a form of widespread blindness.

In geopolitical terms, the invisible abyssal lines created by the mechanistic paradigm have produced a dichotomy between North and South, with the application of the logic of appropriation–violence towards the latter, maintained by means of the coloniality of power. In the field of Economics, the ideas of material accumulation and infinite growth have come to dominate over the global development model, treating the economy as an autonomous, closed system. This has prompted the creation and reproduction of a linear, uni-directional and intensive production model, as well as consumption patterns guided by the values of speed, excess and wastefulness, in total disconnection from Nature's biological cycles.

Finally, it has been highlighted how under the mechanistic paradigm, Law has become disconnected from the more holistic and community-based legal systems of past eras, in selecting individual private property as the most important legal concept, regulated and governed as it is by state sovereignty. Mechanistic Law has also become distanced from the complexity of life, supporting the Economics of material accumulation and growthism as a promise that progress and prosperity may finally be achieved for the improvement in the social well-being. It has been revealed, however, that these models founded on the mechanistic paradigm have in actual fact deepened social and geopolitical inequality, resulting in a decline in the quality of life and the extrapolation of the planet's biophysical limits.

However, the ecological crises, including the waste crisis, have been challenged by the emergence of the environmental movement. Understood as a representation of organized socio-environmental conflicts, this movement is plural, multi-faceted and heterogeneous. In the various streams highlighted, with their respective schools of thought, are points susceptible to criticism alongside positive elements. Particularly in terms of the waste crisis, although the stream of the 'gospel of eco-efficiency' proposes strategies for overcoming it, the 'environmental justice' stream stands out as providing an approach which, apart from being environmental, addresses the social dimension of the problems caused by waste, which is why waste may also be considered as a problem of justice.

This book has chosen to focus on the theme of 'justice' as an important way of assisting in the process of overcoming the waste crisis, mainly because of its potential for strategic diversification, its ability to eschew the chrematistic language of the more economics-focused proposals, while also unifying ecological and social demands. According to this line of thought, and from the dialogue and engagement between the different streams of environmentalism as well as between the different perspectives on justice selected, Chapter 2 proposed the construction of a theoretical framework of justice for socio-ecological issues, naming it 'socio-ecological justice', with the establishment of general assumptions for an *a posteriori* and pragmatic definition of justice.

The 'what' and the 'who' of this justice have been developed with the aim of applying them to different socio-ecological demands, including those related to waste, in a unified although not uniform fashion, adopting the pluralism of ideas in order to avoid excluding valid arguments and strategies. The 'what' of justice, representing its sphere of application, has been proposed in multi-dimensional terms for the inclusion of different demands. The 'who' of justice, in turn, referring to the community of justice, has emerged from the adoption of an ecocentric ethics, including humans and non-humans of present and future generations as well as Nature and its inter-systemic relations. The combination of these two strategies in the same theoretical framework aims to overcome the anthropocentric, liberal–distributive and reductionist (mechanistic) justice model.

A theoretical framework for socio-ecological justice has been proposed, including at least the demands for distribution, recognition and representation (the 'what' of justice) as part of the search to develop the basic capabilities for the full functioning of individuals and communities, allowing for the promotion of the protection of ecological integrity and the dignity of life (the unifying normative core of multi-dimensional ontology) based on the recognition of the intrinsic value both of human beings and of the non-human natural world (ecocentric ethics). Furthermore, the principle of 'all-subjected' has been chosen as the normative principle for the *a posteriori* and pragmatic definition of the 'who' of this justice, with these the minimum general premises of the framework conceived.

In theoretical terms, the ecological crises (including the waste crisis) have sparked the development of new models linked by the aim of overcoming the mechanistic paradigm. In Part One of this book, Chapter 3 emphasized the theories of complex thinking and the ecology of knowledges as epistemological models of the new paradigm sought. In geopolitical terms, the idea of the socialization of power along with elements of the proposal of socio-ecological justice emerged as a possibility for tackling the North–South dichotomy upheld by the coloniality of power. Ecological Economics, Productive Ecology and the Circular Economy have been highlighted in the economic field and Ecological Law in the legal field as theoretical developments expanding on the epistemological models outlined.

Part One of this book was dedicated to investigating the possible origins and causes of the waste crisis, as well as to presenting some theoretical models developed with the aim of instigating a paradigmatic revolution for overcoming the waste crisis and the ecological crises as a whole. Based on these models, Part Two then proposed a more specific assessment of how the legal regulation of waste has been carried out in a general sense, and of cross-border waste transfers in a more specific way, to gauge whether this juridicization has been undertaken in order to push for an ecologization of the Law based on the adoption of a socio-ecologic perspective.

Due to the complexity of the subject, waste has been revealed to represent not only an ecological and a justice problem, but also a legal problem. In a detailed study of the five regulatory instruments selected – Brazilian, European and international – multiple limitations have been revealed as well as a lack of harmonization in the legal regulation of waste across the different global regulatory levels. The ambiguities and uncertainties generated by the regulations studied begin with the very concept of 'waste', going through other elements, such as its classification in terms of hazardousness, and the ideas surrounding its management, thus demonstrating the difficulty of applying and implementing such instruments in order to achieve waste management that is just and ecologically sound.

Neglect of the preventive aspect in such regulatory instruments has also been revealed. While completely absent from international regulations, prevention of waste is touched on by the European and Brazilian regulations, although only in the sense of encouraging it. Although the ban on poor management may be identified (even if indirectly) in all the regulations, the 'duty of best management' has not received due attention in any of them, with waste hierarchies only established under the Brazilian and European regulations, and even these are susceptible to criticism. Furthermore, waste transfer operations are missing from them all, and only the European legislation features provisions on the use of the best available technology (BAT). As a consequence, a lack of systemic control instruments has been revealed in all of the regulations, as well as diffuse control by means of public participation in the international regulations.

With the legal regulation of waste analysed in the instruments selected, Chapter 5 gave particular attention to cross-border waste transfers. Emphasizing the criteria (or conditions), procedures and provisions of solutions to problems emerging from these transfers under each of the regulations studied has led to the conclusion that, while the international instruments adopt a more economics-focused perspective on the subject, with the establishment of some environmental limitations, the European and Brazilian regulations are based on a more socio-ecological perspective.

As factual data and a critical analysis of the international legislation regulating waste transfers reveal that such regulatory instruments are unable (or insufficient) to guarantee the establishment of just and sustainable global waste management, instead treating the subject as an international waste trade, the initial hypothesis of this book has been confirmed.

It may thus be understood that without the creation of international regulations adopting a socio-ecological perspective, the achievement of a global waste management system that respects Nature and its laws and that guarantees socio-ecological justice to all humans and non-humans of present and future generations will be a rather ambitious target. Thus, harnessing the theoretical framework devised in Chapter 2 and the new epistemological, geopolitical, economic and legal models presented in Chapter 3, this book has also sought to reflect on possible strategies for establishing just and ecologically sustainable waste management as an avenue for overcoming the current waste crisis.

Chapter 6 outlined more specific strategies for action (ecology of action) for the establishment of this management, in order for it to meet the premises of socio-ecological justice, promoting ecological integrity (sustainability) and the dignity of life (justice). All of the strategies have been inspired to a greater or lesser extent by the various theoretical developments made as the book developed. The idea thus took shape of the 'governance of material flows' as a wide and expansive notion, in which global waste management is included.

In order to respond to many of the problems and challenges highlighted in Chapters 4 and 5, strategies have been presented on the structure for negotiating and developing the legal regulation of global waste management (featuring the third element of the notion of socio-ecological justice – the 'how' of justice), the general content of this regulation, and specific criteria and procedures for waste transfer requests.

Although the geopolitical approach to the global North and South has been adopted, the strategies presented at the end of the book may be applied even in light of the occurrence of shifts to global (im)balances of power, as long as the minimum criteria developed for waste transfers are always taken into consideration. It may thus be understood that the clear adoption of a socio-ecological perspective oriented towards the promotion of ecological integrity and the dignity of life allows for such criteria to function both in cases of waste transfers classified as hazardous as for those considered to be non-hazardous. Furthermore, they must also be appropriately applied both to transfers made from North to South and from South to North, given that the socio-ecological justice framework developed here does not just reflect the intra-generational aspect, but also encompasses the inter-generational and inter-species aspects.

As highlighted at countless points, all of the proposals made throughout this book should not be considered as singular, complete and definitive. Even if far from being realized in practical terms, such models serve as options for paths within an infinite field of possibilities for overcoming the waste crisis, and ultimately, creating a more just and sustainable future for humans and non-humans.

REFERENCES AND FURTHER READING

Abramovay, R (2016) *Beyond the Green Economy*, Routledge, Oxon

ABRELPE, Associação Brasileira de Empresas de Limpeza Pública e Resíduos Especiais (2017) *Panorama dos Resíduos Sólidos no Brasil 2016* [Online] abrelpe.org.br/pdfs/panorama/panorama_abrelpe_2017.pdf (archived at https://perma.cc/2A4Y-A5JG)

Acselrad, H, Mello, CC do Amaral and Bezerra, G das Neves (2009) *O que é Justiça Ambiental*, Garamond, Rio de Janeiro

Alier, JM (2003) *The Environmentalism of the Poor: A study of ecological conflicts and valuation*, Edward Elgar Publishing Ltd, Cheltenham

Alves, E, Souza, N and Eismann, M (2013) *Bioquímica. Introdução ao metabolismo: guia ilustrado sobre metabolismo humano*, Publicação independente

Amnesty International/Greenpeace (2012) *The toxic truth about a company called Trafigura, a ship called the* Probo Koala, *and the dumping of toxic waste in Cote D'Ivoire* [Online] www.amnestyusa.org/files/afr310022012eng.pdf (archived at https://perma.cc/Z64Y-VMEW)

Anand, R (2004) *International Environmental Justice: A North–South dimension*, Ashgate, Hampshire, England

Aragão, MA de Sousa (2003) *O Direito dos Resíduos*, Almedina, Coimbra

Aragão, MA de Sousa (2006) *O princípio do nível elevado de proteção e a renovação ecológica do direito do ambiente e dos resíduos*, Almedina, Coimbra

Aragão, MA de Sousa (2009) Direito Administrativo dos Resíduos, in *Tratado de Direito Administrativo Especial,* eds P Otero and P Gonçalves, Almedina, Coimbra

Ayala, P de Araújo (2011) *Devido Processo Ambiental*, Lumen Juris, Rio de Janeiro

Bacon, F (2012) *Novum Organum Scientiarum: New Method,* Bottom of the Hill Publishing

Bartl, A (2014) Waste and entanglements of the waste hierarchy, *Waste Management*, (34), pp 1–2

Baudrillard, J (1998) *The Consumer Society: Myths and structures*, Sage, London

Bauman, Z (2007) *Consuming Life*, Polity, Cambridge

Beck, U (1992) *Risk Society: Towards a new modernity*, Sage, London

Benjamin, AH (2010) Constitucionalização do ambiente e ecologização da Constituição brasileira. in *Direito Constitucional Ambiental Brasileiro*, eds JJG Canotilho and JRM Leite, Saraiva, São Paulo

Bosselmann, K (2013) Grounding the rule of law, in *Rule of Law for Nature: New dimensions and ideas in environmental law,* ed C Voigt, Cambridge University Press, New York

Bosselmann, K (2016) *The Principle of Sustainability: Transforming law and governance,* Routledge, New York

Braungart, M and McDonough, W (2009) *Cradle to Cradle: Re-making the way we make things*, Vintage Books, London

Brazil (1993) Decree n. 875, Diário Oficial da União, 20 July. Available from: www.planalto.gov.br/ccivil_03/decreto/D0875.htm (archived at https://perma.cc/J4G7-HAGM)

Brazil (2010) Law no 12.305, Diário Oficial da União, 03 August. Available from: www.planalto.gov.br/ccivil_03/_ato2007-2010/2010/lei/l12305.htm (archived at https://perma.cc/79AN-J8BA)

Bugge, HC (2013) Twelve fundamental challenges in environmental law: an introduction to the concept of rule of law for nature, in *Rule of Law for Nature: New dimensions and ideas in environmental law,* ed C Voigt, Cambridge University Press, New York

Bustamante, LP (2007) *Los Derechos de la Sustentabilidad: Desarrollo, consumo y ambiente*, Colihue, Buenos Aires

Cambridge (2011) Cambridge Business English Dictionary, Cambridge University Press

Capra, F (1996) *The Web of Life: A new synthesis of mind and matter*, Flamingo, London

Capra, F (2002) *The Hidden Connections: A science for sustainable living*, Flamingo, London

Capra, F, and Mattei, U (2015) *The Ecology of Law: Toward a legal system in tune with nature and community*, BK Publishers, Oakland

Charonis, GK (2012) Degrowth, steady state economics and the circular economy: three distinct yet increasingly converging alternative discourses to economic growth for achieving environmental sustainability and social equity, in *World Economic Association Sustainability Conference 2012*. Available from: sustainabilityconference2012.weaconferences.net/papers/degrowth-steady-state-economics-and-the-circular-economy-three-distinct-yet-increasingly-converging-alternative-discourses-to-economic-growth-for-achieving-environmental-sustainability-and-social-eq/ (archived at https://perma.cc/SHJ7-R6KU)

Chrysoulakis, N *et al* (2013) Sustainable urban metabolism as a link between biophysical sciences and urban planning: The BRIDGE project, *Landscape and Urban Planning Journal*, (112), pp 100–117

Cilliers, P, and Nicolescu, B (2012) Complexity and transdisciplinarity: discontinuity, levels of reality and the hidden third, *Futures*, (44), pp 711–18

Cipriano, TARP (2015) A juridificação dos Resíduos no Brasil, in *Direito Ambientale Sustentabilidade*, eds A Philippi Jr *et al*, Manole, Tamboré

Cointreau, S (2006) *Occupational and Environmental Health Issues of Solid Waste Management. Special Emphasis on Middle- and Lower-Income Countries*, World Bank. Available from: siteresources.worldbank.org/INTUSWM/Resources/up-2.pdf (archived at https://perma.cc/8K7H-WWYW)

CONAMA, Conselho Nacional do Meio Ambiente, Brazil (2012) Resolution n. 452, Diário Oficial da União, 04 July. Available from: www2.mma.gov.br/port/conama/legiabre.cfm?codlegi=676 (archived at https://perma.cc/ST3P-58HM)

Daly, HE (1993a) Introduction to essays toward a steady-state economy, in *Valuing the Earth: Economics, ecology, ethics*, eds HE Daly and KN Townsend, MIT Press, London

Daly, HE (1993b) The steady-state economy: towards a political economy of biophysical equilibrium and moral growth, in *Valuing the Earth: Economics, ecology, ethics,* eds HE Daly and KN Townsend, MIT Press, London

Derani, C (1997) *Direito Ambiental Econômico*, Max Limonad, São Paulo

Descartes, R (1998) *Discourse on Method*, Hackett Publishing Co Inc, Indianapolis

EC, European Commission (2015a) *Generation, treatment and transboundary shipment of hazardous waste and other waste in the Member States of the European Union, 2010–2012* [Online] eur-lex.europa.eu/legal-content/EN/TXT/?uri=COM:2015:660:FIN (archived at https://perma.cc/9PLW-YVE9)

EC, European Commission (2015b) *Guidance document on the definition and classification of hazardous waste* [Online] ec.europa.eu/environment/waste/pdf/consult/Draft%20guidance%20document_09062015.pdf (archived at https://perma.cc/BA6T-ZMD8)

EEA, European Environment Agency, European Union (2009) *Waste without borders in the EU? Transboundary shipments of waste* [Online] www.eea.europa.eu/publications/waste-without-borders-in-the-eu-transboundary-shipments-of-waste (archived at https://perma.cc/Y6QV-UURF)

EMF, Ellen MacArthur Foundation (2015) *Growth Within: A circular economy vision for a competitive Europe.* Available from: www.ellenmacarthurfoundation.org/assets/downloads/publications/EllenMacArthurFoundation_Growth-Within_July15.pdf (archived at https://perma.cc/VV97-QUXE)

Esteva, G (2010) Development, in *The Development Dictionary: A guide to knowledge as power*, ed W Sachs, Zed Books, London

EU, European Union (1975) *Council Directive 75/442/EEC*, Official Journal of the European Communities, 25 July. Available from: eur-lex.europa.eu/legal-content/EN/TXT/PDF/?uri=CELEX:31975L0442&from=EN (archived at https://perma.cc/F9FZ-VSB7)

EU, European Union (1991) *Council Directive 91/156/EEC,* Official Journal of the European Communities, 26 March. Available from: eur-lex.europa.eu/legal-content/EN/TXT/PDF/?uri=CELEX:31991L0156&from=EN (archived at https://perma.cc/CZN2-5WZT)

EU, European Union (1994) *Commission Decision 94/03/EC*, Official Journal of the European Communities, 7 January. Available from: eur-lex.europa.eu/legal-content/EN/TXT/PDF/?uri=CELEX:31994D0003&from=EN (archived at https://perma.cc/D22K-CVNA)

EU, European Union (1996) *Council Directive 96/61/EC*, Official Journal of the European Communities, 10 October. Available from: eur-lex.europa.eu/legal-content/EN/TXT/PDF/?uri=CELEX:31996L0061&from=EN (archived at https://perma.cc/RN7Q-GEUS)

EU, European Union (1997) *Council Resolution 97/C 76/01*, Official Journal of the European Communities, 11 March. Available from: eur-lex.europa.eu/legal-content/EN/TXT/PDF/?uri=OJ:C:1997:076:FULL&from=EN (archived at https://perma.cc/SEL7-4NCV)

EU, European Union (2000) *Commission Decision 2000/532/EC*, Official Journal of the European Communities, 6 September. Available from: eur-lex.europa.eu/legal-content/EN/TXT/PDF/?uri=CELEX:32000D0532&from=EN (archived at https://perma.cc/SE48-KLFX)

EU, European Union (2006a) *Directive 2006/12/EC of the European Parliament and of the Council*, Official Journal of the European Union, 27 April. Available from: eur-lex.europa.eu/LexUriServ/LexUriServ.do?uri=OJ:L:2006:114:0009:0021:en:PDF (archived at https://perma.cc/WL7L-V6H2)

EU, European Union (2006b) *Regulation (EC) n. 1013/2006 of the European Parliament and of the Council*, Official Journal of the European Union, 12 July. Available from: eur-lex.europa.eu/legal-content/EN/TXT/PDF/?uri=CELEX:32006R1013&from=EN (archived at https://perma.cc/Z7R4-NVA2)

EU, European Union (2007) *Communication from the Commission to the council and the European Parliament on the Interpretative Communication on waste and by-products, COM(2007)59*. Available from: eur-lex.europa.eu/legal-content/EN/TXT/PDF/?uri=CELEX:52007DC0059&from=EN (archived at https://perma.cc/A7AP-P74P)

EU, European Union (2008) *Directive 2008/98/EC of the European Parliament and of the Council*, Official Journal of the European Union, 22 November. Available from: eur-lex.europa.eu/legal-content/EN/TXT/PDF/?uri=CELEX:32008L0098&from=EN (archived at https://perma.cc/EX4G-5Y4P)

EU, European Union (2014) *Commission Decision 2014/955/EU*, Official Journal of the European Union, 30 December. Available from: eur-lex.europa.eu/legal-content/EN/TXT/PDF/?uri=CELEX:32014D0955&from=EN (archived at https://perma.cc/89K3-EMVZ)

Eurostat, European Union (2019a) *Environmental Data Centre on Waste, Indicators, Municipal Waste Generation and Treatment, By Treatment Method* [Online] ec.europa.eu/eurostat/web/waste (archived at https://perma.cc/EBT8-SW4W)

Eurostat, European Union (2019b) *Statistics Explained: Waste shipment statistics* [Online] ec.europa.eu/eurostat/statistics-explained/index.php/Waste_shipment_statistics (archived at https://perma.cc/X83J-GMPR)

Ewijk, S Van and Stegemann, JA (2014) Limitations of the waste hierarchy for achieving absolute reductions in material throughput, *Journal of Cleaner Production*, (132), pp 122–28

Fischer-Kowalski, M and Huttler, W (1999) Society's metabolism: the intellectual history of materials flow analysis, Part II, 1970–1998, *Journal of Industrial Ecology*, **2** (4), pp 107–36

Flores, JH (2009) *A (re)invenção dos direitos humanos*, Fundação Boiteux, Florianópolis

Fontdeglória, X (2018) China não vai mais reciclar plástico de outros países: gigante asiático proíbe as importações de lixo do exterior para melhorar o meio ambiente, *El País*, 10 January. Available from: brasil.elpais.com/brasil/2018/01/05/internacional/1515145196_165569.html (archived at https://perma.cc/JME5-ME48)

Fraser, N (1996) Social justice in the age of identity politics: redistribution, recognition and participation, *The Tanner Lectures on Human Values*, Stanford University (May)

Fraser, N (2005) Reframing justice in a globalizing world, *New Left Review*, (36) pp 69–88 (December)

Fraser, N (2008) Abnormal Justice, *Critical Inquiry*, **34** (3), The University of Chicago, pp 393–422

Georgescu-Roegen, N (2011) *From Bioeconomics to Degrowth: Georgescu-Roegen´s 'New Economics' in eight essays*, Taylor & Francis

Gharfalkar, M, Court, R, Campbell, C, Ali, Z and Hillier, G (2015) Analysis of waste hierarchy in the European waste directive 2008/98/EC, *Waste Management*, (39), pp 305–13

Ghisellini, P, Cialani, C and Ulgiati, S (2016) A review on circular economy: the expected transition to a balanced interplay of environmental and economic systems, *Journal of Cleaner Production* (114), pp 11–32

Greenpeace (2010) *The Toxic Ships: The Italian hub, the Mediterranean area and Africa* [Online] www.genovaweb.org/doc/Report-The-toxic-ship.pdf (archived at https://perma.cc/8Y5H-JDFN)

Gronemeyer, M (2009) Helping, in *The Development Dictionary: A guide to knowledge as power,* ed W Sachs, Zed Press

Grosfoguel, R (2010) Para descolonizar os estudos de economia política e os estudos pós-coloniais: transmodernidade, pensamento de fronteira e colonialidade global, in *Epistemologias do Sul*, eds BS Santos and MP Meneses, Cortez, São Paulo

Haberl, H, Fischer-Kowalski, M, Krausmann, F, Martinez-Alier, J and Winiwarter, V (2011) A socio-metabolic transition towards sustainability? Challenges for another great transformation, *Sustainable Development*, (19), pp 1–14

Heijungs, R, Huppes, G and Guinée, JB (2010) Life cycle assessment and sustainability analysis of products, materials and Technologies. Toward a scientific framework for sustainability life cycle analysis, *Polymer Degradation and Stability*, (95), pp 422–28

Honneth, A (1992) Integrity and Disrespect: Principles of a conception of morality based on the theory of recognition, *Political Theory*, **20** (2)

Hoornweg, D and Bhada-Tata, P (2012) *What a Waste: A global review of solid waste management,* World Bank. Available from: siteresources.worldbank.org/INTURBANDEVELOPMENT/Resources/336387-1334852610766/What_a_Waste2012_Final.pdf (archived at https://perma.cc/HSA2-FR5Y)

Hornborg, A (1998) Toward an ecological theory of unequal exchange: articulating world system theory and ecological economics, *Ecological Economics*, **25** (1), pp 127–36

IBAMA, Instituto Brasileiro do Meio Ambiente e dos Recursos Naturais Renováveis, Brazil (2013) Normative Instruction n. 12, Diário Oficial da União, 17 July. Available from: www.ibama.gov.br/sophia/cnia/legislacao/IBAMA/IN0012-160713.pdf (archived at https://perma.cc/DRK6-29SM)

IBGE, Instituto Brasileiro de Geografia e Estatística, Brazil (2001) *Pesquisa Nacional de Saneamento Básico 2000*, Ministério do Planejamento, Orçamento e Gestão, Rio de Janeiro

ISWA, International Solid Waste Association (2016) *A Roadmap For Closing Waste Dumpsites: The world's most polluted places* [Online] www.iswa.org/fileadmin/galleries/About%20ISWA/ISWA_Roadmap_Report.pdf (archived at https://perma.cc/23XX-FB64)

Jackson, T (2011) *Prosperity Without Growth: Economics for a finite planet,* Earthscan, Oxon

Kaza, S, Yao, L, Bhada-Tata, P and Van Woerden, F (2018) *What a Waste 2.0: A global snapshot of solid waste management to 2050.* Urban Development Series. Washington, DC: World Bank. doi:10.1596/978-1-4648 -1329-0. Available from: openknowledge.worldbank.org/handle/10986/30317 (archived at https://perma.cc/WMN7-X44Z)

Kellenberg, D (2015) The economics of the international trade of waste, *The Annual Review of Resource Economics.* Available from: www.annualreviews.org (archived at https://perma.cc/6ZQC-7DTP)

Kelsen, H (2009) *Pure Theory of Law*, The Lawbook Exchange, Ltd

Kumar, A, Espinosa, DCR and Holuszko, ME (2017) E-waste: an overview on generation, collection, legislation and recycling practices, *Resources, Conservation and Recycling*, (122), pp 32–42

Latouche, S (2009) *Farewell to Growth,* Polity, Cambridge

Leff, E (2001) *Saber Ambiental: Sustentabilidade, racionalidade, complexidade, poder*, Editora Vozes, Petrópolis

Leff, E (2006) *Racionalidade Ambiental: A Reapropriação Social da Natureza,* Civilização Brasileira, Rio de Janeiro

Leff, E (2010) *Discursos Sustentáveis*, Cortez, São Paulo

Leite, JRM (2015) *Manual de Direito Ambiental*, Saraiva, São Paulo

Leonard, A (2010) *The Story of Stuff: How our obsession with stuff is trashing the planet, our communities, and our health – and a vision for change*, Free Press, New York

Locke, J (2014) *Two Treatises of Government*, CreateSpace Independent Publishing

Maldonado-Torres, N (2010) A topologia do ser e a geopolítica do conhecimento. Modernidade, império e colonialidade, in *Epistemologias do Sul*, eds BS Santos and MP Meneses, Cortez, São Paulo

Meadows, DH, Meadows, DL, Randers J and Behrens, WW (1972) *The Limits to Growth*, Universe Books, New York

Melo, ME (2016) Pagamento por Serviços Ambientais (PSA): entre a proteção e a mercantilização dos serviços ecossistêmicos no contexto da crise ambiental, PhD thesis, Universidade Federal de Santa Catarina, Florianópolis

Miller, MAL (1995) *The Third World in Global Environmental Politics*, Boulder, CO, Lynne Reinner Publishers

Montibeller-Filho, G (2008) *O mito do desenvolvimento sustentável: meio ambiente e custos sociais no moderno sistema produtor de mercadorias*, Editora UFSC, Florianópolis

Moran, EF (2006) *People and Nature: An introduction to human ecological relations*, Blackwell Publishing, Oxford

Moreno, M, Braithwaite, N and Cooper, T (2014) *Moving Beyond the Circular Economy*, Nottingham Trent University. Available from: www.eukn.eu/fileadmin/Files/News/2015/R-Care_Innovation_final_version-libre.pdf (archived at https://perma.cc/7NWV-CW7Y)

Morin, E (2008) *On Complexity*, Hampton Press, New Jersey

Morin, E, and Kern, A-B (1999) *Homeland Earth: A manifesto for the new millennium*, Hamption Press, New Jersey

Moses, K (2013) China leads the waste recycling league: EU legislation is fuelling a multibillion-dollar market. As landfill charges increase, it is often cheaper to send rubbish abroad, *The Guardian*, 14 June. Available from: www.theguardian.com/environment/2013/jun/14/waste-trade-china-recycling-rubbish (archived at https://perma.cc/SJ4V-LQAM)

Muradian, R, Walter, M and Alier, JM (2012) Hegemonic transitions and global shifts in social metabolism: implications for resource-rich countries. Introduction to the special section, *Global Environmental Change*, pp 559–67

Newton, I (2016) *The Principia: Mathematical Principles of Natural Philosophy*, CreateSpace Independent Publishing

Ninni, K (2011) *Lixo 'útil' poupa recursos naturais e energia, mas importação ainda é tabu*, O Estado de São Paulo, 08 October. Available from: sustentabilidade.estadao.com.br/noticias/geral,lixo-util-poupa-recursos-naturais-e-energia-mas-importacao-ainda-e-tabu,782914 (archived at https://perma.cc/E27W-SSSL)

Nixon, R (2011) *Slow Violence and the Environmentalism of the Poor*, Harvard University Press

Nunes, JA (2010) O resgate da epistemologia, in *Epistemologias do Sul*, eds BS Santos and MP Meneses, Cortez, São Paulo

Nussbaum, MC (2004) Beyond 'compassion and humanity': justice for nonhuman animals, in *Animal Rights: Current debates and new directions*, eds CR Sustein and M Nussbaum, Oxford University Press

Nussbaum, MC (2006) The moral status of animals, *Chronicle of Higher Education*, 3 February. Available from: www.chronicle.com/article/The-Moral-Status-of-Animals/25792 (archived at https://perma.cc/QAV7-3JDZ)

Nussbaum, MC (2011) *Creating Capabilities: The human development approach*, The Belknap Press of Harvard University Press

Öberg, IMF (2016) Análise da aplicação da Convenção da Basileia pelo governo brasileiro com foco no trânsito transfronteiriço de resíduos, Masters dissertation, Universidade de São Paulo, São Paulo

OECD, Organisation for Economic Co-operation and Development (2009) *Guidance Manual for the implementation of council decision C(2001)107/FINAL, as amended, on the control of transboundary movements of wastes destined for recovery operations*. Available from: www.oecd.org/env/waste/guidance-manual-control-transboundary-movements-recoverable-wastes.pdf (archived at https://perma.cc/HVG8-YQMS)

Ost, F (1995) *A natureza à margem da lei: a ecologia à prova do direito*, Instituto Piaget, Lisbon

PACE, Platform for Accelerating the Circular Economy (2019) *A New Circular Vision for Electronics: Time for a global reboot*, World Economic Forum [Online] www3.weforum.org/docs/WEF_A_New_Circular_Vision_for_Electronics.pdf (archived at https://perma.cc/SJ92-FUMH)

Pellow, DN and Brulle, RJ (2005) The future of environmental justice movements, in *Power, Justice, and the Environment: A critical appraisal of the environmental justice movement*, eds DN Pellow and RJ Brulle, MIT Press

Peña, DG (2005) Autonomy, Equity, and Environmental Justice, in *Power, Justice, and the Environment: A critical appraisal of the environmental justice movement*, eds DN Pellow and RJ Brulle, MIT Press

Pope, K (2017) *Understanding Planned Obsolescence: Unsustainability through production, consumption and waste generation*, Kogan Page, London

Preston, FA (2012) Global Redesign? Shaping the Circular Economy, *Energy, Environment and Resource Governance*. Available from: www.chathamhouse.org/sites/files/chathamhouse/public/Research/Energy,%20Environment%20and%20Development/bp0312_preston.pdf (archived at https://perma.cc/9MGG-VNMH)

Quijano, A (2010) Colonialidade do poder e classificação social, in *Epistemologias do Sul,* eds BS Santos and MP Meneses, Cortez, São Paulo

Rammê, RS (2012) *Da justiça ambiental aos direitos e deveres ecológicos: conjecturas político-filosóficas para uma nova ordem jurídico-ecológica*, EDUCS, Caxias do Sul

Rammelt, CF and Crisp, P (2014) Systems and thermodynamics perspective on technology in the circular economy, *Real-world Economics Review* (68), pp 25–40

Rawls, J (2005) *A Theory of Justice*, Harvard University Press

Rifkin, J and Howard, T (1980) *Entropy: A new world view*, The Viking Press, New York

Sachs, W (2010) Introduction, in *The Development Dictionary: A guide to knowledge as power*, ed W Sachs, Zed Books, London

Santos, B de Sousa (2010a) Para além do pensamento abissal: das linhas globais a uma ecologia de saberes, in *Epistemologias do Sul*, eds BS Santos and MP Meneses, Cortez, São Paulo

Santos, B de Sousa (2010b) Um Ocidente não-ocidentalista? A filosofia à venda, a douta ignorância e a aposta de Pascal, in *Epistemologias do Sul*, eds BS Santos and MP Meneses, Cortez, São Paulo

Sass, LB and Melo, ME (2014) Governança global ambiental: omissões e contradições no regime internacional da proteção da biodiversidade, *Revista de Direito Ambiental*, 74, pp 337–66

Sbert, JM (2010) Progress, in *The Development Dictionary: A guide to knowledge as power*, ed W Sachs, Zed Books, London

Schlosberg, D (2009) *Defining Environmental Justice: Theories, movements and nature*, Oxford University Press, New York

Sen, A (2001) *Development as Freedom*, Oxford University Press

Silveira, PG (2016) A adoção da melhor tecnologia disponível no licenciamento ambiental brasileiro na perspectiva do Estado Ambiental, Masters dissertation, Universidade Federal de Santa Catarina, Florianópolis

Slade, G (2006) *Made to Break: Technology and obsolescence in America*, Harvard University Press

Smith, A (2014) *The Wealth of Nations*, CreateSpace Independent Publishing

Summers, LH (1991) *The Lawrence Summers Memo*. Available from: www.personal.ceu.hu/corliss/CDST_Course_Site/Readings_old_2012_files/Our%20Words_%20The%20Lawrence%20Summers%20Memo.pdf (archived at https://perma.cc/7P8N-KS2T)

Tamames, R (1983) *Crítica dos limites do crescimento: ecologia e desenvolvimento*, Publicações Dom Quixote, Lisbon

Turner, DA, Williams, IA and Kemp, S (2016) Combined material flow analysis and life cycle assessment as a support tool for solid waste management decision making, *Journal of Cleaner Production*, (129), pp 234–48

Ullrich, O (2010) Technology, in *The Development Dictionary: A guide to knowledge as power*, ed W Sachs, Zed Books, London

UNEP, United Nations Environment Programme (1989) *Basel Convention*. Available from: www.basel.int/TheConvention/Overview/TextoftheConvention/tabid/1275/Default.aspx (archived at https://perma.cc/HGM3-ZCPB)

UNEP, United Nations Environment Programme (1999) *Basel Protocol on Liability and Compensation*. Available from: www.basel.int/TheConvention/Overview/LiabilityProtocol/tabid/2399/Default.aspx (archived at https://perma.cc/CD2V-92K3)

UNEP, United Nations Environment Programme (2009) Secretariat of Basel Convention, *Confirmed cases of illegal traffic*, UNEP-CHW-CAS-TRAF-BrazilUK. Available from: www.basel.int/Implementation/LegalMatters/IllegalTraffic/CasesofIllegalTraffic/tabid/3424/Default.aspx (archived at https://perma.cc/94T5-ZDJX)

UNEP, United Nations Environment Programme (2011a) *Report UNEP/CHW.10/28 of the Conference of the Parties to the Basel Convention on the Control of Transboundary Movements of Hazardous Wastes and Their Disposal on its tenth meeting (COP10)* [Online] www.basel.int/TheConvention/ConferenceoftheParties/ReportsandDecisions/tabid/3303/Default.aspx (archived at https://perma.cc/TS6F-CY8S)

UNEP, United Nations Environment Programme (2011b) *Decision BC-10/3: Indonesian-Swiss country-led initiative to improve the effectiveness of the Basel Convention,* UNEP/CHW.COP.10-BC-10-3, adopted at the Conference of the Parties to the Basel Convention at its tenth meeting (COP10). Available from: www.basel.int/TheConvention/ConferenceoftheParties/ReportsandDecisions/tabid/3303/Default.aspx (archived at https://perma.cc/TS6F-CY8S)

UNEP, United Nations Environment Programme (2011c) Secretariat of Basel Convention, *Confirmed cases of illegal traffic*. UNEP-CHW-CAS-TRAF-Brazil Spain. Available from: www.basel.int/Implementation/LegalMatters/IllegalTraffic/CasesofIllegalTraffic/tabid/3424/Default.aspx (archived at https://perma.cc/94T5-ZDJX)

UNEP, United Nations Environment Programme (2012a) Secretariat of Basel Convention, *Confirmed cases of illegal traffic*. UNEP-CHW-CAS-TRAF-BrazilUSACanada. Available from: www.basel.int/Implementation/LegalMatters/IllegalTraffic/CasesofIllegalTraffic/tabid/3424/Default.aspx (archived at https://perma.cc/94T5-ZDJX)

UNEP, United Nations Environment Programme (2012b) Secretariat of Basel Convention, *Draft report on the implementation of the Basel Convention as it relates to the interpretation of certain terminology*. UNEP-CHW-CLI-REP-Terminology. Available from: www.basel.int/Implementation/CountryLedInitiative/History/Providingfurtherlegalclarity/Reportonterminology/tabid/2674/Default.aspx (archived at https://perma.cc/7FCU-HE6M)

UNEP, United Nations Environment Programme (2015a) *Report UNEP/CHW.12/27 of the Conference of the Parties to the Basel Convention on the Control of Transboundary Movements of Hazardous Wastes and Their Disposal on the work of its twelfth meeting* (COP12) [Online] www.basel.int/TheConvention/ConferenceoftheParties/ReportsandDecisions/tabid/3303/Default.aspx (archived at https://perma.cc/TS6F-CY8S)

UNEP, United Nations Environment Programme (2015b) *Global Waste Management Outlook* [Online] www.unenvironment.org/resources/report/global-waste-management-outlook (archived at https://perma.cc/LP28-N7KX)

UNEP, United Nations Environment Programme (2016) *Follow-up to the Indonesian-Swiss country-led initiative to improve the effectiveness of the Basel Convention. Addendum: Glossary of Terms,* UNEP/CHW.13/4/Add.2. Available from: www.basel.int/Implementation/LegalMatters/LegalClarity/Glossaryofterms/SmallIntersessionalWorkingGroup/tabid/3622/Default.aspx (archived at https://perma.cc/4LUR-FEHV)

UNEP, United Nations Environment Programme (2017) *Report UNEP/CHW.13/28 of the Conference of the Parties to the Basel Convention on the Control of Transboundary Movements of Hazardous Wastes and Their Disposal on the work of its thirteenth meeting (COP13)* [Online] www.basel.int/TheConvention/ConferenceoftheParties/ReportsandDecisions/tabid/3303/Default.aspx (archived at https://perma.cc/TS6F-CY8S)

UNEP, United Nations Environment Programme (2019) *Decision BC-14/16: Providing further legal clarity*, UNEP/CHW.COP.14-BC-14-16, adopted at the Conference of the Parties to the Basel Convention at its fourteenth meeting (COP14). Available from: www.basel.int/TheConvention/ConferenceoftheParties/ReportsandDecisions/tabid/3303/Default.aspx (archived at https://perma.cc/TS6F-CY8S)

da Veiga, JE (2010) *Sustentabilidade: A legitimação de um novo valor*, Senac, São Paulo

Winter, G (2008) A fundament and two pillars: the concept of sustainable development 20 years after the Brundtland Report, in *Sustainable Development in International and National Law: What did the Brundtland Report do to legal thinking and legal development, and where can we go from here?,* eds C Bugge and C Voigt, Europa Law, Groningen

INDEX

a posteriori (pragmatic) justice 103–04, 131–38, 140–41
Aarhus Convention 242–44
abnormal justice times 104–05, 339–43
abyssal lines 45–46, 49, 50, 68–69, 72–73, 75, 150, 370
Acselrad, H 90–92, 94
action, complexity of 336–39
administrative controls 240–44, 245, 261, 346–49
administrative costs 306, 307
Alier, JM 38, 39, 79, 93, 330–31
  cult of wilderness 81–82
  environmental justice 88, 89
  general procedure for waste transfers 361–63
  gospel of eco-efficiency 83, 86
  social metabolism 32, 34–35
all-subjected principle 123–24, 134, 141, 340
Alves, E 28–29
amber control procedure
  EU Regulation 294, 295–96, 297–98
  OECD Decision 281, 282–85, 292
anabolism 328–29
Anand, R 52–54, 275–76, 277, 278
anthropocentrism 126–29
appropriation-violence dichotomy 49–51
Aragão, MA de Sousa 13, 14, 15, 78–79, 242–43, 252–53, 350
  legal concepts of waste 201, 205, 206
  location of waste treatment 353, 354–56
  material flows 325, 326–29, 332
  prohibition of poor waste management 236–38
asymmetry of power 333
autopoiesis 138

Bacon, F 41–42, 68
banned waste 314–15, 319
Basel Convention 3, 19–22, 198–99, 331, 346
  Ban Amendment 269, 277–78, 322, 346
  Brazil's signing of 313–14, 319
  classification of waste 222–25, 260
  Glossary of terms 210–13
  legal concept of waste 209–13, 214
  waste management regime 227–29, 235–36, 237, 239–40, 242, 243
    recovery and disposal 254–55, 256, 257
  waste transfer regulation 265–80, 321–22
    criteria for waste transfers 268–69, 358
    critical analysis 273–80
    general procedure for transfers 269–71
    problems affecting transfers 271–72, 279
Basel Protocol on Liability and Compensation 279–80, 322, 346
basic capabilities 118–20, 132–33, 135
Baudrillard, J 63
Bauman, Z 63
Beck, U 98
best available technology (BAT) 246–47
best management 238
  duty of 244–47, 262, 346, 347, 349–52
  waste hierarchy and identifying 247–59
biological metabolism 28–29
Bosselmann, K 125, 126, 127–29, 133, 136
Braungart, M 29, 60–61, 62
Brazil 6, 16, 87, 372
  classification of waste 215–17, 260
  CONAMA Resolution 216–17, 231–32, 314–15, 317
  IBAMA *see* IBAMA
  waste management regime
    controls 240–41, 242, 243–44, 245
    legal concept of waste 228, 229
    legal subjects and responsibility system 230–32, 237
    waste hierarchy 248, 252, 253, 256, 257–58
  waste transfer regulation 313–21, 323–24
  waste transfers to 1, 19–23, 24, 319
Brazilian National Solid Waste Policy (NSWP) 198, 199
  classification of waste 215–16, 217, 260

legal concept of waste 201–04, 214
waste management regime 228, 229, 230–31, 237, 240–41, 242, 243–44, 245, 248, 252, 253, 256, 257–58
waste transfer regulation 314, 319
Brulle, RJ 97
Brundtland Report 85
Bugge, HC 177–78
Bustamante, LP 63
by-products 206, 208, 209

Canada 20
capabilities 116–21, 132
basic 118–20, 132–33, 135
combined 118, 132
Capra, F 41, 43, 67, 71, 147, 148–49
ecological law 178–90
carrying capacity 35, 65
Cartesian method 42–43
cascading 332
catabolism 328–29
China 28, 169
circular economy 167–74, 192–93
civil society 333–34, 342, 344
classic waste law 327
classifications of waste 215–27, 259–60, 306
Basel Convention 222–25, 260
Brazil 215–17, 260
EU 217–22, 260
global waste management system 345–46
OECD Decision 225–26, 260
Clausius, R 143–44
closed cycles 332
closed systems 34, 170–71
Coase theorem 84
Cointreau, S 27
colonialism 47–48
combined capabilities 118, 132
commensurability of values 361–62
communities 72, 185–86
transformed into sovereign entities 184, 186–87
Community Strategy for Waste Management (CSWM) 249
complex thinking 143–50, 191
complexity 5–6, 180
of action 336–39
CONAMA Resolution 216–17, 231–32, 314–15, 317
concepts of waste management 227–29, 260–61
conflicts, socio-environmental 81–92, 99–100
Consumer Revolution 62–63
consumerism 60–67
consumption 158, 351
de-consumption 332
contextualized hierarchies 152–53
contingent valuation method (CVM) 84–85
control instruments 240–44, 245, 261, 346–49
see also amber control procedure; green control procedure
controlled waste 314–17, 319–20
Copernicus, N 41
cosmetic environmental assessments 361
costs
administrative 306, 307
transfer of socio-ecological costs in waste transfers 35–40, 50
of waste disposal 26–27
counter-hegemonic globalization 154–55, 191
Crisp, P 171
criteria for waste transfers
Basel Convention 268–69, 358
EU Regulation 297–98
global waste management system 357–60
OECD Decision 281–85, 358
critical theories of justice 111–22
cross-border waste transfers *see* waste transfers
cult of wilderness 81–83, 86, 92, 138

Daly, HE 161
Darwin, C 144–45
de-consumption 332
decoupling 85–86, 159, 172
deep ecology 82–83, 92–100, 139
dematerialization 85–86, 159, 172
democratically instituted environmental policies 90, 92
de-production 332
Descartes, R 41, 42–43
destination 202–03, 204
developed (Northern) countries 2, 26–27, 51–54
developing (Southern) countries 2, 18, 26–27, 51–54
development 57–60
different development model 90, 91
developmental agendas 53
dialogism 340–41, 344
dichotomous relationships 48–50
diffuse control 242–44, 245, 261

dignity of life 135–36, 137, 155–56, 363–64
discarding 205, 207
disorder 144, 145–46
disposal
 by GNP of receiving countries 26
 costs of 26–27
 EU waste transfers for 308, 309–10
 global waste management system 351, 352
 legal waste management regimes 248, 250, 252–59
 objections to waste transfers destined for 301, 302
dissipative structures 145–47
distribution 106–16, 121, 135–38, 140–41
duty of best management 244–47, 262
 global waste management system 346, 347, 349–52
dynamic perspective 327–29

ecodesign 182–83
eco-efficiency, gospel of 83–86, 92, 94–95, 138–39
eco-literacy (ecological literacy) 182
ecological advantages of waste transfer 363
ecological debt 101
ecological economics 160–63, 172–73, 192
ecological existential minimum 238–39
ecological integrity 129, 135–38, 155–56, 363–64
ecological justice 128–31
ecological law 178–90, 193
ecological modernization 37, 83–86, 94
ecological pluralism 100–06
ecological rucksack 32, 38
ecologically unequal exchange 38–39
ecology of knowledges 150–54, 191, 334
ecology of the poor 90, 92–100, 121, 139
economic growth 55–60, 64–66
economics
 ecological 160–63, 172–73, 192
 environmental 84–86, 101, 157–60, 192
 mechanistic 55–67, 71, 73, 76
 paradigm shifts in 157–74, 192–93
economism 18
ecosystem perspective 30–31
electronic waste 17
Ellen MacArthur Foundation (EMF) 167–69, 170
emancipation-regulation dichotomoy 49–50
empowerment 334
energy consumption 35
entropy 33, 143–45, 146, 190–91
environmental agendas 53
environmental economics 84–86, 101, 157–60, 192
environmental justice 86–92, 121, 122–31, 139
environmental law 174–78, 193
environmental liabilities 101
environmental movement 79–80, 81–106, 138–39, 154–55, 370
 challenges and limitations of the schools of thought 92–106
 as a socio-environmental conflict 81–92
environmental racism 88–89, 96
environmentally-sound waste management 265–66, 273–75, 311, 312
epistemology
 mechanistic 40–47
 paradigm shifts 143–54, 191
equity 106–08
EU Regulation 198, 199
 waste management regime 233–35, 239–40, 242, 244, 245, 246–47
 waste transfers 293–313, 323, 324
European Economic Area (EEA) 307
European Free Trade Association (EFTA) 308
European Union (EU) 6, 353, 372
 classification of waste 217–22, 260
 EU Regulation *see* EU Regulation
 illegal waste transfers from 23, 24–25
 legal concept of waste 204–09, 214
 Waste Framework Directive *see* Waste Framework Directive
 waste generation 15–16
 waste management regime
  BAT 246–47
  controls 242, 243, 244, 245
  key concepts 228, 229
  legal subjects and responsibility system 232–35, 237
  waste hierarchy 248–51, 252, 253–54, 256, 258
 waste transfer regulation 23, 24–25, 293–313, 323, 324
  criteria for waste transfers 297–98, 358–59
  critical analysis 311–13
  general procedure 298–304
  problems affecting waste transfers 304–11
European Waste Catalogue (EWC) 204–05
 classification of waste 218–19, 220
exergy 165
externalities 36–37

Fischer-Kowalski, M 29, 30–33, 37
Flores, JH 103
forced waste 201
Ford, H 63
Fraser, N 104–05, 111–16, 122–25, 131–32, 339–43
freedom 107
functioning 117–18

Galileo Galilei 41
general procedure for waste transfers
  Basel Convention 269–71
  EU Regulation 298–304
  global waste management system 360–65, 366
  OECD Decision 285–87
    exceptions to general procedure 288–90, 292–93
generative ownership 184, 187–88
geopolitics
  mechanistic 47–54
  post-abyssal 154–57, 191–92
Georgescu-Roegen, N 159–60
Gharfalkar, M 251–52
global economic system 53–54
global waste management system 9, 336–68, 373–74
  classification of waste 345–46
  complexity of action 336–39
  content of legislation 343–57, 368
  drafting and negotiation of international legislation 339–43
  duty of best management and waste management hierarchy 349–52
  legal concepts of waste 343–45
  location of waste treatment 353–57
  responsibility system and control instruments 346–49
  waste transfers 357–67, 368
    criteria for 357–60
    general procedure for 360–65, 366
    problems affecting 365–67
global waste trading market 17, 27
goals of governance of material flows 331–33
gospel of eco-efficiency 83–86, 92, 94–95, 138–39
governance 329–31
  of material flows 331–36, 367
green control procedure
  EU Regulation 294, 295–96, 297
  OECD Decision 281–82, 283, 291–92
green economy 85
Grosfoguel, R 46, 48, 156
growthist economy 55–60, 64–66

Haberl, H 35
hazardous waste 17
  legal classifications of waste 215–27, 260
  transfers 19–21, 24–25
Hobbes, T 68
holographic principle 148
horizontal transfer 35–36
Hornborg, A 39
hunter-gatherers 14

IBAMA 314, 315–16, 317, 319–20
  Regulatory Instrument no 12 231, 232, 317–19
illegal waste transfers 1, 25–26
  Basel Convention 271–72, 279
  to Brazil 19–23, 319
  EU Regulation 304, 305–06, 307
  OECD Decision 287, 288, 289
income inequality 64
India 87
individual private property 67–68, 69, 70–71
individualism 55–56
inductive method 41–42
Industrial Revolution 60–62
inequalities 108
integrated sustainable waste management (ISWM) 356–57
inter-cultural translation 152–53
inter-generational justice 126, 127, 130, 134–36
internal capabilities 118
international institutions 341, 342, 344
International Solid Waste Association (ISWA) 356
international trade 17, 27, 35–40, 75
inter-species justice 126, 130, 131–32, 133–36
intra-generational justice 126, 134–36
intrinsic value 133–36
invisible hand of the market 56
Ivory Coast 23

Jackson, T 64, 171
Jevons paradox 171
just transition 90, 91
justice 8, 80, 106–38, 139–41, 339–43, 370–71
  critical perspectives of 111–22
  ecological 128–31
  environmental 86–92, 121, 122–31, 139
  mechanistic 106–11
  socio-ecological *see* socio-ecological justice
  'what' of 106–22
  'who' of 122–38

Kelsen, H 70
Khian Sea incident 18–19, 51
knowledge
  ecology of knowledges 150–54, 191, 334
  mechanistic epistemology 40–47

Latouche, S 64
law
  classifications of waste *see* classifications of waste
  concepts of waste *see* legal concepts of waste
  ecological 178–90, 193
  environmental 174–78, 193
  mechanistic 67–74, 76, 174–75, 189, 190
  paradigm shifts in 174–90, 193
  regimes for waste management *see* legal regimes for waste management
  social metabolism and 326–29
Laws of Nature 43
laws of thermodynamics 65, 143–44, 146
Leff, E 162–66
legal concepts of waste 9, 200–15, 259, 326
  Basel Convention 209–13, 214
  Brazil 201–04, 214
  EU 204–09, 214
  global waste management system 343–46
  OECD Decision 213–14
legal regimes for waste management 9, 227–59, 260–63
  control instruments 240–44, 245, 261
  duty of best management 244–47, 262
  key concepts 227–29, 260–61
  legal subjects 230–36, 237, 261
  prohibition of poor waste management 236–40, 245
  responsibility systems 230–36, 237, 261
  waste hierarchy 247–59, 262–63
  see also global waste management system
legal subjects 230–36, 237, 261
liberal anthropocentric perspective 126–29
life cycle assessment (LCA) 347–48
linear production 60–67
living systems 145–50
location of waste treatment 257–59, 263
  global waste management system 353–57
Locke, J 55–56

maldistribution 111–12
Maldonado-Torres, N 47, 48
management, and governance 329–30
material flow analysis (MFA) 347, 348
material flows 325–26
  global waste management system 346–49
  governance of 326–36, 367
Mattei, U 41, 43, 67, 71
  ecological law 178–90
McDonough, W 29
mechanistic economics 55–67, 71, 73, 76
mechanistic epistemology 40–47
mechanistic geopolitics 47–54
mechanistic justice 106–11
mechanistic law 67–74, 76, 174–75, 189, 190
mechanistic paradigm 40–74, 75–77, 143, 145, 157, 189, 369–70
Melo, ME 330
metabolism
  biological 28–29
  social *see* social metabolism
Metal Age 14
meta-political justice 114–15, 122
Mexico 21
Middle Ages 14–15
minimizing waste generation 265–66, 273
mirror-entry waste 219, 221–22
misrepresentation (misframing) 114–15, 122–23
mixtures of wastes 288, 289, 290, 292–93
Montibeller-Filho, G 84
Moran, EF 82
Morin, E 43–44, 58–59, 145–50, 337–38
multi-criteria analysis method 362–64
Muradian, R 33, 38, 81, 87, 99

Naess, A 83
Napoleonic Code 69–70
negentropic production 163–66, 172–73
Neolithic period 14
Newton, I 41, 43
Nixon, R 97–98, 99
non-compliance problems
  Basel Convention 271, 272, 279
  Brazil 318–19
  EU Regulation 304–05, 307
  OECD Decision 287–88, 289, 292
non-hazardous waste
  classifications of waste 215–27
  transfers 24–25
non-recognition 112–13
non-regression, principle of 245–46
Northern countries 2, 26–27, 51–54, 60
Nunes, JA 152–54
Nussbaum, MC 116–20, 132

objections to waste transfers 300–03
OECD Decision 198, 199
  classification of waste 225–26, 260
  legal concept of waste 213–14
  waste management regime 228, 229, 235–36, 237, 242, 243, 256–57
  waste transfer regulation 280–93, 322–23
    criteria for waste transfers 281–85, 358
    critical analysis 291–93
    exceptions to general procedure 288–90, 292–93
    general procedure 285–87
    problems affecting waste transfers 287–88, 289, 292
open epistemology 149–50
open systems 34, 146
order 144, 145–46
origin 202–03, 204
orthopaedic thinking 45
Ost, F 67–68, 83, 93, 185, 186–87
  ecological law 180–81
  environmental law 176–77
  patrimony 188–89

participation 119–20
participatory parity 112–16, 119, 131–32
patrimony 181, 188–89
Pellow, DN 97
Peña, DG 96–97, 109–10
photosynthesis 163
Pigovian tax 84
planned obsolescence 61
plastic waste 21–22
pluralism, ecological 100–06
political justice 114–15, 122
poor, ecology of the 90, 92–100, 121, 139
poor waste management, prohibition of 236–40, 245, 346, 347
post-abyssal geopolitics 154–57, 191–92
post-abyssal thinking 150–54, 191
power
  asymmetry of 333
  disconnecting law from violence and 184–86
  socialization of 156, 191–92
precautionary principle 345
prevention 329
Prigogine, I 145
prior consent 360, 364
problems affecting waste transfers
  Basel Convention 271–72, 279
  Brazil 318–19, 321
  EU Regulation 304–11
  global waste management system 365–67
  OECD Decision 287–88, 289, 292
Probo Koala incident 23
production 158, 170–71
  de-production 332
  linear 60–67
  negentropic 163–66, 172–73
productive ecology 163–67, 172–73, 192
products 208, 209
programmes 337–38
progress 57, 58
prohibition of poor waste management 236–40, 245, 346, 347
property
  generative ownership 184, 187–88
  individual private property 67–68, 69, 70–71
proximity, principle of 354–56

Quijano, A 47

Rammelt, CF 171
Rawls, J 106–10
recognition 106, 111–16, 121, 135–38, 140–41
recovery
  EU waste transfers 308–09, 310
  global waste management system 351, 352
  legal waste management regimes 248, 250, 252–59
  objections to waste transfers destined for 301–02, 303
reductionism 43
regulation-emancipation dichotomy 49–50
representation 106, 111–16, 121, 135–38, 140–41
reprocessing 351, 352
ReSOLVE 168–69
resource scarcity 64–65
resources, waste as 329
responsibility systems
  global waste management system 346–49
  legal waste management regimes 230–36, 237, 261
reuse 351–52
Rifkin, J 40, 41, 55–57, 65
risk society 15, 17, 98

Sachs, W 58
Santos, B de Sousa 44–46, 52, 150–52, 154–55
Sass, LB 330
Schlosberg, D 104, 109, 110–11, 130, 136–38
Scientific Revolution 41
self-eco-organization 146–50
self-sufficiency principle 353–54, 355–56
shallow ecology 83–86, 92–100, 139
slow violence 97–100
Smith, A 55, 56
social contract 106–07
social metabolism 28–35, 37–39, 66, 74–75, 369
  and law 326–29
social status 113, 134
socialization of power 156, 191–92
society 55–56
socio-ecological costs, transfer of 35–40, 50
socio-ecological justice 3–4, 8, 80, 129–30, 154–56
  ecological law 183–84
  global waste management system 341–43, 344
  theoretical framework 8, 131–38, 140–41, 371
socio-economic system perspective 30–31
socio-environmental conflicts 81–92, 99–100
soft law 73
sound waste management 238, 311, 312
  Basel Convention 265–66, 273–75
Southern countries 2, 18, 26–27, 51–54, 60
sovereign entities 184, 186–87
space 122–31
Spain 20
state (origin, destination and) 202–03, 204
state sovereignty 67, 68, 69, 187
static perspective on waste 326–27
status 113, 134
steady state theory 161
strategies 337–39
strong questions and weak answers 44–45
subject
  aspects of justice 122–31
  legal subjects 230–36, 237, 261
Summers, L 18
sustainability 85

tacit consent 292, 300
theoretical models 8, 142–94
  circular economy 167–74, 192–93
  complex thinking 143–50, 191
  ecological economics 160–63, 172–73, 192
  ecological law 178–90, 193
  ecology of knowledges 150–54, 191, 334
  environmental economics 157–60, 192
  environmental law 174–78, 193
  post-abyssal geopolitics 154–57, 191–92
  productive ecology 163–67, 172–73, 192
thermodynamics, laws of 65, 143–44, 146
three-dimensional model of justice 111–16
time 122–31
toxic pollution for no one 90–91
Trafigura Beheer BV 23
transboundary agreements on waste transfers 306, 307
transdisciplinary approach 5–6
transfers of socio-ecological costs 35–40, 50
Truman, HS 58

uniformity 104
United Kingdom (UK) 20
United Nations (UN) 13
  Special Rapporteur 19, 349
United States of America (USA) 88–89, 93
unity in diversity 104
universalism 47
  of arrival or confluence 103–04

values, commensurability of 361–62
violence
  appropriation-violence dichotomy 49–51
  disconnecting law from power and 184–86
  slow 97–100

waste crisis 1–2, 7–8, 13–14, 78–79, 197
Waste Framework Directive 198, 199, 323
  classification of waste 218, 220
  legal concept of waste 207–09, 214
  waste management regime 228, 229, 232–33, 239, 241, 242, 243, 244, 245, 246, 248–51, 252, 253–54, 256, 258
waste generation 14–17, 74
  minimizing 265–66, 273
waste management
  legal regimes *see* legal regimes for waste management
  within the governance of material flows 334–36

waste management hierarchy
global waste management system 349–52
legal waste management regimes 247–59, 262–63
waste transfers 1, 9, 74, 99–100, 258–59, 345–46
global waste management system 357–67, 368
illegal *see* illegal waste transfers
legal regulation 264–324, 372–73
Basel Convention 265–80, 321–22
Brazil 313–21, 323–24
criteria for waste transfers 268–69, 281–85, 297–98, 357–60
EU 23, 24–25, 293–313, 323, 324
general procedure 269–71, 285–87, 298–304, 360–65, 366
OECD Decision 280–93, 322–23
solutions for possible problems 271–72, 279, 287–88, 289, 292, 304–11, 365–67
origins and causes of the waste crisis 17–28, 50–51
transfer of socio-ecological costs in 35–40, 50
weak answers and strong questions 44–45
Westphalian framework 122–23
wilderness, cult of 81–83, 86, 92, 138
World Bank 15, 18
worldviews 40–41